PROTEIN QUALITY IN HUMANS:

ASSESSMENT AND IN VITRO ESTIMATION

PROTEIN QUALITY IN HUMANS:

ASSESSMENT AND IN VITRO ESTIMATION

edited by

C.E. Bodwell, Ph.D.

Chief, Protein Nutritional Laboratory
Beltsville Human Nutrition Research Center
Human Nutrition, SEA, USDA, Beltsville, MD

J.S. Adkins, Ph.D.

Chairman, Department of Human Nutrition
and Food
School of Human Ecology, Howard University
Washington, DC

D.T. Hopkins, Ph.D.

Director, Nutritional Biochemistry Laboratory
Ralston Purina Company
St. Louis, MO

AVI PUBLISHING COMPANY, INC.
Westport, Connecticut

Library of Congress Cataloging in Publication Data

Main entry under title:

Protein quality in humans.

　　Includes bibliographical references and index.
　　1. Proteins in human nutrition. I. Bodwell, C.E.
II. Adkins, James S. III. Hopkins, Daniel T. [DNLM:
1. Dietary proteins—Analysis.　2. Dietary proteins—
Metabolism.　QU 55 P96807]
QP551.P6976　1981　　　641.1'2　　　　81-10894
ISBN 0-87055-388-7　　　　　　　　　AACR2

Printed in the United States of America by
The Saybrook Press, Inc.

Preface

Human assays are impractical for the routine evaluation of the large number of traditional and new protein sources available for meeting human needs for dietary protein. Consequently, for the most part, rat bioassays have been used to estimate the protein quality of foods for human consumption. In recent years, however, the scientific community has become increasingly critical of the use of animal bioassays for predicting protein quality for humans. This has been due to the increasing evidence of a lack of agreement between estimates of protein nutritive value obtained in human studies and estimates obtained on the same protein sources by use of animal bioassays. The use of *in vitro* methods provides an alternative approach to that of using either human or animal bioassays for assessing protein nutritive value.

The discussions in this book provide a critical assessment of the precision and accuracy realized in estimates of protein nutritional value in human studies. In view of these assessments, the degree of accuracy and precision in estimates of protein nutritional value which is scientifically acceptable but practical in application is defined. With these considerations as background, the current status and potential applications of *in vitro* measures of protein nutritional quality are assessed.

In Part I, "The Importance of Protein Quality," the significance of the nutritional quality of protein is addressed in relation to global and U.S. needs for providing dietary protein to meet essential amino acid requirements. The special needs of regulatory agencies and of industry in defining protein quality and in assuring the consumer of the level of protein quality of specific sources of protein is also discussed.

In Part II, "Assessing Protein Nutritive Value in Humans," standard long-term conventional procedures and various short-term procedures which use human subjects are described. A critical assessment of the accuracy and precision realized in estimates of protein nutritional value obtained in human studies is given. The effects on protein quality of factors such as energy-protein interrelationships and the importance of nitrogen digestibility in protein quality evaluation are assessed.

In Part III, "In Vitro Methods For Assessing Protein Nutritional Value," *in vitro* methods based on amino acid analyses, estimates of essential amino acid bioavailability, enzymic hydrolyses, microbiological assays and biological or *in vitro* methods for predicting protein digestibility are described. The use of amino acid scores for predicting protein nutritional value for children and for adults is evaluated. Lastly, a critical assessment of the precision and accuracy which can be expected from the use of *in vitro* procedures for estimating protein quality.

In Part IV, "Task Force Reports and Conference Overview," discussions are given which summarize the current "state-of-the-art" in relation to protein quality evaluation and which evaluate our current knowledge about the validity and potential use of *in vitro* assays for predicting protein quality for human consumption.

We believe that this book is unique in that, for the first time, various *in vitro* methods for estimating protein quality have been discussed within the context of a critical awareness of the precision and accuracy of both the human assays which are used for validating these methods and of the *in vitro* methods described. This book thus provides the most critical and up-to-date assessment of protein quality evaluation methods currently available.

C.E. BODWELL, PH.D.

J.S. ADKINS, PH.D.

D.T. HOPKINS, PH.D.

July 1, 1981

Contributors

ARROYAVE, G., Ph.D., Institute of Nutrition of Central America and Panama, PO Box 1188, Guatemala, Central America

ALTSCHUL, AARON M., Ph.D., Georgetown University, Washington, D.C.

BODWELL, C.E., Ph.D., Protein Nutrition Laboratory, Nutrition Institute, Human Nutrition, Science and Education Administration, United States Department of Agriculture, Beltsville, Maryland 20705

BRESSANI, RICARDO, Ph.D., Division of Agriculture and Food Science, Institute of Nutrition of Central America and Panama, PO Box 1188, Guatemala, Central America

BROWN, W.D., Ph.D., Institute of Marine Resources, University of California, Davis, California 95616

CALLOWAY, DORIS HOWES, Ph.D., Department of Nutritional Sciences, University of California, Berkeley, California 94720

CARPENTER, K.J., Ph.D., Department of Nutritional Sciences, University of California, Berkeley, California 94720

CHRISTENSEN, D.A., Ph.D., Department of Animal and Poultry Science, University of Saskatchewan, Saskatoon, Saskatchewan, Canada S7N 0W0

ELÍAS, LUIZ G., Ph.D., Division of Agriculture and Food Science, Institute of Nutrition of Central America and Panama, PO Box 1188, Guatemala, Central America

FINLAYSON, A.J., Ph.D., Prairie Regional Laboratory, National Research Council of Canada, Saskatoon, Saskatchewan, Canada S7N 0W0

FORD, J.E., Ph.D., National Institute for Research in Dairying, Shinfield, Reading RG2 9AT, England

FRIEDMAN, M., Ph.D., Western Regional Research Center, Science and Education Administration, United States Department of Agriculture, Berkeley, California 94710

HACKLER, L.R., Ph.D., Department of Foods and Nutrition, University of Illinois, Urbana, Illinois 61801

HAPPICH, M.L., Eastern Regional Research Center, Agricultural Research, Science and Education Administration, United States Department of Agriculture, Philadelphia, Pennsylvania 19118

HARPER, A.E., Ph.D., Departments of Nutritional Science and Biochemistry, University of Wisconsin, Madison, Wisconsin 53706

HOPKINS, DANIEL T., Ph.D., Ralston Purina Company, Checkerboard Square, St. Louis, Missouri 63188

JANSEN, G. RICHARD, Ph.D., Food Science and Nutrition Department, Colorado State University, Fort Collins, Colorado 80523

JEWELL, D.K., Ph.D., Food Protein Research Group, Room 20, Filley Hall, University of Nebraska, Lincoln, Nebraska 68583

KENDRICK, J.G., Ph.D., Food Protein Research Group, Room 20, Filley Hall, University of Nebraska, Lincoln, Nebraska 68583

MARABLE, NINA L., Ph.D., Virginia Polytechnic Institute and State University, Blacksburg, Virginia 24061

MARGEN, SHELDON, M.D., School of Public Health, University of California, Berkeley, California 94720

MITCHELL, GERALDINE V., Ph.D., Division of Nutrition, Bureau of Foods, Food and Drug Administration, Washington, D.C.

MURPHY, ELIZABETH W., Food Ingredient Assessment Division, Science Program, Food Safety and Quality Service, United States Department of Agriculture, Washington, D.C. 20250

NAVARRETTE, DELIA A., Ph.D., Division of Agriculture and Food Science, Institute of Nutrition of Central America and Panama, PO Box 1188, Guatemala, Central America

NESHEIM, ROBERT O., Ph.D., The Quaker Oats Company, Merchandise Mart Plaza, Chicago, Illinois 60654

PELLETT, P.L., Ph.D., Department of Food Science and Nutrition, University of Massachusetts, Amherst, Massachusetts 01003

PHILLIPS, J.G., Ph. D., Biometrical and Statistical Services, NER, AR, Science and Education Administration, United States Department of Agriculture, Eastern Regional Research Center, Philadelphia, Pennsylvania 19118

PINEDA, O., Ph.D., Institute of Nutrition of Central America and Panama, PO Box 1188, Guatemala, Central America

RAND, WILLIAM M., Ph.D., Laboratory of Human Nutrition, Department of Nutrition and Food Science and Clinical Research Center, Massachusetts Institute of Technology, Cambridge, Massachusetts 02139

RITCHEY, S.J., Ph.D., Department of Human Nutrition and Foods, Virginia Polytechnic Institute and State University, Blacksburg, Virginia 24061

SAMONDS, K., Ph.D., Department of Food Science and Nutrition, University of Massachusetts, Amherst, Massachusetts 01002

SANZONE, GEORGE, Ph.D., Virginia Polytechnic Institute and State University, Blacksburg, Virginia 24061

SARWAR, G., Ph.D., Bureau of Nutritional Sciences, Food Directorate,

Health Protection Branch, Department of National Health and Welfare, Tunney's Pasture, Ottawa, Ontario, Canada K1A 0L2

SATTERLEE, L.D., Ph.D., Food Protein Research Group, Room 20, Filley Hall, University of Nebraska, Lincoln, Nebraska 68583

SCRIMSHAW, NEVIN S., Ph.D., Laboratory of Human Nutrition, Department of Nutrition and Food Science and Clinical Research Center, Massachusetts Institute of Technology, Cambridge, Massachusetts 02139

TAPER, L.J., Ph.D., Department of Human Nutrition and Foods, Virginia Polytechnic Institute and State University, Blacksburg, Virginia 24061

TKACHUK, R., Ph.D., Grain Research Laboratory, Canadian Grain Commission, Agriculture Canada, Winnipeg, Manitoba, Canada R3C 3G9

TORÚN, BENJAMÍN, Ph.D., Program of Physiology and Clinical Nutrition, Institute of Nutrition of Central America and Panama, PO Box 1188, Guatemala, Central America

VANDERVEEN, JOHN E., Ph.D., Division of Nutrition, Bureau of Foods, Food and Drug Administration, Washington, D.C.

VARGAS, EMILIO, Ph.D., United Nations University Fellow, Institute of Nutrition of Central America and Panama, PO Box 1188, Guatemala, Central America

VITERI, F., Ph.D., Institute of Nutrition of Central America and Panama, PO Box 1188, Guatemala, Central America

WILLIS, B. WELLS, Food Ingredient Assessment Division, Science Program, Food Safety and Quality Service, United States Department of Agriculture, Washington, D.C. 20250

YOUNG, VERNON R., Ph.D., Laboratory of Human Nutrition, Department of Nutrition and Food Science and Clinical Research Center, Massachusetts Institute of Technology, Cambridge, Massachusetts 02139

Contents

Conference Organization

Chairman: C.E. Bodwell, Chief, Protein Nutrition Laboratory, Nutrition Institute, Human Nutrition, SEA, USDA, Beltsville, MD. 20705

Co-Chairmen: J.S. Adkins, Chairman, Program In Human Nutrition and Food, School of Human Ecology, Howard University, Washington, D.C. 20059

D.T. Hopkins, Director, Nutritional Biochemistry Laboratory, Ralston Purina Company, Checkerboard Square, St. Louis, MO. 63188

Task Force I

Chairman: S. Margen, University of California

Group Leaders: R.P. Abernathy, Purdue Univ.
D.A. Benton, Ross Laboratories
R. Bressani, INCAP
G.R. Jansen, Colorado State Univ.
R.O. Nesheim, Quaker Oats
N.S. Scrimshaw, MIT
M.E. Swenseid, UCLA

Task Force II

Chairman: A.E. Harper, Univ. of Wisconsin

Group Leaders: A.M. Altschul, Georgetown Univ.
L.R. Hackler, Univ. of Illinois
H.D. Hurt, Quaker Oats
W. Martinez, USDA
P.L. Pellett, Univ. of Mass.
H.L. Wilcke, Ralston Purina Co.
V.R. Young, MIT

Conference Staff: Ms. Debbie Wamser, Ralston Purina
Mr. Robert Staples, USDA
Mr. Gregory Foy, Howard Univ.
Ms. Sarah Ngundam, Howard Univ.

Program Advisory Committee:

R.H. Anderson, General Mills, Inc.
M. Brown, Food & Nutrition Board
M.A. Burnett III, GMA
K.J. Carpenter, Univ. of Calif.
I. Fried, USDA
L.R. Hackler, Univ. of Ill.
A.E. Harper, Univ. of Wisc.
D.M. Hegsted, USDA
G.R. Jansen, Colo. State Univ.
P.A. Lachance, Rutgers Univ.
K. McNutt, Nutrition Consortium
M. Milner, Amer. Inst. of Nutrition
R.O. Nesheim, Quaker Oats Co.
P.L. Pellett, Univ. of Mass.
I.I. Rusoff, Nabisco, Inc.
L.D. Satterlee, Univ. of Neb.
N.S. Scrimshaw, MIT
J.E. Vanderveen, FDA
V.R. Young, MIT

Acknowlededgments

This book is the result of a conference and workshop which was held at Airlie House, Warrenton, Virginia, March 23–26, 1980, and which was made possible, in part, by a grant from the Division of Problem Focused Research, National Science Foundation, Washington, DC, to the Program In Human Nutrition And Food, School of Human Ecology, Howard University, Washington, DC. Appreciation is expressed to the authors of the various chapters, the two Task Force Chairmen, the leaders of the discussion groups, the conference staff, and the members of the Program Advisory Committee. The able secretarial assistance of Ms. Debbie Wamser and Mrs. Ethel C. Cole is gratefully acknowledged. Contributors from various Federal Agencies are expressing their own opinions and not those of the Federal Government.

Part I The Importance of Protein Quality

Nutritional Significance of Protein Quality: A Global View

Nevin S. Scrimshaw

The 1960s in the UN system were years of concern for protein quantity and quality in the diets of developing countries. The Protein Advisory Group (PAG), later the Protein-Calorie Advisory Group, was formed, the UN Advisory Committee on Science and Technology (ACAST) investigated ways to avert the impending protein crisis, and the Secretary General of the UN convened a special panel on this subject (DESA/UN, 1971).

During this period, the success of Green Revolution research on cereals and government incentives for increased cereal production were already leading to a decline in the *per capita* availability of legumes, sometimes an absolute decrease, and to steep price increases. Populations do not thrive on cereal alone, but on a cereal staple plus some more concentrated sources of proteins, usually one or more legumes supplemented with very small amounts of fish or other animal protein. It was foreseen that population pressures forcing governments to focus on producing more of the cereal staples, and the greater profitability for farmers to grow them instead of legumes, would seriously affect poor populations and vulnerable groups such as preschool children and pregnant or nursing mothers.

In fact, a reduced availability of legumes, continuing social inequity, and population trends have resulted in an increase in the number of malnourished individuals worldwide. Even kwashiorkor has returned in some populations from which it had largely disappeared.

Global perception of world food and nutrition problems abruptly changed, however, after the meeting in 1971 of a joint FAO/WHO Expert Committee on Energy and Protein Requirements that lowered estimates of safe allowances for dietary protein for human adults by about 30% without appreciably altering estimates of mean dietary energy requirements (FAO/WHO, 1973). Re-evaluation of average dietary intakes in developing country populations using these new standards frequently indicated that calories were slightly more deficient than protein. It became fashionable to emphasize the deficiency of energy in the diets of low-income populations and to point out that if this were corrected, protein needs, as indicated by the 1971 committee report, would be met.

Economists and policy makers concluded mistakenly, but understandably, that it was not necessary to be concerned with problems of either protein quantity *or* quality, but only with seeing that people obtained enough of their usual diets. Concentration on cereal yields and on the increased production of roots and tubers, including cassava, to meet caloric requirements was assumed to be sufficient. In recent meetings to discuss nutritional goals for plant breeders, or to develop national food and nutrition policy, voices have been raised with great confidence stating that the focus should be on raising yields or increasing consumption of cereals because protein needs would also be taken care of in this manner.

Such a conclusion, although it has an element of truth, ignores a number of important factors:

First: Individuals can adapt to a wide range of dietary energy intakes so that their physiological requirements are met, even when the intake is much lower than that suggested by current requirement estimates. This adaptation involves, to a limited extent, a reduction in basal metabolic rate (BMR), but mainly it is due to a decrease in physical activity. There is simply no other way to explain why populations averaging 60 to 80% of their supposed mean dietary energy requirements do not continuously lose weight and die. Some societies even show evidence of cultural adaptation to the necessity for lower levels of activity than characteristic of those whose food supply is greater. On the other hand, an increase in the caloric intake of populations for whom dietary energy is limiting results in a prompt increase in activity (Viteri and Torún, 1975).

Even children adapt equally quickly to reduced dietary energy. Torún *et al.* (1979) have shown that children receiving adequate protein, but faced with a 20% reduction in dietary energy, promptly come back into energy balance by reducing their activity and increasing their sleeping and resting time while continuing to grow at a normal rate. For children, this enforced compensatory reduction in activity could have consequences for stimulation of learning and behavior. For adults, the consequences may be both economic and social. But in neither case is the necessary consequence a weight loss indicative of physiological deficiency. Rutishauser and Whitehead (1972) have described children whose activity levels are so low that they grow on diets only 1.5 × BMR. Of course, they live in near thermoneutral environments.

Second: The socio-economic and environmental circumstances likely to require an adaptation to low dietary energy intakes are generally characterized by factors that can increase protein needs. Enteric infections and parasites can impair intestinal absorption. A major concern for populations of developing countries, especially for preschool children, is the high prevalence, duration, and severity of infections that condition growth, that precipitate acute nutritional deficiency diseases in individuals whose diet is borderline, and that are a frequent cause of death. Moreover, there is a need for higher than normal concentrations of protein relative to calories for optimum recovery from acute infectious episodes (Table 1.1). In children, this is important for catch-up growth (Rutishauser and Whitehead 1972; Viteri *et al.* 1979).

TABLE 1.1 ADDITIONAL PROTEIN AND CALORIES FOR CATCH-UP GROWTH FOR A CHILD WEIGHING 7 KG*

Gm Protein Retained	10	30	50	70
Gm Protein/kg Required	1.96	2.86	3.23	3.96
Calories/kg in Diet	113	123	133	144
% Protein/Calories	7.0	9.3	9.7	11.0

*Adapted from R.G. Whitehead (see Viteri *et al.* 1979).

There is a close correspondence between a reduction in measures of cell-mediated immunity and the degree of growth retardation and impairment of protein status. For repletion and catch-up, more of the usual diet may simply not provide sufficient protein for optimum recovery.

Nutritionists as well as economists must learn to appreciate that the dietary standards for normal, healthy individuals cannot be applied indiscriminately to underprivileged groups containing large numbers of vulnerable and unhealthy individuals.

Far from an adaptation to low protein intakes under these circumstances, the reverse occurs. Growth failure in children is, of course, a kind of adaptation, but one that has lasting consequences. There is an "adaptation" to actual protein deficiency by an increase in nitrogen retention. However, this occurs only while the individual is depleted and is not the matching of long-term requirements to long-term intakes that occur when calories are mildly or moderately deficient.

May there not be genetic adaptation to low protein intake among some groups? Preliminary examination of such data as are available does not suggest genetic differences of a magnitude that can be detected in small samples. The apparently greater efficiency of nitrogen retention in Nigerian laborers reported by Nicol (1959) and Nicol and Phillips (1976) can probably be explained by a protein-depleted state when they entered the study (see Scrimshaw, 1977). The small difference in obligatory nitrogen excretion between Taiwanese and MIT University students described by Huang was balanced by higher fecal N losses, so that total endogenous N losses were similar (Huang *et al.* 1972; Scrimshaw *et al.* 1972).

The United Nations University (UNU) World Hunger Programme is sponsoring metabolic balance studies in Brazil, Chile, Colombia, Egypt, Guatemala, Korea, Mexico, and Thailand to compare with data from similar research protocols from Berkeley, California; Cambridge, Massachusetts; Moscow, USSR; and Tokushima, Japan. A workshop under the auspices of the International Union of Nutritional Sciences (IUNS) Committee 6 of Commission I on Protein-Energy Requirements will analyze the results, and a report of the findings will appear in a forthcoming supplement to the UNU Food and Nutrition Bulletin.

Third: For diets low in both caloric and protein density, calculations of the possibility of meeting protein needs when enough is consumed to meet estimated calorie requirements may be irrelevant if the sheer bulk of the diet makes this virtually impossible. They become even more unrealistic when these calculations assume a quality of diet based on *per capita* averages without recognizing that the proportion of biologically available protein decreases as the protein quality of family diets becomes lower with decreasing income. Moreover, those family members who are already most vulnerable frequently receive less than their quantitative and qualitative share of the family diet.

Fourth: Of particular relevance to this conference is the method proposed in the FAO/WHO (1973) report for converting protein requirements based on egg or milk as reference proteins to the protein of actual diets. The conversion utilizes values for net protein utilization (NPU) and the report assumes that 30% more nitrogen is required for nitrogen balance than would be predicted from the NPU value of proteins, as conventionally measured in rats or at 0.3 to 0.4 gm of protein per kg in adult human subjects. Our own work and that of others reported at this meeting show that this factor increases as protein quality decreases so that even more protein from poor diets is required for N balance than predicted from relative NPU values. This leads to an overestimation of the capacity of poor-quality diets to meet human needs.

It is not the purpose of this paper to dwell further on the importance of taking into account human needs for protein as well as energy in agricultural planning, and in the formulation and implementation of food and nutrition policies and programs (Scrimshaw 1977, 1978). It is appropriate, however, to emphasize that there are large populations of the world—and their numbers are steadily increasing—for whom available protein intake may be sufficiently limiting to interfere with resistance to infection, catch-up growth, recovery from infection, the course and outcome of pregnancy, and possibly other aspects of health. For them, differences in the quality of the protein they ingest can be of critical importance.

Fifth: The FAO/WHO (1973) allowances were intended for normal, healthy populations. Unfortunately, insufficient data were available to give quantitative instructions as to how they might be adjusted for populations who are not normal or healthy by Western standards.

Sixth: Economists and planners, as well as nutritionists, should be aware that the estimates of protein requirements appearing in the 1973 FAO/WHO report were based on very limited data, and may have been underestimated for a series of methodological reasons that are only now becoming evident. The recommendations for adults were based mainly on U.S. university students and on only short-term balance studies. A series of papers from our laboratory suggest that the allowance of 0.57 gm of protein per kg/day for healthy adults is adequate for one month, but after two or three months is associated with loss of lean body mass, decreases in serum albumin and hemoglobin levels, and abnormal liver enzyme activity (Garza *et al.* 1977a,b;1978). No adverse changes were observed in a study of similar duration in which beef protein was fed at a level of 0.8 gm/kg/day. Our data from this and other MIT studies agree with the conclusion of Calloway in these proceedings (Chapter 10) that the FAO/WHO safe level of egg or milk protein meets average protein needs, but not the mean plus two standard deviations as intended.

At the FAO/WHO (1973) protein allowance level, excess calories can produce positive N balance, but only at the cost of weight gain, an obviously undesirable characteristic for an allowance intended for the entire world. Moreover, as Calloway states, at each level of protein there is an energy level beyond which increasing calories without increasing protein intake does not continue to improve N retention. In fact, it can lead to kwashiorkor if protein levels are low.

These results should serve as a warning to treat the 0.57 gm recommendations with reservation until it can be reviewed and revised at the FAO/WHO/UNU meeting scheduled for 1981. Whitehead has voiced similar concern that the protein and energy requirement estimates for the growth of children are based on a daily growth rate that is assumed to be 1/365th of the annual growth rate, despite the fact that normal children often grow at two or three times this rate for extended periods (see p. 34, Viteri *et al.* 1979).

During catch-up growth following infections or severe malnutrition under village conditions, the rate may be as much as nine times the average daily rate, and up to 18 times under hospital conditions (see p. 38, Viteri *et al.* 1979; Whitehead, 1977). Since currently estimated protein requirements for growth are about 10 to 20% of the total requirement, depending on age, this would imply periods when protein needs would be considerably greater than recommendations, if full catch-up were to occur. As innumerable authors have pointed out, very little catch-up occurs among underprivileged children in developing countries and they remain permanently stunted. In effect, the peaks or spurts in growth are cut off.

Of course, catch-up growth rates require more calories as well, but the proportion of total calories required for growth is much smaller than that of protein, as Table 1 illustrates. Hence, for both catch-up and recovery from depletion there is a disproportionate increase in the need for protein relative to energy. Two recent studies, one of Graham *et al.* (1977) in Peru and one unpublished of Margen using Guatemalan data, conclude that growth failure and variations in growth among preschool children of low-income families are associated with deficiencies of dietary protein and not of dietary energy as has been postulated. The protein calories required are a function of diet quality since in any situation where protein becomes limiting, protein quality counts.

Seven: The argument that traditional diets provide sufficient protein if only people can get enough to eat has both historical justification and biological necessity to support it, but it may not be true for the diets of some of today's low-income populations. Economic and social changes, population pressures, political actions, and natural disasters can alter traditional patterns adversely. Diets based on cassava are particularly likely to become protein-deficient whenever circumstances preclude their containing sufficient legume or animal protein. This is also true for maize- and sorghum-based diets. In these circumstances, of course, protein quantity and quality matter.

In principle, if everyone consumed average diets and had average protein and energy needs, we could forget about protein, but the real world is not that simple. Some individuals have higher protein needs for physiological or pathological reasons, and these do not necessarily coincide with higher energy needs. Moreover, the average diet may be satisfactory for maintenance, but far from optimal for recovery from infections and other causes of stress, or for catch-up growth. It is essential that diets have adequate margins of safety, not only for the variation in protein requirements among healthy individuals, but also for those who, in some populations, are unhealthy most of the time and in all populations some of the time. It is plain foolishness to take the average per cent of net dietary protein calories, based on average calorie requirements and "safe allowance" for protein intended for normal, healthy individuals as a reference point for adequacy, but some have done so. Each of the components of this figure—energy, protein, and NPU—have their own variances, and the combined variance is large. Even with the 1973 FAO/WHO figures, the average calorie requirement minus 2 SD and average protein requirement plus 2 SD, the resulting figure of 7% is too low. No one would, for long, voluntarily accept a diet of less than 10% protein calories, despite its theoretical adequacy.

GLOBAL ENERGY AVAILABILITY

Much is sometimes made of the fact that the existing food supply would be adequate to meet all human needs if it were adequately distributed, but clearly it is not. In fact, dietary energy intakes are so closely related to income that *per caput* figures give no reliable indication of the proportion of the population whose caloric intakes are deficient (Scrimshaw, 1977). Figure 1.1 based on family dietary intake and income data for Northeast Brazil, illustrates the extent to which upper-income families consume more than their supposed requirements and lower-income families proportionately less. Since the amount apparently used by the former exceeds feasible intakes, even if obesity is developing, some of this must represent waste, and in theory should be subtracted before *per caput* consumption is calculated. At the other end of the socio-economic scale, caloric intakes that are only 70 or 80 per cent of suggested requirements can only mean that adaptation occurs, primarily through an involuntary reduction of physical activity. The relationship between income and food intake distorts the meaning of all average values for food availability within countries. Table 1.2 gives another example—the large differences in food availability among income groups in India.

TABLE 1.2. AVERAGE CALORIE AND PROTEIN AVAILABILITY PER CONSUMER UNIT OF HOUSEHOLDS BY EXPENDITURE GROUPS, INDIA, 1971–1972[1]

Expenditure (Rupees per Caput per month)	Urban			Rural		
	Percent of Households	Calories (per Consumer Unit per Day)	Proteins (Grams per Consumer Unit per Day)	Percent of Households	Calories (per Consumer Unit per Day)	Proteins (Grams per Consumer Unit per Day)
0–15	0.9	1,228	37	3.9	1,493	46
15–21	3.7	1,582	46	10.5	1,957	60
21–24	3.6	1,821	54	7.1	2,287	69
24–28	6.0	1,970	58	10.2	2,431	73
28–34	10.2	2,130	62	15.2	2,734	82
34–43	14.9	2,343	69	17.7	3,127	93
43–55	15.4	2,622	76	14.4	3,513	105
55–75	16.9	2,872	82	11.5	4,016	121
75–100	11.3	3,190	91	5.2	4,574	139
More than 100	17.0	3,750	110	4.2	6,181	182

[1] Source: The Fourth World Food Survey, FAO, Rome, 1977, Food and Agriculture Organization of the United Nations, FAO Statistics Series No. 11; FAO Food and Nutrition Series No. 10.

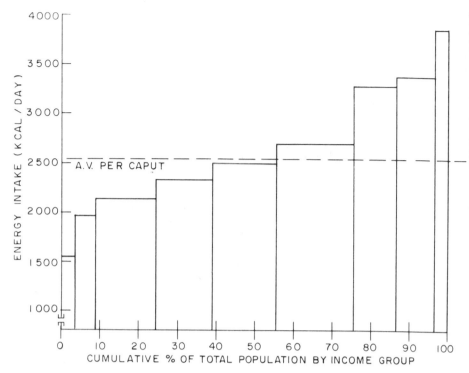

FIG. 1.1.   EFFECT OF INCOME ON ENERGY INTAKE IN BRAZIL; MEAN REQUIRE-
MENT = 2450 KCAL PER DAY; FROM THE GETULIS VARGAS FOUNDATION (1970) AND
JANSEN *ET AL.* (1977)

GLOBAL PROTEIN AVAILABILITY AND DISTRIBUTION WITH INCOME

Existing protein sources for human consumption are even more capable of meeting the needs of everyone, if equitably distributed, than are those for calories, but the capacity of populations that are better off to command not only greater quantities of protein *per caput*, but also protein of higher quality, is well known. Figure 1.2 illustrates this for the population of Northeast Brazil. The variation in distribution of *total* protein with family income looks similar to that for calories. When the distribution of animal protein, and hence of protein quality by income, is examined, it is similarly skewed. Since protein availability to meet physiological needs is a multiple of both quality and quantity, it is not surprising that the lower shaded bars representing *available* protein are even more skewed with income than those for total protein. This fact is rarely taken into account in the calculations of economists and planners.

Figure 1.3 shows another way of looking at this phenomenon. The vertical bars show the distribution of mean family protein intakes within income groups. At requirement level, egg protein is about 60% utilized according to

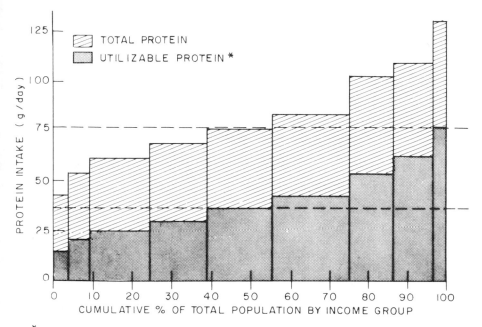

* ASSUMING PROTEIN
UTILIZATION RANGING
FROM 35-60%

FIG. 1.2.   EFFECT OF INCOME ON PROTEIN INTAKE IN BRAZIL (SEE FIGURE 1 FOR SOURCES OF DATA)

Inoue's (Inoue et al., 1973), our own (Young et al. 1973), and other data, and not the nearly 100% observed at the levels of intake conventionally used to determine NPU. Assuming an NPU at requirement level to be approximately 60% for the highest quality diets, a diagonal line rather than a horizontal one must be used for comparison of total protein intake with protein requirement or allowances. This kind of representation shows clearly that protein is not limiting in the upper half of the population by the standards used, and in fact, most of these groups are consuming considerably more than they need, but it is marginal or deficient for those in the lowest income group. It is for these that protein quality can make a difference and for whom more of the same diet may be neither feasible nor adequate.

Any efforts to increase the availability of protein foods or the quality of dietary protein must be directed at these groups and not at those with adequate purchasing power. Somehow this point is missed again and again in criticisms of past emphasis on protein quantity and quality. The impending protein crisis foreseen in the 1960s, and sometimes referred to as an increasing protein gap, was never considered by knowledgeable people to be a global or even national lack of protein. Rather, it was a further reduction

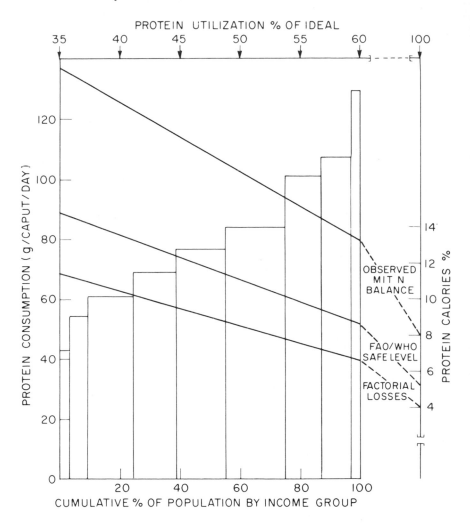

FIG. 1.3.   BRAZILIAN PER CAPUT PROTEIN CONSUMPTION BY INCOME GROUP RE-
LATED TO PROTEIN NEEDS AND UTILIZATION (SEE FIGURE 1 FOR SOURCES OF
DATA)

in the available protein in the diets of the poor, because production of
legumes and animal protein was expected not to keep pace with the increase
in cereals, roots, and tubers that could fill stomachs and meet apparent
energy needs more cheaply.

GLOBAL SIGNIFICANCE OF PER CENT OF DIETARY
PROTEIN CALORIES

Since there are wide variations in physiological requirements for calories
as the result of social and economic factors—with a decrease common when

these are unfavorable, while protein requirements either remain the same or are increased under adverse socio-economic and environmental circumstances—further consideration must be given to the per cent of available protein calories in diets for low-income groups. Where this percentage is low or marginal, protein quality becomes important.

Assuming that current FAO/WHO standards are applicable, and that milk, meat, or eggs are the sole sources of protein, the per cent of protein necessary to cover the needs of nearly all of the population are easily calculated. As suggested above, the calculations indicate that if the FAO/WHO recommendations are applicable, 7% of protein calories from meat, milk, or eggs should be sufficient for nearly everyone., Our actual observations on MIT students suggest a value of 10% for N balance studies of ten days duration, as shown in Figure 1.4, using the observed means and calculated 97.5% confidence limits. This value will, of course, increase progressively as the quality of dietary protein decreases when measured by NPU. For lower socio-economic groups for whom the quality of protein is poor, the increase required is difficult to achieve without improvement in quality as well as quantity of the dietary protein consumed daily.

The goal in improving dietary protein quality should be to bring the per cent of available protein calories up sufficiently that protein needs of nearly everyone are met when the diet meets energy requirements. The nature of the effect of protein quality on the per cent of available protein calories required can be readily visualized from Figure 1.4. The data for quantitative guidelines should come from the studies currently sponsored by the UNU, FAO, WHO, and others, and should be available for the international meeting scheduled for 1981 to review protein-energy needs.

EVIDENCE FOR PROTEIN QUALITY EFFECTS

Studies on both children and adults consistently confirm the effects on nitrogen balance of improving protein quality under appropriate test conditions. Without changing N intake, negative balances become less negative or positive, and positive balances become more so. In view of the many such studies, it is surprising to hear even some nutritionists saying that protein quality does not matter in adults. It does not if you are relatively well-off, and already eating far more protein than you need. It may matter if you are poor, elderly, sick, or even pregnant or lactating. For children in a period of growth spurt or catch-up growth, protein quality can be even more important.

It is also frequently stated that there is no point in improving the protein quality of diets unless apparent caloric deficits are also corrected; otherwise, you are giving expensive protein to be used for energy when cheap calorie sources would do as well. The answer to this is like an answer to the question: "Why are you starving your dog?"—"I am not starving my dog, and besides, I do not have a dog." Low-calorie intakes by comparison with FAO/WHO (1973) standards do not usually indicate physiological deficiency of calories. If the individual is in energy balance even at comparatively low-calorie intakes, the argument that protein is pre-empted for

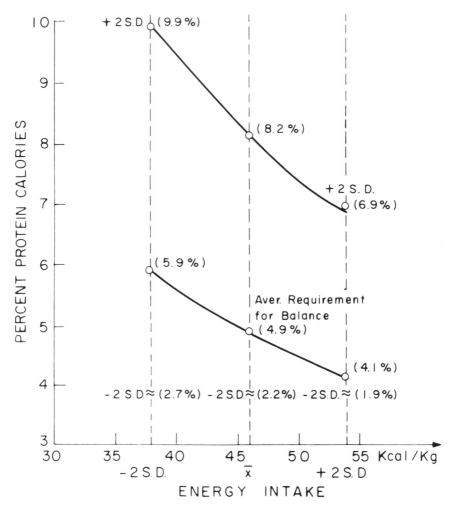

FIG. 1.4.   NET DIETARY PROTEIN CALORIES PERCENT FOR NITROGEN BALANCE
WITH HIGH QUALITY PROTEIN (EGG, BEEF) WHEN ENERGY NEEDS OF MIT STU-
DENTS (N = 7) ARE MET

energy is not applicable. However, individuals who are calorie- and protein-
deficient *do* respond to changes in protein quality about as well as they
would if calories were not deficient.

Table 1.3 illustrates the results of giving young men 0.27 grams of wheat
protein per kg at their normal calorie intakes with and without 2.25%
lysine addition (Scrimshaw *et al.* 1973). The study was then repeated with
calorie intake reduced 20% and the subjects' activity maintained so that
they lost weight at the expected rate. Nevertheless, despite the genuinely
deficient dietary energy, the improvement in N retention with lysine was

TABLE 1.3. URINARY UREA AND NITROGEN BALANCE IN YOUNG MEN STUDIED FOR THE EFFECTS OF LYSINE SUPPLEMENTATION (2.25% OF TOTAL PROTEIN INTAKE) OF WHEAT GLUTEN AT ADEQUATE AND RESTRICTED CALORIC INTAKES

Level of Protein, g/kg/day	Adequate Energy				Restricted Energy			
	No Lysine		Lysine		No Lysine		Lysine	
	Urea	N Balance	Urea	N Balance	Urea	N Balance	Urea	N Balance
0.27	2.21 ±0.44[1]	−1.29 ±0.33[1]	2.04 ±0.59	−1.05 ±0.48	2.93 ±0.58	−2.02 ±0.49	2.76 ±0.99	−1.70 ±0.96
0.73	6.97 ±0.80	−0.72 ±0.73	6.43 ±0.92	−0.09 ±0.69	7.53 ±0.91	−1.40 ±0.54	7.31 ±0.98	−0.97 ±0.61

[1] Mean (gms/day) ± SD for 6 and 8 subjects studied at the 0.27 and 0.73 g protein levels, respectively; from Scrimshaw et al. 1973.

essentially the same. At a level of 0.73, the subjects were in balance when dietary energy was adequate and the diet was supplemented with 2.25% lysine. There was no significant difference in the effect of lysine supplementation when subjects received either 100 or 80% of required dietary energy under the same conditions. We have repeated this study using the same design with chickpea as the protein source and methionine as the supplementary amino acid and obtained the same results. Vargas and Bressani of INCAP have excellent unpublished data showing equally clear protein quality differences among various mixtures of maize, beans, and milk, at 100% and 80% of usual dietary energy.

CONCLUSIONS

Energy deficits have their own social and economic consequences that must be corrected. Protein deficiencies have health consequences that are serious. When protein deficiency causes growth failure or death from low resistance to infection, its effects are permanent. Protein quantity and quality can indeed limit the nutrition, health, and functional performance of vulnerable populations wherever they are found, whether in developing or in industrialized countries.

Given all of the above considerations, the statement that, for populations deficient in both calories and protein, calories should receive first priority, must be evaluated critically according to the circumstances. Comparison of dietary intake with dietary standards often suggests that calories are limiting for the reasons I have discussed, but this is likely not to be true physiologically. We need to strive for a balanced improvement in both the protein and energy value of the diets of low-income groups in developing countries.

REFERENCES

DESA/UN (Department of Economic and Social Affairs, United Nations). *Strategy Statement on Action to Avert the Protein Crisis in the Developing Countries*. United Nations, New York, ST/ECA/144 E/5018/Rev. 1, 1971.

FAO/WHO (Food and Agriculture Organization/World Health Organization). 1973.   Joint FAO/WHO Ad Hoc Expert Committee: *Energy and Protein Requirements*. World Hlth. Org. Tech. Rep. Ser. No. 522, WHO, Geneva.; FAO Nutr. Mtgs. Rep. Ser. No. 52, FAO, Rome.

GARZA, C., SCRIMSHAW, N.S., and YOUNG, V.R.   1977a.   Human Protein Requirements: Evaluation of the 1973 FAO/WHO Safe Level of Protein Intake for Young Men at High Energy Intakes. Brit. J. Nutr. *37*:403.

GARZA, C., SCRIMSHAW, N.S., and YOUNG, V.R.   1977b.   Human Protein Requirements: A Long-Term Metabolic Nitrogen Balance Study in Young Men to Evaluate the 1973 FAO/WHO Safe Level of Egg Protein Intake. J. Nutr. *107* (2): 335.

GARZA, C., SCRIMSHAW, N.S., and YOUNG, V.R.   1978.   Human Protein Requirements: Interrelationships between Energy Intake and Nitrogen Balance in Young Men Consuming the 1973 FAO/WHO Safe Level of Egg Protein, with Added Non-Essential Amino Acids. J. Nutr. 108 (1): 90.

GETULIO VARGAS FOUNDATION."Food Consumption in Brazil," Brazilian Institute of Economics. United States Department of Agriculture, Economic Research Service and Israel Program for Scientific Translations, November, 1970.

GRAHAM, G.G., MORALES, E., PLACKO, R.R., and MacLEAN, W.C., JR. 1979.   Nutritive Value of Brown and Black Beans for Infants and Small Children. Am. J. Clin. Nutr. 32: 2362.

HUANG, P.C., CHONG, H.E., and RAND, W.M.   1972.   Obligatory Urinary and Fecal Nitrogen Losses in Young Chinese Men. J. Nutr. 102:1605.

INOUE, G., FUJITA, Y., and NIIYAMA, Y.   1973.   Studies on Protein Requirements of Young Men Fed Egg Protein and Rice Protein with Excess and Maintenance Energy Intakes. J. Nutr. 103: 1673.

JANSEN, G.R., JANSEN, N.B., SHIGETOMI, C.T., and HARPER, J.M.   1977. Effect of Income and Geographic Region on the Nutritional Value of Diets in Brazil. Am. J. Clin. Nutr. 30: 955.

NICOL, B.M.   1959.   The Protein Requirements of Nigerian Peasant Farmers. Brit. J. Nutr. 13: 307.

NICOL, B.M. and PHILLIPS, P.G.   1976.   The Utilization of Dietary Protein by Nigerian Men. Brit. J. Nutr. 36: 337.

RUTISHAUSER, I.H.E. and WHITEHEAD, R.G. Energy Intake and Expenditure in 1-3-Year-Old Ugandan Children Living in a Rural Environment. 1972.   Brit. J. Nutr. 28: 145.

SCRIMSHAW, N.S.   1977.   Through a Glass Darkly: Discerning the Practical Implications of Human Dietary Protein-Energy Interrelationships. W.O. Atwater Memorial Lecture. Nutr. Rev. 35 (12): 321.

SCRIMSHAW, N.S.   1978.   Relationship of Improved Food Production, Processing, and Distribution to Malnutrition in Developing Countries. Proceedings of the Symposium on Canada and World Food, Royal Society of Canada and The Agricultural Institute of Canada, Ottawa, Canada, pp. D-5-1-D-5-19.

SCRIMSHAW, N.S., HUSSEIN, M.A., MURRAY, E., RAND, W.M., and YOUNG, V.R.   1972.   Protein Requirements of Man: Variations in Obligatory Urinary and Fecal Nitrogen Losses in Young Men. J. Nutr. 102 (12): 1595.

SCRIMSHAW, N.S., TAYLOR, Y.S.M. and YOUNG, V.R.   1973.   Lysine Supplementation of Wheat Gluten at Adequate and Restricted Energy Intakes in Young Men. Am. J. Clin. Nutr. 26: 965.

TORÚN, B., VITERI, F.E., ARROYAVE, G., PINEDA, O., ARAYA, H., and GARCIA, S.   1979.   Interacción de Proteinas y Energia en la Dieta: Efecto de la Ingesta Energetica sobre los Requerimientos Proteinicos de Niños con Dietas Basadas en Maiz y Frijol Negro. Instituto de Nutrición de Centro

America y Panama, *Informe Anual*, 1° de enero − 31 de diciembre de 1978. Documento C INCAP 30/2.

VITERI, F.E. and TORÚN, B. 1975. Ingestion Calorica y Trabajo Fisico de Obreros Agricolas en Guatemala. Bol. Ofic. Sanit. Panam. *LXXVIII* (1): 58.

VITERI, F.E., WHITEHEAD, R.G., and YOUNG, V.R. (Eds.) 1979. *Protein-Energy Requirements under Conditions Prevailing in Developing Countries: Current Knowledge and Research Needs.* The United Nations University World Hunger Programme. *Food and Nutrition Bulletin* Supplement 1, July.

WHITEHEAD, R.G. 1977. Protein and Energy Requirements of Young Children Living in the Developing Countries to Allow for Catch-up Growth After Infections. Am. J. Clin. Nutr. *30*: 1545.

YOUNG, V.R., TAYLOR, Y.S.M., RAND, W.M., and SCRIMSHAW, N.S. 1973. Protein Requirements of Man: Efficiency of Egg Protein Utilization at Maintenance and Sub-maintenance Levels in Young Men. J. Nutr. *103* (8): 1164.

2

Importance of Protein Quality in the United States Diet

A. E. Harper

It is almost 100 years since Voit observed that different food and feed proteins were not equivalent in their ability to support growth and nitrogen retention in young animals and nitrogen balance in adults (McCollum, 1957). These and similar observations made before the end of the 19th century served as the stimulus for development of methods for determining the nutritional quality of proteins. The basic methods were developed over 50 years ago by Osborne, Mendel and Ferry (1919); Thomas (1909); Mitchell (1924); and McCollum (1939). These methods, based on measurements of growth responses or nitrogen retention of animals fed various sources of protein under standardized conditions, made it possible to rank proteins in a general way for their potential in meeting nitrogen and amino acid needs before much was known about the amino acid composition of proteins and amino acid requirements of either animals or human subjects. Nevertheless, Osborne and Mendel, McCollum, and Mitchell all recognized in the 1920s that the biological methods for evaluation of protein quality were indirect measures of the potential effectiveness of different proteins in meeting the amino acid needs of the test organism (Mitchell, 1954).

In any discussion of evaluation of protein *quality*, it is critical at the outset that a distinction be made between the two major objectives of protein *quality* evaluations. The first objective is to permit ranking of proteins according to their potential nutritive value and to permit detection of changes in nutritive value that may occur during processing, storage or preparation of foods. The other is to permit prediction of the contribution a food protein or mixture of food proteins makes toward meeting nitrogen and amino acid requirements for growth or maintenance of people of various ages. Many discussions of methods for measurement of protein *quality* founder because of failure to distinguish clearly between these objectives.

The approaches that can be used effectively in achieving the objective of comparing proteins for their potential nutritive value are usually, although they need not be, distinctly different from those used in assessing the *efficacy* of proteins for improving the protein nutriture of human subjects. For the first, all that is needed is a simple standardized method for determining the ability of proteins to promote growth or maintain nitrogen retention of a test organism. It must rank proteins consistently and reproducibly. To achieve the second objective, that of assessing the *efficacy* of proteins for improving protein nutriture of human subjects, requires knowledge of human amino acid requirements, of the amino acid composition of both the diet and the supplement under consideration, together with knowledge of any unavailability of critical amino acids in both the diet and the supplement (Harper, 1977).

IMPORTANCE OF PROTEIN QUALITY

Before going farther, we must deal with the question of whether it is important to evaluate foods and diets for protein *quality* in the United States or other countries where average per capita consumption of protein and specific amino acids exceeds considerably the accepted requirements. I am tempted to respond to this question by asking: "Is there a need for dietary standards and knowledge of the nutrient content of foods in countries where nutritional deficiency is rare?" If we believe that it is important to monitor the nutritional *adequacy* of the food supply and the *adequacy* of nutrient intake of the population, then it is necessary to have information about the nutritional *quality* of proteins in terms that permit prediction of the *efficacy* of different food proteins in meeting human nutritional needs.

I would then hasten to *qualify* this statement by asserting that classical protein *quality* measurements are of only limited value for these purposes. The objective of monitoring the food supply and its nutrient content is to assess the *adequacy* of nutrient intakes of the population and to detect trends in intake that may indicate the possibility of nutritional problems developing. Before human requirements for water soluble vitamins had been *quantified*, it was valuable to know that some foods were better sources of these nutrients than others. Once the requirements had been *quantified*, however, *adequacy* of the food supply could be estimated much more accurately by determining how closely the *quantities* of nutrients in the food supply or in the diet corresponded with the *quantities* required. The need for protein is a need for a specific *quantity* of α-amino nitrogen and for specific *quantities* of individual amino acids. Therefore, for assurance of the *adequacy* of the food supply, and to assess the *adequacy* of intakes of individuals, we need to know how closely the amounts of α-amino nitrogen and of the specific essential amino acids in the food supply or in diets correspond with the amounts needed (NAS, 1974A, pp. 167–183). Knowledge of the average amount of protein consumed by the population in relation to Recommended Dietary Allowances is not an appropriate substitute for this type of information.

The information we have about protein intakes and protein status in the United States indicates that most people consume more than an adequate amount of protein. There are individuals with low protein intakes but observations showing that low concentrations of serum albumin are observed less frequently than low protein intakes indicate that inadequate protein nutrition is not a health problem that is of general concern.

Individuals most likely to have inadequate protein intakes are those whose food intake is low owing to illness or debility associated with aging, or to mental or physical defects. It is unlikely that persons who are consuming foods in quantities that provide an adequate amount of energy (calories) will have an inadequate intake of protein. Nevertheless, as energy expenditure and energy intake fall, as they have been doing in the U.S., it becomes increasingly important to ensure that the quality of protein in the diet is high and that knowledge of the relationship between protein quality measurements and their use in ensuring that amino acid needs have been met, is accurate.

PROTEIN QUALITY AND HUMAN AMINO ACID REQUIREMENTS

If we wish to monitor the *adequacy* of diets and the *adequacy* of protein nutriture of the population, we must also give some thought to the appropriateness of the dietary standards. To use "protein" as the standard for meeting α-amino nitrogen and amino acid needs is not much more specific than using a liver extract that has been evaluated for its *adequacy* in meeting vitamin needs as the standard for evaluating the vitamin status of the food supply.

It is between 20 and 30 years since estimates of human amino acid requirements were made and almost 25 years since an expert committee of FAO proposed the use of an amino acid scoring procedure for evaluating diets for their ability to meet human amino acid needs (FAO, 1957). The information on this subject has been reviewed quite thoroughly twice within the past few years (FAO/WHO, 1973; NAS, 1974A). It would therefore seem to be time to give serious consideration to dealing with amino acids in dietary standards exactly as we deal with other essential nutrients. A small step in this direction was made in the RDA bulletin (NAS, 1974B) by including in the protein section a tentative standard amino acid pattern against which it was proposed that diets or food proteins could be evaluated.

There have been criticisms that amino acid requirements are not accurate, that amino acid analyses of proteins leave much to be desired and that we know little about amino acid availability. I would suggest that from a practical viewpoint, these criticisms are red herrings and that: 1) the estimated amino acid requirements of infants, upon which the amino acid scoring patterns are based, are more accurate than estimates of requirements for several nutrients, such as vitamins A, D and E and minerals such as calcium and zinc; 2) the analytical methods for amino acids, despite the limitations imposed by destruction of amino acids during hydrolysis of food proteins, are at least as reliable as those for folic acid, some of the trace minerals and vitamin E; 3) knowledge of the availability of amino acids is no less than that of the availability of iron, folic acid and vitamin B_6. The accuracy of predictions of the amounts of proteins required to meet amino acid needs of children bear this out (Arroyave, 1974; Pineda et al. 1981, Chapter 3). Establishment of RDA's for methionine, lysine and tryptophan, and possibly threonine, the few amino acids that are likely to be low in human diets, should serve as a stimulus for research on availability and analysis of amino acids and should provide an acceptable basis for estimating the ability of mixtures of proteins to meet human requirements for amino acids.

At this point I would conclude, firstly, that the time has come in evaluating diets for protein *adequacy*, to establish RDA's for the amino acids that are most likely to be limiting in human diets; and, secondly, that in assessing the *adequacy* of protein nutriture, the amount of nitrogen and the amounts of the most critical amino acids being consumed in the diet should be measured.

PROTEIN QUALITY AND REGULATORY OR OTHER PURPOSES

This brings us to the question of the use of protein *quality* measurement for nutrition labeling, for regulatory purposes and for ensuring that protein *quality* has not deteriorated during processing. I would begin by assuming that the major reason for regulation and labeling is to provide consumers with assurance about the *quality* of the protein in food products and to provide the basis for corrective action if the protein does not meet some acceptable standard. Secondly, I would recommend that the label not be viewed as an effective tool for nutrition education and would reiterate that classical protein *quality* measurements do not provide information that is of much use in assessing or in attempting to improve protein nutriture.

The implication of these assumptions is that any simple reliable method for assessing protein *quality* will serve to provide quality assurance for consumers and for regulatory purposes. The protein *quality* measurement used should provide values that are readily understandable and the values reported for different proteins should, within narrow limits, be directly proportional to each other. These criteria would eliminate from consideration immediately the "protein efficiency ratio" method as it is currently used. I would agree with McLaughlan (1979) that modification of the current PER method by including adjustment for the weight loss of a group of animals fed a protein-free diet, so that "net protein ratio" can be calculated is the simplest procedure. It should satisfy the criterion of providing values that are directly proportional to each other within acceptable limits. I would go one step farther, and propose that values should be related to those of the best *quality* proteins so that they can be expressed in units between zero and 100. This is essentially the "relative net protein ratio" procedure. For the standard of comparison I would use methionine supplemented casein, as it approaches in nutritional *quality* the highest *quality* food proteins and is a protein that is generally of uniform nutritional *quality*. This method could be used for foods, purified proteins or food analogues. The values reported would permit direct comparison of the *quality* of proteins relative to the highest *quality* proteins in simple easily understood terms. It would not, however, permit comparisons of proteins for their *efficacy* in meeting human nutritional needs unless the product represented the entire diet. Even then the value would not permit calculation of the amount of protein required to meet a standard such as the RDA because protein *quality* measurements using animal assays are made under conditions in which protein is the limiting component of the diet. Under such conditions protein is used with greater efficiency than when it is consumed in an amount that meets the requirement for growth or maintenance.

CONCLUSIONS

In conclusion, I would reiterate that it is crucial to distinguish clearly between the two major purposes of protein *quality* evaluation: one, that of ranking proteins according to their potential nutritive value; and, two, that of assessing the *efficacy* of proteins in meeting human requirements for amino acids.

A simple, reliable bioassay will serve to detect deterioration of a product during processing and to provide assurance of the *quality* of proteins for most regulatory purposes. Methods used for these purposes should provide values for protein *quality* that are directly proportional to each other over the range expected in foodstuffs, with the value for the highest *quality* proteins being set at 100. The simplest method that meets these criteria is the "relative net protein ratio" method. Such methods cannot be used to predict the value of proteins in meeting human protein requirements.

The effectiveness of food proteins or diets in meeting human requirements for amino acids and total protein can be determined only by amino acid scoring procedures. For this, it is necessary to have a reliable standard amino acid scoring pattern based on human amino acid requirements against which food proteins can be evaluated, and to know the nitrogen content and amino acid composition of the proteins. Satisfactory amino acid scoring patterns have been developed. Results obtained so far in evaluating diets by scoring procedures suggest that low availability of amino acids is not a serious limitation of these methods with many foods. Nevertheless, simple, reliable methods for assessing the bioavailability of the amino acids in foodstuffs are needed to provide for the widest application of amino acid scoring methods.

REFERENCES

ARROYAVE, G.   1974.   Amino acid requirements by age and sex. *In* Nutrients in Processed Foods—Proteins, P.L. White and D.C. Fletcher (Editors). Publishing Sciences Group, Inc., Acton, MA., pp. 15–28.

FAO (FOOD AND AGRICULTURE ORGANIZATION). 1957. Protein requirements. FAO Nutr. Stud. No. 16, 52 pp. FAO, Rome.

FAO/WHO (FOOD AND AGRICULTURE ORGANIZATION/WORLD HEALTH ORGANIZATION). 1973. Energy and protein requirements. WHO Technical Report Series No. 522, FAO Nutrition Meetings Report Series No. 52. FAO/WHO, Geneva.

HARPER, A.E. 1977. Human amino acid and nitrogen requirements as the basis for evaluation of nutritional quality of proteins. In Food Proteins, J.R. Whitaker and S.R. Tannenbaum (Editors). AVI Publishing Co., Inc., Westport, CT., pp. 363–386.

McCOLLUM, E.V. 1957. A History of Nutrition. Houghton Mifflin Co., Boston, pp. 220–223.

McCOLLUM, E.V., ORENT-KEILES, E. and DAY, H.G. 1939. The Newer Knowledge of Nutrition, 5th Edition. The Macmillan Co., New York.

McLAUGHLAN, J.M. 1979. Critique of methods for evaluation of protein quality. In Soy Protein and Human Nutrition, H.L. Wilcke, D.T. Hopkins and D.H. Waggle (Editors). Academic Press, New York, pp. 281–297.

MITCHELL, H.H. 1924. A method of determining the biological value of protein. J. Biol. Chem. 58:873.

MITCHELL, H.H. 1954. The dependence of the biological value of food proteins upon their content of essential amino acids. Deutschen Akademie der Landwirtsch. zu Berlin, Vol. V/2: 279–325.

NAS (NATIONAL ACADEMY OF SCIENCES). 1974A. Improvement of Protein Nutriture. Committee on Amino Acids, Food and Nutrition Board. NAS/NRC, National Academy of Sciences, Washington, D.C.

NAS (NATIONAL ACADEMY OF SCIENCES). 1974B. Recommended Dietary Allowances, 8th edition. Committee on Dietary Allowances, Food and Nutrition Board. NAS/NRC, National Academy of Sciences, Washington, D.C.

OSBORNE, T.B., MENDEL, L.B. and FERRY, E.L. 1919. A method of expressing numerically the growth-promoting value of proteins. J. Biol. Chem. 37:223.

PINEDA, O., TORUN, B., VITERI, F. and ARROYAVE, G. 1981. Protein quality in relation to estimates of essential amino acid requirements. In Protein Quality in Humans: Assessment and In Vitro Estimation, C.E. Bodwell, J.S. Adkins and D.T. Hopkins (Editors). AVI Publishing Co., Westport, CT.

THOMAS, K. 1909. Ueber die biologische Wertigkeit der Stickstoffsubstanzen in verschiedenen Nehrungsmitteln; Beitrage zur Frage nach dem physiologischen Stickstoffminimum. Arch. Physiol. 219–302. (German)

C. l

DISCUSSION

DR. FRIEDMAN: I would like to ask for comments from anyone whether consideration has been given to nutritional or health-related consequence of excess protein intake. Would someone care to comment?

DR. BODWELL: Dr. Linkswiler's group and the Berkeley group have found some evidence that there's a deleterious effect of high intakes of protein on calcium retention in humans, but this is not confirmed by all studies and I think this is somewhat undecided. Other than that, if you go to the rat, of course, there could be renal damage and so forth but I don't think that's been demonstrated in humans. Dr. Scrimshaw might want to comment on that later.

The question I had for Dr. Harper is, I thought in the middle of your talk, you were rolling along very nicely toward using amino acid composition data in some way, in reference to nutritional labeling, as opposed to rat growth assays but then you ended up right back with the rat assay. Would you care to comment? Where are you headed?

DR. HARPER: I am still concerned about the different uses that are made of protein quality measurements. I see no other way of assessing the value of a food for human nutriture accurately than by determining alpha-amino nitrogen content and amino acid content. For many labeling regulations and regulatory actions, we need not be concerned with that much detail about protein nutriture but only with the comparative value of different products. If you have a series of heat-treated products that may have suffered damage, all you need is a method that will permit you to detect an effect and take action to correct it. A method that gives values between 0 and 100, such as relative net protein ratio, so people can understand that the protein is 100% or 70% of the value of the best quality proteins, is probably all that is needed for many purposes. I wonder how many people calculate their amino acid intake. They may want amino acid data to see if certain proteins complement each other but that's done mainly by dietitians rather than by the general public who would not have available to them the amino acid analyses. We need something comprehensible for the public and at the same time, valuable for nutritionists and dietitians.

DR. EDWARDS: I indicated I wanted to make a comment, not to ask a question. I take issue with the philosophy that just because the majority of Americans exceed the recommended allowance for protein, it is not a problem that we should be concerned about. And I'll give you about four reasons that I take issue with that philosophy.

First, the Hanes data cover the period from 1971−74 and there's definitely a trend that there is an increase in teenage pregnancies. As you noted, the females on the slide below the Hanes cut-off point and the RDA requirements was 30% in one case, 60% in one case, as the age went from 15 to 17 and on out. Indeed, the number of teenage pregnancies is increasing in 1978, 1979, 1980, and probably in the next decade because our mores are changing. The new generation is a little different from our generation, and we can expect that this will have definite and serious repercussions, particularly because unemployment is also increasing among teenaged youth.

Now, the other thing that's very important, I think, is to recognize that when inflation hits families, it hits the low-income families worst of all. Therefore, persons who are using the food dollar will change purchases from protein such as meat and eggs to beans and peas and spaghetti and macaroni and you have the wheat products and the lysine deficiencies and so on and so forth coming into play in greater proportion than you would normally assume to be present. I have a feeling that this has a definite correlation to the high incidence of nutritional anemia because many of our protein foods such as meat and eggs are excellent carriers of iron. We're running into a situation now that may not have existed to the same extent when the Hanes survey was made.

Now, definitely there is a correlation between dietary protein of a mother and intellectual development and here again, we see some definite consequences of inadequate protein intake at the level of the mother.

I would say, finally, that the reason that there is so much obesity among low-income groups, I said obesity, is because low-income groups use their food dollar for high carbohydrates that are usually low in protein and so this shouldn't fool us. It's a different kind of obesity and it's called the obesity of poverty. If this group would focus on these kinds of issues and see that the vulnerable groups are the groups that we need to address our concerns to, rather than those of us who are well off, then I think the objectives of the conference can be well served.

DR. SCRIMSHAW: As to the question of excess protein, it would seem that if there were any dangers of damage to normal, healthy people, this would have been demonstrated long ago. Furthermore, the records of individuals in societies with very high protein intakes fail to reveal problems. These include not only traditional Eskimos and populations in Argentina and Uruguay, but hunters and gatherers who gorge themselves on a large kill and eat enormous quantities of meat before carrying the rest back to the village. The suggestion that dietary protein excess can be a practical problem for populations is simply not supported by evidence. Of course, with kidney disease or the congenital metabolic disorder, phenylketonuria, protein intakes must be limited.

DR. MARGEN: I want to take issue with Dr. Scrimshaw if I might. The first thing that I'd like to mention is the fact that as you stated there are populations who have been on high protein diets and demonstrated no deleterious effects. I think that the populations that you are referring to are generally, the hunters and gatherers, and these are not necessarily consistent high protein eaters. They usually will go on protein splurges. I think, however, that the work in Dr. Linkswiler's and our laboratory at least calls for some caution in the matter of a high protein diet, with respect to the effects of protein on calciuria as Dr. Bodwell mentioned. That is, even on a low- or a high- calcium diet, an increment of protein of about 75 grams leads to about a 100% increase in the amount of calcium excreted. If an individual is on a low-calcium diet, the only place from which this can come is bone. In view of the relatively low-calcium diets which are prevalent in our country, I think that we have to be rather alert to this problem. The reason I state this is, that the amount of protein which is consumed by many of our adults, in the presence again of a relatively low-calcium diet may be significant. But there's another factor, and I don't want to get involved with that this evening, and that has to do with a different way of looking at protein requirements. That is, looking at it in a rather dynamic way that Dr. Sukhatme and I have been doing (Amer. J. Clin. Nutr. 31:1237, 1978), and examining the mathematical relationships, if you like, and the pattern of protein regulation. If one examines this, one finds that there is a regulation of protein metabolism within a certain range. And this we call a stochastic regulated process. If the amount of protein falls too low or the amount of protein increases, this regulated pattern starts undergoing disorganization. From the physiological point of view, it indicates the lack of the organism to adapt and indicates again a condition of stress and serious potential threat.

I would like to comment also, if I may for a moment, on Dr. Edwards' remarks because I think that what she has pointed out is something which is extremely important. That is, that we have to examine the problems of nutrition in our society, particularly in the deprived and poor in our society, and the potential impact which both inflation and our social policies may have. The potential impact which the alteration in dietary pattern induced by socio-economic factors has among these people today is virtually un-known. However, I do think it is very important that, as we examine them, we also be very careful in trying to separate, in part, the scientific facts from the social implications and then to bring them together and merge them. I think that we as scientists have the right to debate this issue entirely in the scientific sense but the social policy decisions must be made by our policy makers. We must be extremely cautious in what we say in this arena, although in our democracy we may freely criticize their conclusions. We have to be extremely scientific birds, but we have to meld with some aspects of social concern.

DR. SCRIMSHAW: Just to clarify the preceding point regarding possible calcium imbalance: The point I was making referred to the lack of problems with high protein intakes when diets are adequate in other nutrients. With high levels of any food or any nutrient, you run the risk of imbalance. The possibility of protein-imbalance is no exception.

DR. VANDERVEEN: I would like to focus attention on the point that Dr. Harper had made. Indeed, a majority of the protein labeling presently deals with nutrition labeling. However, in some cases PER measurements are used for other regulatory functions. For example, in infant formulas such measurements are used for a quality cut-off point. Either the quality is there or it is not there. If the quality cut-off is not attained, then the product cannot be called suitable for infants' diets. Similarly, in the case of the tentative final order for vegetable protein products, there are cut-offs. In both cases, quality standards are theoretically based on human requirements. However, casein is used as the standard and rat growth is used as the measure. In both cases, the PER value is used to determine whether the product is marketable or whether it can be called a substitute or an imitation food. So present protein quality measurements do both. In some cases, it's used as a comparison such as for nutrition labeling, but in other cases protein quality measurements are used for standards.

3

Protein Quality in Relation to Estimates of Essential Amino Acids Requirements

O. Pineda, B. Torún, F. E. Viteri and G. Arroyave

According to the FAO/WHO Expert Committee on Energy and Protein Requirements convened in 1971 (FAO/WHO, 1973) the quality of a protein can be estimated from its essential amino acid composition and the safe protein intake at any age with the aid of a scoring pattern derived from amino acid requirement values. Since the requirements for total nitrogen and essential amino acids decline with age (Cohn *et al.*, 1976; Waterlow and Stephen, 1967), the "quality" of a protein must also be related to the age of the subject who consumes it (Arroyave, 1975a; 1975b). This emphasizes the need for precise estimates of amino acid requirements at different ages. Until now, experimental data on amino acid requirements existed only for infants, 10–12 year-old children and adults. No experimental data were available for children of preschool age.

For infants, the data of Holt and Snyderman (1967) and of Fomon and Filer (1967), who used synthetic diets or a variety of infant formulas, provided information on the amounts of essential amino acids required to maintain expected growth rates and nitrogen retentions. The FAO/WHO Committee decided to combine the lower values in the two sets of data to estimate the amino acid requirements for infants under 6 months of age. From these data, an amino acid scoring pattern was calculated in terms of mg of amino acid per g of protein. A protein with an equivalent amino acid composition to that of the scoring pattern would satisfy the infant's amino acid needs if ingested at a "safe level of intake" of 2 g/kg/day (FAO/WHO, 1973).

For children 10–12 years of age, the essential amino acid requirements used are those proposed by Nakagawa *et al.* (1960, 1961a, 1961b, 1962, 1963). These values were derived from nitrogen balance data and were based on the lowest intake required to produce a positive nitrogen balance in all subjects studied (3 to 4 children per amino acid). The children ate mixtures of pure essential amino acids with glycine or mixtures of non-essential amino acids as the source for non-essential nitrogen. However, there is some uncertainty about whether to consider the proposed figures as requirement values since the intake increments used in the studies were large and, therefore, the levels which gave positive nitrogen balances were not necessarily the lowest possible.

Information on amino acid requirements for adults is available from several sources. Rose (1957) determined the amounts necessary to maintain positive nitrogen balance, using diets composed of pure essential amino acids with glycine or urea as non-essential N sources. Since the variability among subjects was large, the level selected was that of the man with the highest requirement. Leverton's studies with young women (1959) used as the discriminating criterion the attainment of nitrogen equilibrium by all subjects. Using the data from women available until 1963, Hegsted (1968) published values derived by regression analysis. Although these figures have been used extensively, the FAO/WHO Committee (1973) suggested figures which were lower than Rose's but higher than Hegsted's.

In summary, experimental data related to essential amino acid requirements were available only for certain age groups. Furthermore, the criteria used in the interpretation of these data were based only on nitrogen balance studies or on adequate weight gain of infants. There was no information from studies in which other indicators were evaluated simultaneously to attain a more precise estimate of amino acid requirements.

The FAO/WHO Expert Committee concluded that a standard amino acid scoring pattern could be established to assess a protein's nutritive quality relative to the pattern. It was accepted that breast milk was the appropriate food for infants; and, therefore, their amino acid requirements "should be excluded from the application of any tentative guide to protein scoring that might be developed for older children and adults" (FAO/WHO, 1973). The Committee also agreed that any such reference pattern should satisfy the needs of preschool children since a protein that was adequate for them would also be adequate for adults, whereas the reverse might not be true. With these points in mind, a provisional reference amino acid scoring pattern was devised for application to all ages beyond infancy, based on *theoretical estimates* of the amino acid requirements of preschool children which were derived from the information available for other age groups and a safe level of protein intake of 1.2 g/kg/day. It is important to emphasize that, until now, there was no direct experimental evidence on the amino acid requirements of young growing children beyond infancy (i.e., children of preschool age).

Since the provisional pattern is still used without change, it is important to establish whether the estimates of amino acid requirements used to recommend that pattern are adequate. The similarity of the provisional pattern with that recommended to fulfill infants' requirements (FAO/WHO, 1973) suggests that the needs for certain essential amino acids may be overestimated in the former. If this were the case, the result would be an underestimation of protein quality based on the reference amino acid scoring pattern.

MEASUREMENT OF AMINO ACID REQUIREMENTS FOR CHILDREN OF PRESCHOOL AGE

Investigations were carried out at the Institute of Nutrition of Central America and Panama (INCAP), designed to provide *experimental* information on essential amino acid requirements for 2-year old children using several criteria of evaluation. We studied a total of 42 children ranging in age from 21 to 27 months. These children had been malnourished but they had fully recovered at least one month prior to the beginning of the investigation. During that time they were healthy and growing at adequate rates. At the beginning of the studies, their weight was over 95% of that expected for their height, their creatinine height index was above 0.90 (Viteri and Alvarado, 1970) and they had normal serum total proteins, albumin and amino acid patterns and normal hematologic indices. During the course of the study, the children were maintained under close medical supervision to ensure that their health and growth were not impaired.

Throughout the study, all children received dietary nitrogen (N) in amounts equivalent to 1.2 g of protein/kg/day. The diets consisted of a core of 0.3 g of cow's milk protein per kg per day plus a mixture of purified amino acids in the same proportions and amounts found in 0.9 g of milk protein/kg/day. Table 3.1 shows the amino acid contents of this semi-synthetic diet.

TABLE 3.1.   CALCULATIONS FOR THE PREPARATION OF THE EXPERIMENTAL DIETS

			Amino acid composition of the basal semi-synthetic diet	
Amino acid	mg/g of milk protein[1]	mg/1.2 g of milk protein	Milk core: mg/0.3 g of milk protein	+ mg of purified amino acids equivalent to 0.9 g milk protein
Isoleucine		47	56	14	42
Leucine		95	114	28	86
Lysine		78	94	23	71
Total sulfur a.a.[2]		33	40	10	30
Methionine	25		30	8	22
Cystine	8		10	2	8
Total aromatic a.a.[3]		102	122	31	91
Phenylalanine	54		65	16	49
Tyrosine	48		58	14	44
Threonine		44	53	13	40
Tryptophan		14	17	4	13
Valine		58	70	17	53
Histidine		27	32	8	24
Alanine		35	42	10	32
Aspartic Acid		77	92	23	69
Glutamic Acid		222	266	67	199
Glycine		20	24	6	18
Proline		91	109	27	82
Serine		58	70	17	53

[1]FAO/WHO (1973) and FAO (1970).
[2]Met/Cys = 3.05
[3]Phe/Tyr = 1.13

The composition of the basal diet is that shown in the last two columns. The essential amino acid under study was partially replaced by glycine in order to provide the essential amino acid at five different levels of intake. These levels varied between "maximum" (equivalent to that in 1.2 g of milk protein/kg/day) and "core" (equivalent to that in 0.3 g of milk protein/kg/day). For example, in the case of lysine, the following levels were provided by the semi-synthetic diets:

Milk	Lysine		Total Lysine

1) Core + 71 mg = 94 mg, equivalent to that in 1.2 g milk protein/kg/day
2) Core + 43 mg = 66 mg, equivalent to that in 0.85 g milk protein/kg/day
3) Core + 30 mg = 53 mg, equivalent to that in 0.68 g milk protein/kg/day
4) Core + 16 mg = 39 mg, equivalent to that in 0.50 g milk protein/kg/day
5) Core + 0 mg = 23 mg, equivalent to that in 0.30 g milk protein/kg/day

A similar procedure was followed with five other essential amino acids. Table 3.2 shows the amounts of each amino acid fed at the five levels of intake. The amount of total nitrogen was kept constant with changes in the dietary glycine using isonitrogenous equivalents as shown in Table 3.3.

All diets provided 100 kcal/kg/day with 30% of the energy from vegetable oil. They were complemented with vitamins, minerals, and electrolytes to satisfy the children's needs. The diets were fed as 5 identical meals per day.

TABLE 3.2.   ESSENTIAL AMINO ACID INTAKES FOR EACH AMINO ACID UNDER STUDY

Amino acid	mg of amino acid per g of milk protein	Amino acid intake levels, mg/kg/day[1]				
		A	B	C	D	E
Isoleucine	47	14	24	32	40	56
Lysine	78	23	39	53	66	94
Methionine + Cystine	33	10	16	22	28	40
Threonine	44	13	22	30	37	53
Tryptophan	14	4	7	10	13	17
Valine	58	17	29	39	49	70

[1] Equivalent to the amino acid contents of 0.3 (A), 0.5 (B), 0.68 (C), 0.85 (D) and 1.20 (E) g of milk protein/kg/day.

TABLE 3.3.   ISONITROGENOUS EQUIVALENTS OF GLYCINE TO COMPENSATE FOR CHANGES IN THE DIETARY ESSENTIAL AMINO ACIDS

Amino acid	Nitrogen content, %	mg of glycine used to substitute for each mg of the essential amino acid
Isoleucine	10.68	0.57
Lysine	19.16	1.03
Methionine	9.39	0.50
Cystine	11.66	0.62
Threonine	11.76	0.63
Tryptophan	13.72	0.74
Valine	11.96	0.64
Glycine	18.66	1.00

Experimental Periods and Sequences

The requirements for total sulfur amino acids, isoleucine, lysine threonine, tryptophan, and valine were studied, each with six children. In the investigation of each amino acid, three children followed a "descending" and three an "ascending" experimental design as described below. Prior to the beginning of the study, all the children ate a milk-based diet which provided between 2 and 3 grams of milk protein and 100 kcal per kg/day.

Descending Design

The children received a milk formula which provided 1.2 g protein/kg/day for nine days. They were then progressively adapted to the milk plus amino acid diet (basal diet) described above. This adaptation period lasted 6 days during which the milk diet was gradually replaced by the basal diet; thus, by the 6th day, the children ate exclusively the basal diet. The children remained with this diet for 9 days. During the last 4 days of the 9-day period, excreta were collected for analyses of nitrogen and determination of nitrogen balance and for measuring other biochemical components in urine (see below). After that, the level of the specific amino acid under study was decreased every 9 days in four sequential steps until the content of the amino acid was equivalent to that of 0.3 g of milk protein/kg/day. Studies were conducted during the last 4 days of each period. In total, five 9-day periods were involved per essential amino acid.

Ascending Design

This was similar to the descending design but the children started with the lowest level of intake of the amino acid being studied (i.e., the level contained in 0.3 g of milk protein/kg/day) which was then increased every 9 days until the diet contained an amount of the amino acid under study equivalent to that of 1.2 g milk protein/kg/day. Except for the specific amino acid under study, the concentration of all other essential and non-essential amino acids was kept constant based on the composition shown in Table 3.1.

Evaluation Variables

1. Nitrogen Balance: Nitrogen excretion was measured in a 4-day fecal collection and in individual 24-hour urine collections throughout the same 4 days. The nitrogen content of the diet was also measured in each balance period. Balance was calculated as intake minus urinary and fecal excretions and minus 8 mg N/kg/day to account for integumental and other miscellaneous losses (Viteri and Martinez, 1981).

2. Urinary excretion of urea and creatinine: These were determined in the four 24-hour specimens collected during each balance period.

3. Creatinine-Height Index (CHI): CHI was calculated from the child's

height, which was measured at 14-day intervals (Viteri and Alvarado, 1970), and from the mean daily creatinine excretions.

4. Blood amino acids: Free amino acids in plasma and erythrocytes were determined after 12 hours of overnight fast and again 4 hours after breakfast, by gas liquid chromatography of the trimethylsilyl derivatives (Gehrke and Leimer, 1970; Gehrke and Takeda, 1973).

RESULTS AND DISCUSSION

A detailed assessment of the evaluation variables, which included the analysis of various ratios of the free amino acids in plasma and erythrocytes and the changes from fasting to post-prandial values, indicated that only the plasma concentration of the free amino acid under study after an overnight fast, the nitrogen balance and the urinary urea/creatinine ratio changed in relation to the amino acid intake. The criteria used to determine amino acid requirements were:

1. Free amino acid concentration in plasma after an overnight fast: the lowest level of amino acid intake *before* the plasma concentration of that amino acid showed a consistent tendency to decrease, which suggests a decrement in dietary amino acid adequacy.

2. Nitrogen balance:

a) Lowest level of amino acid intake which allowed *all* children to retain at least 16 mg N/kg/day, which is the nitrogen requirement for growth in this age group (FAO/WHO, 1973).

b) Lowest level of amino acid intake *before* the mean N balance showed a consistent tendency to decrease, which suggests a decreased utilization of dietary amino acids and/or an increased catabolism of endogenous protein.

3. Urea/creatinine ratio in urine collected over the 4-day period.

a) Lowest level of amino acid intake which allowed *all* children to maintain a urinary urea/creatinine ratio lower than 2 standard deviations above the mean value observed in all children consuming the basal diet (i.e., lower than 2.49 + 2 (0.90) = 4.29).

b) Lowest level of amino acid intake before the mean urinary urea/creatinine ratio showed a consistent tendency to increase, which suggests an increment in deamination of endogenous amino acids.

These criteria allowed a safety margin above the mean dietary amino acid requirements since they identified the level of intake before deleterious changes took place. That margin of safety was increased by the fact that when the various indicators suggested two different levels as the lowest adequate intake, the higher of the two intake levels was selected on the basis that when dietary requirements are established, it is safer to err by excess than by deficit.

The criteria described under headings 1, 2a, and 3b, which involved the beginning of a definite pattern of change in plasma amino acid concentration, nitrogen balance and urinary urea/creatinine ratio, respectively, might in some instances depend on the subjective assessment of the person who interpreted them. Therefore, the four investigators involved in the study interpreted the results independently. They all agreed almost always and in the few occasions when they did not, the judgement of three prevailed over that of a lone dissenter.

The results obtained with lysine will be presented in detail as an example of the selection criteria. Figure 3.1 shows N balance corrected for integumental and other miscellaneous losses. Only 5 children were studied at the lowest level of intake. Also, the data of one child who had diarrhea during the balance period with an intake of 66 mg lysine/kg/day were not included in the analysis. The figure shows that with lysine intakes below 53 mg/kg/day some children retained less than 16 mg N/kg/day. The trend toward a decrease in N retention could be interpreted to begin with lysine intakes below 66 mg/kg/day based on the mean values. If the individual variability is taken into account, the trend begins with intakes below 53 mg/kg/day. Figure 3.2 shows that the mean concentration of free lysine in plasma began decreasing with dietary intakes below 66 mg/kg/day. Figure 3.3 shows that the mean of the urea/creatinine ratios in urine began increasing with intakes below 66 mg/kg/day, coinciding with the appearance of ratios above 4.19 in some children.

The nitrogen balance data suggested that an intake of 53 mg lysine/kg/day was adequate for all the children studied, whereas the other indicators suggested that some children may need higher intakes. As mentioned above, the criteria which prevailed were those that allowed a greater safety margin, even at the risk of overestimating the requirement level. Therefore, the dietary intake recommended for lysine was defined as 66 mg/kg/day.

The same procedures were followed to interpret the results obtained for the other amino acids under investigation. Table 3.4 shows the recommended levels of intake for the six essential amino acids that were studied, expressed as mg of each amino acid or as milk protein equivalents (i.e., the amount of milk protein which contains the corresponding amount of the essential amino acid). Based on obligatory fecal N losses of 20 mg/kg/day (Torún et al., 1981a) the mean "true" digestibility of the semi-synthetic diet was 97.3%. Table 3.4 also shows the amounts of essential amino acids absorbed at the recommended levels of intake, assuming that all amino acids were absorbed in the same proportion of 97%.

In the case of threonine and tryptophan, the various indicators suggested that the lowest adequate level of intake was either 0.85 or 1.20 g, of milk protein/kg/day. Since the difference between these two levels of intake is large (34%) while the other discrete intervals of intake below 0.85 g/kg/day were smaller, it was decided to use a range rather than a specific value as a first approximation of the recommended intake level for threonine. This range is shown in Table 3.4.

However, studies on protein requirements of similar children recently carried out at INCAP and discussed in another section of this Conference (Torún et al., Chapter 20) supported our conclusions and allowed us to decide upon a specific value instead of a range of values for the threonine recommendation. Table 3.5 shows the amino acid contents of the safe levels

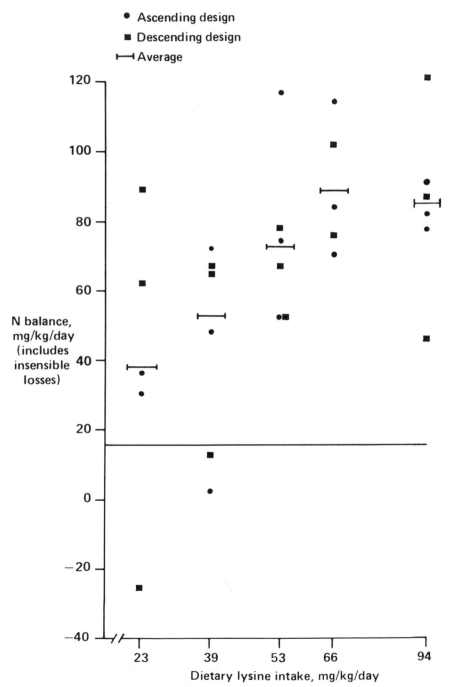

FIG. 3.1.   NITROGEN BALANCE DETERMINED IN CHILDREN FED DIFFERENT LEVELS OF LYSINE; SEE TEXT FOR EXPLANATION OF ASCENDING OR DESCENDING DESIGN

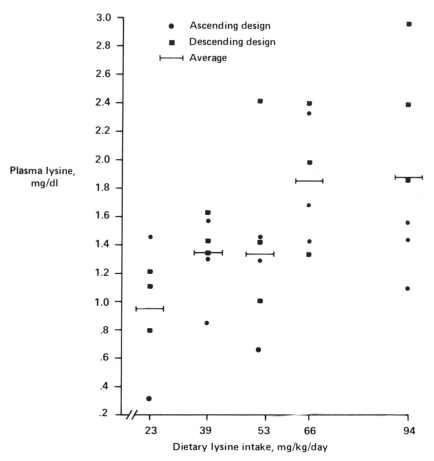

Incap 80–331

FIG. 3.2.   FASTING LEVELS OF PLASMA FREE LYSINE IN CHILDREN FED DIFFERENT LEVELS OF LYSINE

TABLE 3.4.   AMINO ACID INTAKES RECOMMENDED FOR PRE-SCHOOL CHILDREN (2-YEARS OLD), INCAP 1980

Amino acid	Amino acid intakes expressed as milk protein equivalents, g/kg/day	Amino acids (mg/kg/day) ingested	absorbed[1]
Isoleucine	0.68	32	31
Lysine	0.85	66	64
Methionine + Cystine	0.85	28	27
Threonine	>0.85, <1.20	>37, <53	>36, <52
Tryptophan	0.85	13	12.5
Valine	0.68	39	38

[1]Corrected for "true" digestibility of 97%.

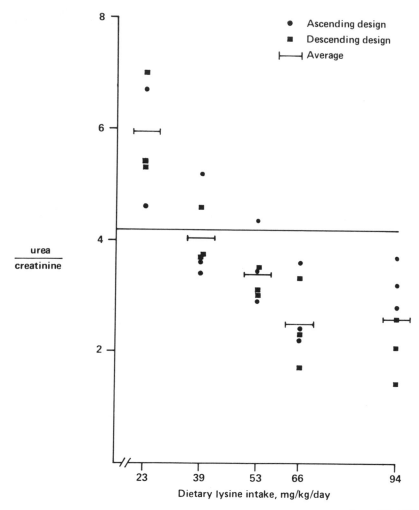

FIG. 3.3.   UREA/CREATININE RATIO VALUES CALCULATED FROM ANALYSES OF 96-HOUR URINE COLLECTIONS (4-DAY BALANCE PERIODS) FROM CHILDREN FED DIFFERENT LEVELS OF LYSINE

of protein intake determined with cow's milk and a soybean protein isolate. They are similar to the lower limit of the threonine range. Therefore, it was decided to use 38 and 14.4 mg/kg/day, respectively, as the recommended levels of intake. These figures must be corrected for the "true" digestibility of the dietary protein sources, because if it is not the limiting amino acid in either milk or soy protein, the recommended level might be closer to 12 mg/kg/day.

TABLE 3.5.   COMPARISON OF RECOMMENDED AMINO ACID INTAKES (mg/kg/day) WITH THE CONTENTS OF SAFE LEVELS OF PROTEIN INTAKES FOR 2-YEAR OLD CHILDREN

Amino acid	Amino acids provided by		Recommended amino acid intakes
	0.79 g of milk protein/kg/day	0.98 g of soy protein isolate/kg/day	
Isoleucine	46	51	32
Lysine	66	61	66
Methionine + Cystine	30	26	28
Threonine	35	36	38
Tryptophan	13.8	16.3	13
Valine	52	52	39

The recommended levels of intake of the six essential amino acids investigated in this study and supported by the studies on protein requirements are 21–35% lower than the currently accepted recommendations of isoleucine, sulfur amino acids, threonine, and valine for 2-year old children, which are based on the FAO/WHO (1973) amino acid scoring pattern and a safe intake level of egg or milk protein of 2 g/kg/day (Table 3.6). The recommendation for tryptophan, on the other hand, is 8% higher although this may be an overestimation. These differences are important in terms of protein quality assessment since they may lead to changes in the amino acid scoring pattern for proteins, as discussed in detail in another section of this Conference (Torún et al., Chapter 20). Some proteins which are considered to be limited by their methionine or threonine contents may, in fact, have a better nutritive quality than present assessments suggest, provided that other essential amino acids do not limit their quality to the same extent as the currently accepted methionine (or threonine) related limitations. In contrast, many vegetable protein sources will still be considered of a low quality since lysine or tryptophan may be their limiting amino acid. The results of the present study does not indicate that the FAO/WHO estimates for these amino acids are excessive. Furthermore, the nutritional value of such proteins is impaired by relatively low digestibilities which implies that larger amounts of these proteins must be ingested to secure adequate levels of total nitrogen absorption, which is independent of their essential amino acid composition.

TABLE 3.6.   COMPARISON OF ESSENTIAL AMINO ACID INTAKES (mg/kg/day) RECOMMENDED BY FAO/WHO (1973) AND BY THIS STUDY FOR 2 YEAR-OLD CHILDREN

Amino acid	(A) FAO/WHO[1]	(B) This Study	(B) as % of (A)
Isoleucine	48	32	67
Lysine	66	66	100
Methionine + Cystine	42	28	67
Threonine	48	38	79
Tryptophan	12	13	108
Valine	60	39	65

[1] Based on the FAO/WHO (1973) amino acid scoring pattern and a safe level of protein intake equivalent to 1.2 g of milk or egg protein/kg/day.

SUMMARY

The FAO/WHO essential amino acid scoring pattern was based on the requirements of preschool children. Those requirements, however, were theoretical estimates derived from information on amino acid requirements of infants under six months of age, 10–12-year old children and non-pregnant adults. That information was based on nitrogen balance studies and on weight gains of infants. The present study investigated the requirements of children 21–27 months-old for isoleucine, lysine, methionine + cystine, threonine, tryptophan, and valine. Each amino acid was studied with six children who ate a mixture of cow's milk and synthetic amino acids which provided increasing (3 children) or decreasing (3 children) amounts of the specific amino acid under study at 9-day intervals. Among the various indicators explored, only the plasma concentration of the free amino acid under study after an overnight fast, the nitrogen balance and the urinary urea/creatinine ratio changed in relation to the amino acid intake. Daily requirements of all the children studied were satisfied by the absorption of 31 mg isoleucine, 64 mg lysine, 27 mg methionine + cystine, 36–52 mg threonine, 12.5 mg tryptophan and 38 mg valine/kg body weight. The corresponding requirements suggested by FAO/WHO in 1973 were 48, 66, 42, 48, 12 and 60 mg amino acid/kg. Further studies (Torún et al., Chapter 20) suggested 37 mg/kg/day as the requirement for threonine absorption, and confirmed that the FAO/WHO requirements for isoleucine, methionine + cystine, threonine, and valine are too high. Those studies also supported the conclusion that the values established in our investigations are similar to "recommended" or "safe levels" of intake rather than mean requirements. These new recommendations may lead to changes in amino acid scoring patterns and in the protein quality assessment of some protein sources.

ACKNOWLEDGEMENTS

These studies were conducted partly through the financial support of the United States' National Institutes of Health (NIH Grant R22 AM 17086).

REFERENCES

ARROYAVE, G.   1975a.   Nutritive value of dietary proteins: For Whom? Proc. 9 Int. Cong. Nutr., Mexico 1972, Vol. 4, p. 43. Karger Basel, Switzerland.

ARROYAVE, G.   1975b.   Amino acid requirements and age. In Protein Calorie Malnutrition. R.E. Olson (Editor).   Acad. Press, N.Y., p. 1.

COHN, S.H., VASWANI, A., ALOIA, J.F., ROGINSKY, M.S., and ELLIS, K.Y.   1976.   Changes in body composition with age measured by total body neutron activation. Metab., 25: 85.

FAO (Food and Agriculture Organization). 1970. Aminoacid content of foods and biological data on proteins. Rome, Italy (FAO Nutritional Studies, No. 24).

FAO/WHO (Food and Agriculture Organization/World Health Organization). 1973. Energy and protein requirements: Report of A Joint FAO/WHO Ad Hoc Expert Committee on Energy and Protein Requirements. WHO Technical Report Series No. 522. World Health Organization, Geneva.

FOMON, S.J., and FILER, L.J. 1967. Aminoacid requirements of normal growth. In Amino Acid Metabolism and Genetic Variation, W.L. Nyan (Editor), McGraw Hill, N.Y., p. 391.

GEHRKE, C.W., and LEIMER, K. 1970. Trimethylsilylation of amino acids. Effect of solvents on derivatization using Bis (trimethylsilyl) Trifluoroacetamide. J. Chrom., 53: 201.

GEHRKE, C.W., NAKAMOTO, H., and ZUMWALT, R.W. 1969. Gas-liquid chromatography of protein amino acid trimethylsiyl derivatives. J. Chrom., 45: 24.

GEHRKE, C.W., and TAKEDA, H. 1973. Gas-liquid chromatographic studies of the twenty protein aminoacids: a single column separation. J. Chrom. 76: 63.

HEGSTED, D.M. 1963. Variation in requirements of nutrients amino acids. Fed. Proc. 22: 1424.

HOLT, L.E., and SNYDERMAN, S.E. 1967. The amino acid requirements of children. In Aminoacid metabolism and genetic variation.W.L. Nyan, (Editor). McGraw Hill, N.Y., 0. 381.

LEVERTON, R.M. 1959. Aminoacid requirements of young adults. In Protein and aminoacid nutrition. A.A. Albanese (Editor), p. 407. Acad. Press N.Y.

NAKAGAWA, I., TAKAHASHI, T., and SUZUKI, T. 1960. Aminoacid requirements of children. J. Nutr., 71: 176.

NAKAGAWA, I., TAKAHASHI, T., and SUZUKI, T. 1961a. Aminoacid requirements of children: Isoleucine and Leucine. J. Nutr., 73: 186.

NAKAGAWA, I., TAKAHASHI, T., and SUZUKI, T. 1961b. Aminoacid requirements of children: Minimal needs of Lysine and Methionine based on Nitrogen Balance Method. J. Nutr., 74: 401.

NAKAGAWA, I., TAKAHASHI, T., SUZUKI, T., and KOBAYASHI, K. 1962. Aminoacid requirements of children: Minimal needs to Threonine, Valine and Phenylalanine based on nitrogen balance method. J. Nutr. 77: 61.

NAKAGAWA, I., TAKAHASHI, T., SUZUKI, T., and KOBAYASHI, K. 1963. Amino acid requirements of children: Minimal needs of tryptophan, arginine and histidine based on nitrogen balance method. J. Nutr. 80: 305.

ROSE, W.C. 1957. The aminoacid requirements of adult man. Nutr. Abs. Rev., 27: 631.

TORÚN, B., CABRERA-SANTIAGO, M.I., and VITERI, F.E. 1981a. Protein requirements of preschool children: obligatory N losses and N balance measurements using cow's milk. Arch Latinoam. Nutr. (in press).

TORÚN, B., PINEDA, O., VITERI, F.E., and ARROYAVE, G.   1981b.   Use of Aminoacid composition data to predict protein nutritive value of children with specific reference to new estimates of essential aminoacid requirements. *In* Protein Quality in Humans: Assessment and in vitro estimation, C.E. Bodwell, J.S. Adkins and D.T. Hopkins (Editors), AVI Publishing Co., Westport, Conn., p. 000.

VITERI, F.E., and ALVARADO, J.   1970.   The Creatinine Height Index: Its use in the estimation of the degree of protein depletion and repletion in protein calorie malnourished children. Pediat., 46: 696.

VITERI, F.E., and MARTINEZ, C.   1981.   Integumental nitrogen losses of preschool children with different levels of dietary sources of protein intake. Food Nutr. Bull. (in press).

WATERLOW, J.C., and STEPHEN, J.M.L.   1967.   The assessment of total lysine turnover in the rat by intravenous infusion of L-(U-^{14}C) lysine. Clin. Sci., 33: 489.

YOUNG, W.R., RAND, W.M., and SCRIMSHAW, N.S.   1977.   Measuring protein quality in humans: a review and proposed method. Cereal Chem. *54*: 929.

4

Protein Quality in Relation to FDA Regulatory Needs

John E. Vanderveen, Ph.D. and Geraldine V. Mitchell, Ph.D.

The need to set quality standards and to measure protein quality for the protection of the consumer is based on both public health and economic needs. In this age of technology in which a high proportion of foods in our food supply are completely fabricated, the consumer could be deceived about the protein quality should there be no quality standards. We need only to look back on the recent "liquid protein" fad weight reduction diet which swept the country two years ago to demonstrate this point. During that incident, the Food and Drug Administration received several comments from professionals which revealed their amazement that the so-called "liquid protein" product was largely hydrolyzed collagen. In that instance, we are not sure whether the quality of protein had a major role in the cause of death for so many young women: however, it seems evident that such a product could not provide body needs for maintenance. Perhaps less obvious but still within the realm of possibility is the potential for taking very low quality proteins and fabricating imitation foods which look like, taste like, and function like the foods which they mimic and replace. In those cases where such foods are consumed as the sole source of nutrients, the problem can be very significant. The technology now exists to turn even very low quality vegetable protein into analogs for meat, fish, poultry, dairy foods, and even high quality vegetable proteins. What perhaps is most troublesome is that once protein isolates have been incorporated into foods, it is very difficult, at present, to identify the vegetable source from which they were made. Thus from a quality point of view, compositional information and growth performance information become vital for assessing whether a product will meet consumer needs.

Even if the manufacturer used what was reputed to be a quality protein for the manufacture of a product, there is the chance that the protein was mistreated or poorly manufactured and, as a result, protein quality was compromised. Although some technologists claim that damaged protein would be organoleptically unacceptable, it is possible to mask unacceptable flavors through the use of flavoring agents or combining with other ingredients having more dominant organoleptic qualities. In considering other public health needs, it is most important for some consumers to have information of the source of proteins. A growing number of our population have been found to have allergies or sensitivities for one or more proteins. In most foods today the ingredient statement would clearly indicate the source of a protein produce used as an ingredient. However, with the difficulty of detecting different protein isolates, there is a chance that a mislabeled product could go undetected.

Many food technologists and nutritionists have stated in recent years that protein quality is no longer a consideration in the American diet. They point out that most children, teenagers, and adults consume a variety and an abundance of proteins each day which meet their requirement for all the amino acids. They question, therefore, the need to set protein quality standards for these segments of the population. There are, of course, some small minority who do not eat a variety of foods each day because of medical, physical or economic conditions or because of personal beliefs or traditions. In some instances, the introduction of imitation foods have created a risk for these individuals. However, even if there was no risk, the consumer has a right to know that protein quality from different foods is variable and that substitute foods have a minimum quality standard which is comparable to that of the traditional food. To ignore the quality differences between proteins would be at least an economic issue.

In order to begin to establish quality standards for protein, the range of potential effects on the individual must be considered. Traditionally, the nutritionist centered his concerns on growth, development, and body maintenance. They can be characterized as parameters of quantity of life. Today, however, there is both sentiment and justification to also consider parameters of quality of life. It is our concern that consideration be given to maintenance of physical performance and mental acuity, freedom from diet induced behavioral changes, freedom from non-mortal diseases, and freedom from degenerative diseases which do not result in death throughout the entire life span. What this implies is that we must not only have our standards and criteria for these standards on short term growth and maintenance data, but we should give some consideration of the impact of life long consumption of proteins having different quality. As one example, we should not only be satisfied that a protein provides adequate growth potential in infants but we must also be assured that the immune competence of the child is not compromised. For years, epidemiologists have cited incidents where growth rates of a population appeared to be normal on what was considered poorer quality proteins but the incidence of mortality during contraction of childhood diseases was significantly elevated.

Clearly, the most critical needs among our population are those of the infant and young growing child. We must not only be concerned that the amino acid needs of this group be adequate to support optimum growth and mental development: we must also be sure that their needs are met for maintaining a viable immune response for protection against viral, bacterial and other microbiological diseases. In addition, we must insure their ability to ward off the effects of a variety of toxic substances with which the individual routinely comes in contact in our modern society. An adequate supply of quality protein frequently provides protection against these substances.

Our emphasis on the infants and children does not mean that we regard the needs of the rest of the population as unimportant. Certainly, the needs during pregnancy and lactation of women and the declining years of the elderly are equally as important to our goals for quality of life. I would venture to say that continued needs for quality protein is also more important than presently documented for the rest of the population as well.

In considering the needs of the population, the FDA must focus regulatory proposals on the needs of the most vulnerable groups who could conceivably consume the products being regulated. Obviously, if the products were consumed by infants and toddlers in a significant way, then the quality standards must meet their needs. If, on the other hand, the products were not used for infants or toddlers, then somewhat less restrictive quality standards would be appropriate. However, in keeping with economic considerations, if a product is used to replace a higher quality product in the market place and the product is indistinguishable from the traditional product and marketed in such a way as to imply comparable nutritional value, then regulations should at least provide a quality standard which would assure the consumer of no loss in meeting his or her individual body needs.

In recent years, there has been continuing effort in FDA to assess the most appropriate way to measure protein quality. The current approved methodology, the protein efficiency ratio (PER), has been criticized for not accurately assessing the biological value of proteins with respect to human requirements. The data derived from human studies have provided some support to this criticism.

The preamble to the tentative final regulation on vegetable protein products responded to this situation in this way:

"The Commissioner recognized the limitations of the PER method. The applicability to human nutrition of the PER assay and other rat growth assays commonly used to estimate the protein quality of foods has been questioned. It is recognized that there is a lack of adequate data regarding the physiological amino acid and nitrogen requirements of both man and animals and the factors which may affect these requirements. Scientific information on the protein quality of foods in humans is of a highly variable nature because of methodological uncertainties and the lack of standardization between experiments. The development and/or improvement of methods to be used to more adequately evaluate the protein quality of foods in human nutrition depends greatly on the availability of more precise and extensive data on the amino acid requirements of humans and animals and on animal research techniques which can be used to compare with human data. Since many of the protein foods included in this regulation will be consumed by children, it is appropriate to use an assay that uses a growing animal for test purposes."

"The PER assay is the only official method currently available for determining protein quality. There are other methods being used to measure protein quality, but the merits of these methods have not yet been established through the collaborative studies which are necessary for them to be made official. The Commissioner therefore concludes that for the purposes of this regulation, it is appropriate to use the PER assay as the basis for determining protein quality."

The PER assay is also used for all other protein quality requirements in FDA regulations (Vanderveen, et al., 1977). Up to this time there is no indication that PER has in any way compromised the safety of the consumer; and if there is an error, it is in favor of the consumer's health. This is not to say that PER should not be changed or replaced. It simply means that in the absence of improved methodology which safeguards human needs, continued use of PER is appropriate from a regulatory point of view.

The FDA is prepared, however, to take a realistic view to the regulation of protein quality when there exists good evidence that providing an exemption from the PER requirements will not have any impact on public health of any population segment who might consume the food. The preamble to the tentative final regulation on vegetable protein products responded to this issue in this way:

"The Commissioner recognized that there is some evidence that under some conditions vegetable proteins having a PER significantly less than that of casein may be nutritionally equivalent to some major protein sources in the diet. However, the data on human subjects available to the Commissioner at this time are insufficient to permit him to conclude with confidence that foods which contain as protein sources vegetable protein products with a lower PER value are nutritionally equivalent to the animal- derived protein foods that they replace. While there are some data on protein quality needs for humans showing that vegetable proteins with a lower PER value are nutritionally valuable, most of the studies to date involve adult test subjects. Since many of the protein foods included in this regulation will be consumed by children, it is appropriate to continue to use the PER assay, which is conducted on growing animals, for test purposes. Because of the importance of ensuring that growing children receive adequate nutrition, the Commissioner believes he must not alter a test criterion for protein quality traditionally used to make nutritional determinations until he is satisfied that the adequacy of any new criterion has been established. The Commissioner will consider exceptions to the PER level criterion for a specific protein used in a specific manner, upon receipt of a petition therefor, supported with adequate data. Such data should include studies on healthy growing children based on acceptable methods and should substantiate the protein quality equivalency of the vegetable protein and the usual major protein source being substituted for in part or in whole. The main concern is for the nutritional equivalence of the protein source for the growth, maintenance, and well-being of children. Data on adult humans will be considered in conjunction with adequate data obtained from feeding studies with children. The petition must correlate the human data submitted with data obtained using the official PER assay or a more appropriate and acceptable method, if available."

It is obvious from the above discussion that establishing a new official method for measuring protein quality is urgently needed. Hopefully, the scientific community, the regulatory agencies, and others who are concerned with consumer needs together with the regulated industry can reach a consensus of scientific opinion on this issue. It appears that an adequate data base now exists to make a start in this area. Briefly then, I will attempt to outline some of what FDA considers essential elements of a new official method:

1. The method must accurately estimate protein quality with respect to the population segments who consume the product.

2. It must be possible for an assay to be accomplished on the finished product.

3. The time to complete the assay should not exceed two weeks. (Preferably less than two days)

4. The assay should not require the use of human subjects but can be verified through the use of clinical studies.

5. The assay must include an assessment of digestibility comparable to that which would be found in human digestion.

6. The assay would utilize a representative aliquot of the sample without extensive preparation. At the same time, the assay should not require amounts in excess of 12 servings.

7. The procedure should have endorsement by AOAC or other appropriate body.

8. The cost of the assay would not exceed two hundred dollars if accomplished in a cost effective way.

In summary, I would like to repeat a statement I made at the Keystone Conference nearly two years ago. In my opinion, the scientific community has to make a meaningful effort to resolve this deficiency in knowledge and to agree on interpretations so that the process of public rule making can be initiated. The continual absence of needed data and consensus of scientific opinion will likely result in legal or political actions to dictate the future use of protein products in the market place.

REFERENCES

VANDERVEEN, J.E. and ADKINS, J.S. 1977. The use of protein quality measurements in Federal Regulations. Cereal Foods World (May).

DISCUSSION

DR. SAMONDS: I wonder if Dr. Vanderveen could expand upon his comment that some in the population depend upon intakes of very poor quality protein. There is a story, perhaps apocryphal, of a student at the University of Massachusetts who exists solely on grass, the green kind, which he grows himself hydroponically. But except for a few bizarre examples like this, how many people really are consuming poor quality proteins, what proteins, how long and why?

DR. VANDERVEEN: Well, admittedly, data are scarce in this area and the last question, I can't answer. We do have information from surveys which would definitely indicate, particularly among the very elderly, that there are some people who consume products that are marginal in nutritional value, particularly protein quality, but also in total quantity of protein. Some of them are very marginal.

There is also the problem associated with thousands of people who took the liquid protein for weight reduction. There is no doubt about the fact that some of them went for a period of six months on hydrolized collagen. Admittedly it was a situation which was unique. Certainly you wouldn't anticipate that this would occur very often. However, we did find people who after completing their weight reduction continued to consume liquid protein together with a carbohydrate source and thought they were getting an adequate diet.

DR. CHENG: Dr. Vanderveen, you mentioned that for nutrition labeling purposes, it is the protein quality that we have to take into consideration. Of course, when we talk about nutritional quality we are using the U.S. RDA factor of either 45 or 65, depending on the PER value of that specific product. On the other hand, I think the quantity of the protein is also important when we deal with the labeling of each individual finished product. I especially want to bring to the attention of this group that the AOAC is proposing the use of different nitrogen to protein conversion factors by the food and feed industries. For example, the conversion factor for a soy product would be 5.7 and for milk products 6.38. I would like to know what FDA's attitude is toward this new proposal, because if it is endorsed by regulatory compliance agencies, it will have a tremendous impact on the food and feed industries regarding nutrition labeling procedures. I personally believe that with so many experts here tonight from industry, universities, and governmental agencies, it would be worthwhile for us to put this question on tomorrow's task force topic.

DR. VANDERVEEN: As you quickly pointed out, that was an AOAC action and not something that we initiated. At the moment we have not come to a conclusion as to exactly what our position should be relative to those values. We agree that they are old values which were created, established in the 1930s. We would also agree that they should be looked at again. As most of you know the AOAC has created a task force to again review their action. FDA will participate in this effort. To my knowledge, the agency has no position on this matter at this time.

DR. BODWELL: On the nitrogen conversion factor, we'd prefer to keep that out of this conference. It was published in the AOAC Handbook but they have withdrawn it. They do have the task force that Dr. Vanderveen mentioned and that task force is going to hold hearings. Certainly AOAC would like very much to have all the written comments they could obtain prior to the middle of June when the task force meets for the third time. There are at least four members of that task force in the audience.

5

Protein Quality in Relation to USDA Regulatory Needs

Elizabeth W. Murphy and B. Wells Willis

The Food Safety and Quality Service (FSQS) of the U.S. Department of Agriculture has responsibility to the American public for assuring that meat and poultry products available for human consumption are wholesome, not adulterated, and properly labeled. Under these legal requirements, producers are able to market a wide range of meat and poultry products which will meet the varied economic, ethnic, and cultural needs of different groups and individuals. However, the Federal Meat Inspection Act and the Poultry Products Inspection Act, under which much of the FSQS's work is done, require us to set product standards and review product labels. These activities are required in order to uphold the nutritional quality of the food supply and assure that products reasonably meet consumer expectations.

Another major responsibility of FSQS is to manage the Government-wide Food Quality Assurance Program. This Program became operational on October 1, 1979, and is the focal point in the Federal Government for coordinating and managing food specifications and product descriptions for foods procured with appropriated funds. The Program addresses document development and quality assurance policies and procedures. Active management is intended to reduce duplication, complexity, and restrictiveness of government procurement documents as well as to lead toward the procurement of more commercial-type food items. Federal agencies such as DOD, VA, Bureau of Prisons, and FNS will be participating in this Program. FSQS also purchases food for the Federal School feeding programs and is responsible for ensuring that the foods purchased conform to the specification and the quality assurance provisions.

Because many of these foods are intended for growing children, their nutritional quality, including protein quality, must be considered carefully in preparing specifications and making purchases. Requirements for protein quality for military personnel, who are largely adults, are less stringent than requirements for children. Thus, problems arise in trying to establish purchase specifications that are similar for foods targeted for these different population groups.

The development in recent years of alternate sources of protein, for use as full or partial replacements for meat, poultry, fish, cheese, or milk, leads to problems for FSQS in assuring the quality of protein in products the agency monitors or purchases. With diminishing supplies and increasing costs of meat, it is highly likely that both consumers and government food programs will be using more alternate proteins as replacements for part of the traditional animal protein foods in meals. At present, most of the alternate protein in the market place comes from the soy bean. However, developmental work currently underway with other proteins leads us to expect that we may someday need to monitor the protein quality of peanut, pea, alfalfa, cottonseed, tobacco, or blood proteins. Evaluations of new technologies, such as mechanical deboning, may also require information on protein quality. We therefore have an increasing need for a method of measuring protein quality that is suitable to use in monitoring compliance with product standards, label claims, and purchase specifications.

Three years ago, at a University of Nebraska conference on Rapid Determination of Protein Quality, an FSQS spokesman outlined our needs for protein quality tests (Fried, 1977):

1. *Tests should be reproducible*, both within the same laboratory over a period of time, and between laboratories.
2. *Tests should be rapid*. It was noted by our former spokesman that long tests add to costs and problems of work flow if product must be held until tests are complete. This is not a problem to industry alone—FSQS needs rapid procedures for specification testing, so that product purchased for the military or USDA feeding programs can be released for distribution as rapidly as possible. Furthermore, extended storage of products in refrigerators or freezers might affect shelf life and costs to consumers.
3. *Tests should be reasonable in cost*. Federal agencies, especially regulatory agencies, are being expected to hold the line on budgets. We thus find ourselves unpleasantly situated between the Scylla of inadequate surveillance and the Charybdis of unacceptable costs if we must use bioassays to monitor protein quality.
4. *Test results should be equivalent to results that would be obtained from the current benchmark procedure*. At present, the Protein Efficiency Ratio (PER) method is the official AOAC procedure for measuring protein quality. An acceptable test must give results in constant relationship to the bioassay PER, so that protein quality as determined will be capable of comparison with historically determined values.

This three-year old outline of requirements for rapid tests of protein quality is still completely applicable today. To this list, however, our chemists have added one more requirement:

5. *Tests should be rugged.* They should be simple and not require a highly skilled professional chemist for adequate performance. For regulatory control, we need routine analyses capable of being performed by a qualified para-professional.

Unfortunately, we have not yet identified a test for protein quality that meets the above five requirements. However, we have added slightly to our experience in dealing with problems of protein quality. The rest of this discussion will pertain to our recent findings and activities.

BIOASSAYS FOR PER

As part of a comprehensive evaluation of health and safety of use of mechanically processed (species) product (MP(S)P, formerly known as mechanically deboned meat), PER's were run on test samples of casein and samples of MP(S)P (Kolbye *et al.* 1977). In all, 63 tests using 10 animals per test were made. Using the data from all tests, measurements of reliability were calculated. The 95 percent confidence interval for an observed PER was found to be ± 0.28 units. If one observed a PER of 2.70, one would be reasonably sure that the true PER value lay somewhere in the interval of 2.42 to 2.98. Our statisticians pointed out that this is only the error due to variability of rats with the same genetic make-up in the same laboratory. It is not a small error for small PER values.

PREDICTING PER WITH EQUATIONS

Several years ago, scientists from FSQS and the USDA Eastern Regional Laboratory jointly developed three equations for predicting PER's of meat and poultry and products containing them (Alsmeyer *et al.* 1974). The equations were then tested on amino acid—PER data supplied by the food industry. The authors concluded that the third equation, which was based on contents of methionine, leucine, histidine, and tyrosine in the product, effectively predicted PER for 66 of the 93 food products—those containing proteins primarily of meat, poultry, grain, or yeast origin. "Effectiveness" was defined as falling within ± 0.2 of the bioassay value for PER. The equations are given in Table 5.1.

TABLE 5.1.   EQUATIONS FOR PREDICTING PER FROM AMINO ACID ANALYSES[1]

Equation 1: $PER = -0.684 + 0.456 \, (LEU) - 0.047 \, (PRO)$
Equation 2: $PER = -0.468 + 0.454 \, (LEU) - 0.105 \, (TYR)$
Equation 3: $PER = -1.816 + 0.435 \, (MET) + 0.780 \, (LEU) + 0.211 \, (HIS) - 0.944 \, (TYR)$

[1]As reported by Alsmeyer *et al.* (1974).

In our own work, we have compared the three equations with bioassay values for PER of beef MP(S)P (Kolbye et al. 1977) and mechanically deboned poultry (MDP) made from cooked fowl frames (Murphy et al. 1979). For these products, equation 1 was a better predictor of PER than equation 2; and equation 3 was completely unacceptable. However, for mechanically deboned poultry, equations 1 and 2 were found to inaccurately predict compliance with a postulated minimum PER of 2.5 in, respectively, 42 and 31 percent of 26 samples. We therefore concluded that further work was needed to develop equations based on amino acids that would be suitable for use with MP(S)(P) and MDP.

ESTIMATING PER BY PERCENTAGE OF ESSENTIAL AMINO ACIDS

The standard for MP(S)P sets a minimum PER of 2.5 (Code of Federal Regulations, 1978). However, a product with an essential amino acid (EAA) content of at least 33 percent of the total amino acids present is accepted as being in compliance with the PER requirement. The seven amino acids included in calculating this percentage are isoleucine, leucine, lysine, methionine, phenylalanine, threonine, and valine.

This alternate measure for acceptable protein quality was included in the regulation in hopes that new data would be generated that would enable us to establish the reliability of the percentage of essential amino acids as an indicator of protein quality. However, little MP(S)P has been produced, and additional data have not been forthcoming. Furthermore, studies made on mechanically deboned poultry did not reveal any relationship between percentage of EAA and PER. Some samples of MDP had high PER's but low percentages of total EAA's, while for other samples the PER's were low and the EAA's were high. We have thus concluded that percentage of EAA's as an indicator for PER is not a good regulatory tool.

Our laboratories are currently making a limited study of the protein digestibility procedure of Satterlee et al. (1977) using samples of mechanically deboned poultry on which PER's were previously determined. Although data from this study have not yet been calculated, our chemists believe this to be a method with some promise.

We find ourselves in a position little changed from three years ago, as we still do not have a practical and acceptable test for protein quality.

REFERENCES

FEDERAL MEAT INSPECTION ACT, 21 U.S.C. 601 et seq.; 1906; amended 1967.

POULTRY PRODUCTS INSPECTION ACT, 21 U.S.C. 451 et seq.; 1957; amended 1968.

FRIED, I. 1977. What the USDA is looking for and will accept in a method for protein quality evaluation. Food Tech. 31 (6): 85.

KOLBYE, A., NELSON, M.A., and MURPHY, E.W. 1977. "Health and Safety Aspects of the Use of Mechanically Deboned Meat." Vol. 1 and 2. Food Safety and Quality Service, USDA, Washington, D.C.

ALSMEYER, R.H., CUNNINGHAM, A.E., and HAPPICH, M.L. 1974. Equations predict PER from amino acid analysis. Food Tech. *28*: 34.

MURPHY, E.W., BREWINGTON, C.R., WILLIS, B.W., and NELSON, M.A. 1979. "Health and Safety Aspects of the Use of Mechanically Deboned Poultry." Food Safety and Quality Service, USDA, Washington, D.C.

CODE OF FEDERAL REGULATIONS. 1978. Title 9, Chapter III, Subchapter A. Mandatory Meat Inspection, Part 319, Definitions and Standards of Identity or Composition, ¶319.5.

SATTERLEE, L.D., KENDRICK, J.G., and MILLER, G.A. 1977. Rapid in-vitro assays for estimating protein quality. Food Tech. *31* (6): 78.

6

Protein Quality in Relation to Industry's Needs

Robert O. Nesheim

PRIMARY FACTORS

There are two primary factors that influence the concern of the food industry with protein quality. The first is the obligation to maintain the quality of our food supply. The second relates to the accuracy of claims made for the protein contribution of a product.

With few exceptions, protein intakes of individuals in the United States are more than adequate to meet the recommended dietary allowance for this nutrient. Under these circumstances it would be easy to be complacent about the protein quality of products introduced in the market place. After all, most Americans consume more protein than they require. This is not a valid argument, however, for not being concerned about the protein quality of our products.

When a new product is introduced into the diet it should supply a protein level at least equivalent in quantity and quality to the product it is expected to replace. Quantity is easily measured by standard analytical procedures. Quality measurements are not as simple, and that essentially is what this conference is all about.

Products may be formulated to be equal in protein quality on the basis of amino acid composition. However, various processes employed in the production of a product may have an adverse effect on protein quality. The protein quality of the ingredients used in formulating the product may be below expectations. Its formula, type of processing, its duration, temperatures employed, and storage conditions and in-home preparation are all factors that determine the protein quality of the product as ultimately consumed. The effect of reducing sugars on the availability of lysine, for example, is well known. While much is known concerning the effect of these various factors on protein quality, ultimately the protein quality of the end product must be measured, particularly when the product is a significant source of protein in the diet or when claims are made. Specifications must be developed for the ingredients and processing, and appropriate controls instituted to ensure that deviations do not occur which will adversely affect product quality.

INDUSTRY NEEDS

During the development of a product estimates of protein quality can be made by various techniques. These may include enzyme digestibility, available lysine, amino acid analyses, and other techniques which are thoroughly familiar to this group and do not require discussion on my part. However, the only officially recognized determination of protein quality by the regulatory agencies is Protein Efficiency Ratio, or PER. This test is time consuming and expensive and not well suited for routine use in either product development or for quality control.

Protein quality is of particular importance in products designed to comprise the entire meal or to supply the total diet for a period of time. These products must meet the RDA for protein for the targeted segment of the market. The quantity and quality of the protein are both of primary importance. Proper formulation, process controls, and quality measurements are essential to ensure that the product delivered meets the appropriate criteria.

In addition to the obligation of industry to provide products of appropriate product quality, product labeling as required in labeling regulations also affects industry's need for satisfactory protein quality measurements. The quality of the protein in a product can have a significant impact on the claim that can be made. For example, nutrition labeling regulations specify how protein claims relative to the U.S. RDA may be made. If the PER of the protein in the product is equivalent to casein, then the claim for the percent of the RDA for protein supplied by a serving of the product is based on a U.S. RDA of 45 grams. If the PER is not equal to casein, then the standard is 65 grams. If the quality of the protein is less than 20 percent of the PER of casein, it cannot be claimed as a source of protein. The significance of this regulation is illustrated by the following example. A product providing 4.5 grams of protein in a serving could claim to provide ten percent of the U.S. RDA if the protein quality was equal to casein but only six percent if the PER was not equal to casein but was better than 20 percent of casein.

This regulation and others have been discussed previously (Chapters 4 and 5) and therefore further elaboration is not required. It is sufficient to state that there are regulations which require that food manufacturers have rapid and efficient procedures for determination of protein quality to determine and document label claims.

REGULATORY ISSUES

While the regulatory requirements certainly cause industry to focus on the need for improved and more rapid methods of protein quality evaluation, many food and ingredient manufacturers would find improved methods highly desirable even in the absence of regulations. Manufacturers concerned with producing quality food products are seeking methods which will enable greater control over the production of products to ensure uniform product quality. To do this it is important to make measurements at certain critical points in the process which will ensure that a product and a process are in control with its specifications during manufacturing.

Therefore, the most desirable type of procedure for evaluation of protein quality is one that has application on line. This requires a test whereby a quality assurance technician could, in a matter of minutes, determine whether a purchased ingredient met its protein quality specifications or a product being produced was within protein quality specifications. No such direct short-term test is available today, and therefore it has been necessary to develop procedures which permit rapid testing and which correlate well with the more time-consuming methods for protein quality assessment. Frequently this involves defining processing parameters which provide reasonable assurance that the end product will meet the protein quality specifications.

Ingredients will likely be used in the manufacturing process before a check can be made on protein quality using PER. The possibility exists that the resulting product will not meet specifications.

VOID IN MEASURING PROTEIN QUALITY

Accepting the fact that it is not possible today nor in the foreseeable future to have a direct protein quality measurement that can be used on line, the next most desirable test is one that can be completed before the product is shipped from the manufacturing plant or the ingredient used in manufacturing a product. This usually means a period of not more than 24 hours. During this period the product can be kept within the tight control of the manufacturer and not released for shipment from the plant until there is assurance that the product has met its specifications. This time frame also puts very strict requirements on the test and is not likely to be met in the near future.

A third level of desirability in terms of a rapid test procedure is one that can be completed before the product passes the manufacturer's control. The product may have left the manufacturing plant but is still in the manufacturer's distribution system so that if the product does not meet the protein quality requirements it can be withheld from general distribution and not be in violation of the label or manufacturer's specification. The manufacturer can then take whatever steps may be necessary to solve the problem.

The least desirable of the procedures for assessing protein quality are those that exist today, where PER is the only officially recognized test and test results usually cannot be obtained until after a product is moved into commercial distribution channels. If tests were run before shipping, the product would have to be retained over a month, necessitating expensive warehousing and inventory costs. In view of this, manufacturers must rely on other systems of on-line control to maintain protein quality. These systems usually have been checked to determine that they provide good correlation with PER test results.

THE NEED FOR CHANGE

Industry is interested in an official method of determining protein quality which will move the time frame of the testing to the shortest period consistent with accuracy, cost, and practicality. To accomplish this it appears to me that we must move away from the biological determination of protein quality, as in PER, to a more direct method. With today's knowledge of human amino acid requirements and our ability to measure amino acid by automated procedures, we should be able to move from antiquated biological determinations to appropriate chemical procedures. I believe we are past the point in our technology where we must depend upon biological tests using a growing rat to measure protein quality of products for humans. In the course of this conference I hope we reach agreement that we can move ahead to improved procedures and define the steps necessary to reach that goal.

PART II ASSESSING PROTEIN NUTRITIVE VALUE IN HUMANS

7

Conventional ("LONG-TERM") Nitrogen Balance Studies for Protein Quality Evaluation in Adults: Rationale and Limitations

William M. Rand, Nevin S. Scrimshaw and Vernon R. Young

A schematic outline of protein metabolism in the human, together with an indication of the various stages in the utilization and metabolism of the products of protein digestion and absorption that offer potential for evaluation of dietary protein quality, is given in Figure 7.1. However, the safety and ethical constraints in studies with human subjects restrict the types of measurements and approaches that can be undertaken to assess the differential, quantitative effects of food proteins on protein and amino acid metabolism in the human body. These measurements (Table 7.1) include determinations of amino acids in body fluids, as indices of dietary protein adequacy, or of the end-products of amino acid metabolism, such as urea and sulfur compounds, in addition to N compounds in the excreta (urine and feces). Furthermore, the incorporation into body protein of ingested nitrogen (N) may be determined by measuring changes in body composition or predicting them by the chemical balance technique. Measurement of plasma metabolite levels, including amino acids, urea, albumin, other transport proteins and/or blood enzymes, has not reached the stage required for routine application in human studies concerned with quantitative assessment of dietary protein quality. Furthermore, techniques for estimation of body composition are not sufficiently precise to allow accurate determination of relatively small changes in body protein content during the relatively brief experimental diet periods characteristic of metabolic balance studies in healthy subjects. Therefore, it is appropriate to focus specific attention on the N balance technique as the current, most widely used

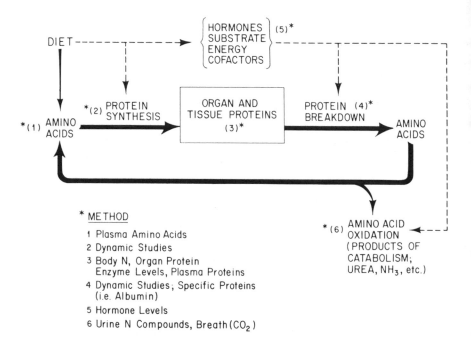

FIG. 7.1.   ILLUSTRATION OF THE MAJOR FLOW OF AMINO ACIDS WITHIN THE BODY, TOGETHER WITH AN INDICATION OF THE METHODS AND MEASUREMENTS THAT MAY BE USED TO ASSESS THE STATUS OF THE VARIOUS PHASES OF BODY PROTEIN AND AMINO ACID METABOLISM
Dotted lines indicate that, in addition to the provision of amino acids and N, the diet influences the regulation of protein metabolism via its effects on the endocrine system (hormones) and energy metabolism.

approach for measurement of dietary protein quality in adult human subjects.

For purposes of this review, we accept the view that direct studies in human subjects serve as a means for hypothesis testing and as a final confirmatory approach for establishing the nutritive value of a food protein source(s) intended for direct human feeding.

Current estimates of amino acid requirements and utilization under various nutritional and physiological states are not sufficiently secure to allow a precise, quantitative prediction of protein quality in humans on the basis of data from *in vitro* methods or bioassays in experimental animals alone, despite their potential or utility. Thus, clinical studies form an important component of the comprehensive assessment of the nutritional value of human diets. We will address a series of issues that concern the design, analysis and interpretation of conventional (long-term) human

TABLE 7.1.   SOME METHODS OTHER THAN N BALANCE THAT HAVE OR MAY BE
USED TO ASSESS PROTEIN QUALITY IN ADULTS

Approach	Comment
Measurement on Blood Free Amino Acids	Further development needed Potential for estimating limited amino acid
Urea Albumin	Of possible value (children)

Proteins (enzymes,
    transport proteins)
	For a quantitative index, more
	    investigation required

Measurements on Urine N compounds (urea, NH_4, 3-Mehis)	Perhaps qualitative indexes
Sulfur compounds	
Body Composition ^{40}K, in vivo neutron activation, Body Water (^2H_2O; $H_2^{18}O$)	Precision a limiting factor

metabolic N balance experiments. By conventional, or long-term, we refer
to studies lasting more than a few weeks and involving one or more experimental dietary periods. In order to restrict the discussion, we will depend
largely on our own findings involving studies in healthy young adult men.
In addition, access to detailed data is necessary to address some of the
problems and these data are not readily accessible in the published literature.

RATIONALE FOR "LONG-TERM" METABOLIC STUDIES

Various considerations need to be taken into account in developing an
appropriate rationale for conventional (long-term) N balance studies in the
evaluation of protein quality in humans. Firstly, it is important to consider
the rate of change in N balance and establishment of a new equilibrium
with an altered level of protein intake or change in source of protein. This
factor also influences the choice of design of N balance study. Secondly,
there is the question of variability of the N balance within an individual
subject for a given N and protein intake. Thirdly, the suggestion has been
made that a rhythm occurs in N balance and retention and this deserves
comment. Finally, the appropriate level of test protein intake should be
discussed. Each of these issues will be examined in the following sections
before an account is given of the analysis and interpretation of data obtained.

Adaptation to Altered N Intake

It is well known that the human body responds to a change in the level
and source of nitrogen intake, or in the adequacy of the dietary essential

amino acid supply, by adjusting the output of urinary N and of fecal N to achieve a new state of balance between total N intake and total N excretion (Martin and Robison, 1922; Munro, 1964; Waterlow, 1968). The rate at which these N losses are adjusted to altered dietary conditions will determine, in part, the minimum length of the experimental period necessary to obtain a reliable estimate of N balance for a given, new level or source of protein intake. We have explored previously this aspect of body N metabolism by determining the time necessary to achieve a relatively steady state in urinary N output when subjects are given a protein-free diet (Rand *et al.*, 1976). As shown in Fig. 7.2, essentially all healthy subjects adapt within 10 days under these dietary conditions. For test levels of protein intake that are closer to either the physiological requirement for protein or usual intake level, the new steady state in urinary N output may be achieved within a shorter period of time, and for some subjects this is likely to take

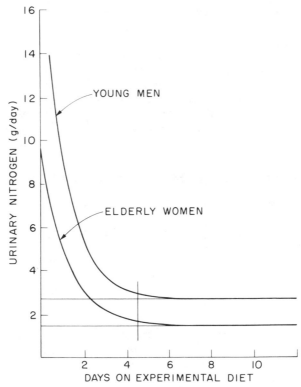

FIG. 7.2. COMPARISON OF CHANGES IN URINARY NITROGEN EXCRETION WHEN YOUNG MEN AND ELDERLY WOMEN ARE GIVEN A VERY LOW-PROTEIN OR PROTEIN-FREE DIET
Vertical line indicates average time to achieve stability for both groups. Taken from Rand *et al.* (1976). Range in time to achieve stability for young men was 3 to 10 days and for elderly women was 3 to 8 days.

only one or two days. In the case of marked increases in N intake, however, the time for adaptation may be longer, since Oddoye and Margen (1979) stated that it took about 9 to 12 days for young adult men to adjust urinary N output to an increase in N intake from an initial lower level of 12 g to a high intake of 36 g N/day. Similarly, they observed that a decrease in N intake from 36 g down to the 12 g N daily level required an adaptation period of 9–12 days to achieve a steady state in urinary N output. Thus, the precise length of period necessary to achieve a new steady state rate of urinary N excretion in adults after an alteration in protein intake appears to depend, in part, upon the magnitude of the change in protein intake relative to that necessary to meet the minimum physiological requirement. On the basis of a long-term metabolic N balance study in which young men received a good quality protein, at a level of 0.8 g protein/kg/day, N balances during the second week had achieved a relative steady state but, as shown in Table 7.2, the first week did not provide N balance data that predicted the long-term status of N balance in these subjects. For these reasons and because there is interindividual variation in the rate of adaptation in urinary N output to changes in protein intake, it is desirable to use about the first four to seven days of a test diet period as an "adjustment" or "adaptation" phase and to incorporate into the experimental design an initial period of one or two days of very low protein intake to promote a rapid adaptation in N metabolism in the population of test subjects. Moreover, it can be seen from the summary of experimental diet conditions used by various investigators (Table 7.3) that this view is reflected in the design of many balance studies with adults. However, it is apparent that there is no standard or set specific design for this adaptation phase. This will depend, in part, upon the specific objectives of the experiment.

TABLE 7.2. DIFFERENCE IN N BALANCE BETWEEN FIRST, SECOND AND AVERAGE OF SUCCEEDING WEEKS IN A 3-MONTH N BALANCE STUDY[1]

	Difference for:	
Subject	Week 1	Week 2
	— mg N/kg/day —	
D.K.	− 2.9	15.2
T.D.	−18.6	1.3
T.K.	−25.4	− 0.3
J.P.	−17.8	13.0
B.F.	−25.7	3.2
D.T.	−16.3	0.2
R.P.	−37.6	− 2.0
A.K.	−17.2	− 4.3

[1]Protein intake of 0.8 g soy isolate protein/kg/day (Supro-710; Ralston Purina).

Daily Variation in Balance

The day-to-day variation in N balance within an individual subject receiving a constant N intake also will determine the length of the experi-

TABLE 7.3.   A SELECTIVE SURVEY OF EXPERIMENTAL DIET CONDITIONS USED TO
ASSESS N RETENTION AND VARIOUS ASPECTS OF DIETARY PROTEIN UTILIZATION
IN ADULTS

Diet Conditions	Author
Non-protein diet for 2 days; test periods 5 days	Hoffman and McNeil (1949)
Non-protein diet for 3–10 days; test periods 8–10 days	
Variable test periods: average 8 days	Bricker et al. (1945)
7 Days preliminary (2 days low N, 5 days 13 g N)	Kies and Fox (1970)
Experimental periods 5 days each	
3 days at low N; 10 days experimental,	Kies et al., (1965)
4 or 5 day test periods	Romo and Linkswiler (1969)
12 day adaptation; 6–8 day test periods	Clark et al. (1963)
7 day adaptation; 7 day test periods	Thomas et al. (1979)
3 day low N; 10 day experimental	Kies and Fox (1972)
10 day standardizing; 11 day experimental (7 days balance)	Clark et al. (1972)
15 day diet periods (N balance days 11–15)	Young et al. (1973)
1 day low N; 15 day experimental (N balance 11–15)	Young et al. (1975)
2 day low N; 12 day standardization: 9 day tests	Zezulka and Calloway (1976)
7 day standard diet: 24 day test	Kishi et al. (1978)
12 day test (last 6 days balance)	Nichol and Phillips (1976)
12 or 15 day test periods	Calloway and Margen (1971)

mental period that is required to obtain a reliable estimate of the N balance for a specific test protein intake level. The precise nature and extent of this variation has not been extensively reported for subjects adapted to constant N intakes for various diets but in our earlier studies we observed an intraindividual variation equal to that between subjects receiving an essentially protein-free diet (Scrimshaw et al., 1972). An analysis of data obtained in two recent long-term metabolic N balance studies confirms and extends these earlier findings, as shown in Table 7.4. These data indicate that the day-to-day intraindividual variation in N balance is often as large as that observed within a group of subjects receiving adequate or generous, but constant, protein intakes. From the standpoint of N balance estimations, the variation in urinary nitrogen excretion makes the major quantitative contribution to fluctuations in actual balance, although the amount of variance in fecal N output can be quite marked. This may be due to

TABLE 7.4.   RANGE FOR INTRAINDIVIDUAL VARIATION AND MEAN INTERINDIVID-
UAL VARIATION IN BALANCE PARAMETERS FOR SUBJECTS ADAPTED TO CON-
STANT, ADEQUATE PROTEIN INTAKES FOR 2 OR 3 MONTHS

| | Study No. (Protein Intake) | |
Parameter	1 (1.5 g/kg/day)[1]	2 (0.8 g/kg/day)[2]
N Balance (mgN/kg/day)		
SD_W[3]	12.1	18.9
SD_B	29.8	22.4
Urinary N: Mean (mgN/kg)	201	109
CV_W[4]	4.8−8.4	7.1−26.1
CV_B	18.2	29.9
Fecal N: Mean (mgN/kg)	27	12.5
CV_W	11.2−35.1	22.9−62.3
CV_B	47.2	35.0

[1] Based on 16 subjects receiving mixed protein sources.
[2] Based on 8 subjects receiving soy isolate (Supro-710; Ralston Purina).
[3] SD_W = within standard deviation; SD_B = between standard deviation.
[4] CV_W = within coefficent of variation; CV_B = between coefficient of variation.

variations in bowel movement and because of the inadequacy of most fecal
markers (e.g., carmine red, F.D. and C. Blue No. 2, charcoal) to serve as a
precise means of separating the fecal N output according to the specific
dietary period. From the data summarized in Table 7.4 it can be seen that,
in the case of some individuals, one third of the time a measured value for a
single balance day may differ from the mean N balance by 20 mg/kg or
more. Clearly, when there is an intraindividual variation of this magnitude
an adequate measure of the average N balance requires an observation
period of greater than one day. For example, a 9-day balance period would
increase the confidence of the estimate of N balance by a factor of 3.

    The intraindividual variation in N balance is due, in part, to experimen-
tal errors introduced during collection and analysis of excreta and we have
explored this problem by measurement of daily urinary creatinine and its
relationship with urinary total N excretion; the latter was used as an index
of the stability of body N balance. Because endogenous creatinine output is
assumed to be a measure of muscle mass (Chinn, 1967; Muldowney et al.,
1957; Graystone, 1968; Meador, et al., 1968) creatinine excretion should
show relatively small daily variations when body composition is unchanged
and should alter only slowly when exogenous sources of creatinine change
(Garza et al., 1977). Thus, for the present purpose, the determination of
urinary creatinine output is considered to provide a valuable estimate of
the accuracy of total 24-hr urine collections in metabolic balance studies. It
should be pointed out here that studies of variation in 24-hr urinary
creatinine excretion within individuals have given conflicting results (Anker, 1954; Applegarth, et al., 1968; Bleiler, et al., 1962; Edwards, et al.,
1969; Curtis and Fogel, 1970; Paterson, 1967; Chattaway, et al., 1969) and
some investigators have concluded that measurement of daily creatinine
excretion is of little or no value in assessing the completeness of urine

collections. However, we consider these investigations inadequate and in agreement with others (Jackson, 1966; Epstein and Schriever, 1970; Cramer *et al.*, 1967) we accept the measurement of daily creatinine excretion as a useful tool for evaluation. of the variation in N balance within subjects. Thus, an analysis of results obtained in 60 healthy male MIT students who participated in eight separate N balance experiments showed, as summarized in Table 7.5, that the coefficient of variation of daily creatinine excretion within most individual subjects was less than 10%. This possibly overestimates the true daily endogenous variation in creatine output because subjects may occasionally delay completion of the 24-hr urine collection by failing to adhere precisely to the requirements of the experimental protocol. When marked variations in creatinine excretion were observed during the initial phases of a study, further explanation about the experimental requirements to the subject usually resulted in a dramatic reduction of the variability of creatinine and N excretion (Figure 7.3).

TABLE 7.5.   COEFFICIENTS OF VARIATION WITHIN SUBJECTS FOR 24-HOUR CREATININE EXCRETION DURING THE ENTIRE METABOLIC BALANCE STUDY PERIODS[1]

Exp. No.	Major Source of Dietary Protein[2]	Protein Intake (g/kg)	No. of Subjects	Coefficient of Variation	
				Mean	Range
1	Wheat gluten	0.27	6	7.0	4.7−9.5
2	Wheat gluten	0.73	8	6.2	2.5−9.2
3	Chick pea[3]	0.27	7	5.7	4.8−7.8
4	Chick pea and rice[4]	0.65	8	4.8	3.5−6.9
5	Dried skim milk	0.5	7	5.1	4.3−5.8
6	Dried skim milk	0.4	7	7.2	5.8−8.7
7	Mixed[5]	0.2 − 0.4	6	10.6	4.7−16.7
8	Egg	0.2 − 0.5	11	5.9	2.6−10.9

[1] Total length of experimental period was 60 days for Experiments 1−6 and 45 days for Experiments 7 and 8.
[2] In Experiments 1−4 the protein was supplemented with the limiting amino acid [lysine (Exp. 1 and 2) and methionine (Exp. 3 and 4)] during half of the experimental period.
[3] If one subject who initially misunderstood instructions was included, the mean would be 7.0 with a range of 4.8−14.9.
[4] Seventy percent of dietary protein was from chick-peas and the remainder from rice.
[5] A mixture of wheat gluten, chick-pea and dried skim milk provided the total protein intake.

These data may then be examined in reference to the coefficients of variation for creatinine excretion within individuals for the last 5 days of a 15-day metabolic balance period and the results compared with those for urinary total N output.

As shown in Table 7.6 for 4 separate studies the intraindividual variation in total urinary N excretion was greater than that for creatinine excretion, showing that there is a daily variation in N balance that exceeds the variation in creatinine output. From these results it is apparent (based on considerations of urinary N excretion) that the estimation of N balance during the last five days of a 15-day metabolic period can be made with a

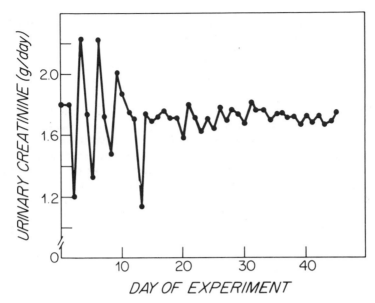

DAY OF EXPERIMENT

FIG. 7.3. VARIATION IN DAILY URINARY CREATININE EXCRETION DURING THE INITIAL AND SUBSEQUENT PHASE OF A METABOLIC BALANCE STUDY IN A HEALTHY YOUNG MALE SUBJECT
Coefficient of variation was 15% and 3.5% during the first and second 15 day periods, respectively (see text).

TABLE 7.6. COEFFICIENTS OF VARIATION FOR 24-HR URINARY TOTAL N AND CREATININE EXCRETION VALUES FOR INDIVIDUAL YOUNG MEN STUDIED IN A SERIES OF METABOLIC N BALANCE EXPERIMENTS

		Urinary N		Urinary Creatinine	
Exp.	Diet[a]	Mean	Range	Mean	Range
1	adeq.	11.8	3.4–19.6	4.1	1.5–10.6
	adeq.[b]	11.6	4.7–21.7	7.5	1.7–21.6
	restr.	11.1	4.8–15.2	5.3	0.7–14.4
	restr.[b]	6.9	3.3–10.3	5.2	3.1–11.4
2	adeq.	6.6	2.4–9.8	5.3	2.3–14.1
	adeq.[b]	6.2	2.5–11.4	3.7	2.4–5.1
	restr.	6.6	3.1–10.6	4.1	1.0–10.3
	restr.[b]	5.2	2.9–7.4	3.4	0.9–5.8
4	adeq.	12.0	7.0–17.2	3.6	1.5–5.1
	adeq.[b]	8.1	2.5–12.7	2.9	0.7–5.2
	restr.	7.6	3.5–12.2	2.8	0.8–4.1
	restr.[b]	10.0	4.0–24.4	3.2	2.3–4.8
8	0.2g/kg	8.2	2.7–11.6	3.3	1.3–6.4
	0.3g/kg	10.7	5.5–18.6	4.2	0.6–14.7
	0.4g/kg	7.7	4.5–12.2	3.9	1.5–7.9
	0.5g/kg	11.1	4.6–21.9	3.9	1.4–8.9

[a] "adeq." and "restr." refer to adequate and restricted energy intake, respectively. For Experiment 8 the diets provided only the adequate level; see Table 5 for numbers of subjects in each experiment.
[b] Amino acid supplementation of dietary protein as described in Table 5.

precision approximating ± 10%. For these various reasons, an experimental diet period extending into and covering about the second week appears rational in studies with adult humans.

Rhythms in N Balance

Another possible reason for the choice of an experimental diet period of about two weeks or even longer is raised by the observations made by Sukhartme and Margen (1978) that day-to-day fluctuations in N balance are not random but are serially correlated as an autoregression process. Accordingly, we (Rand et al., 1979) have examined our data on urinary N excretion in healthy young adult subjects receiving a constant diet supplying either generous or low-protein intakes for periods lasting 8 to 11 weeks. Two statistical models were tested as possible descriptions of daily urinary nitrogen output and these models are shown in Figure 7.4. For the 21 subjects studied, only two showed a statistically significant serial correlation (Model 2) in daily urinary nitrogen output. Hence, variation in the latter can be considered to be due to random experimental and biological variations. We conclude that existence of a rhythm in N metabolism and N output does not appear to impose an additional need to further extend the length of N-balance periods beyond that necessary to achieve an acceptable estimate of the new steady state output of N excretion and, thus, mean N balance after initial adaptation to the diet is achieved.

Effect of Test Protein Level

Choice of the design of conventional metabolic N balance studies for protein quality evaluation should also take into account the known effect of level of protein intake on the efficiency of utilization and retention of dietary protein. We (Young and Scrimshaw, 1978) and others (Calloway and Margen, 1971; Inoue et al., 1973; Kishi, et al., 1978; Hegsted, 1974) have discussed this aspect in detail and so only a limited treatment of the topic will be made here. Basically, experiments with human subjects, including children (Bressani and Viteri, 1971) and adults (Inoue, et al., 1973), indicate that the efficiency of N utilization for high quality proteins is not constant within the submaintenance range of protein intake but changes with the specific test intake level (Inoue et al., 1974). Although the metabolic basis for these changes is not well established, it is worth examining briefly this problem.

As graphically depicted in Figure 7.5, the efficiency of N retention declines as the total protein intake approaches that required for maintenance. Above this level the retention of each further increment in N intake is usually quite low. Although it is still debated whether the N balance response to changes in N intake within the entire submaintenance range is rectilinear (Kishi, et al., 1978) or curvilinear (Young et al., 1973) [although the latter is anticipated on biochemical grounds (Krebs, 1972)], it is clear

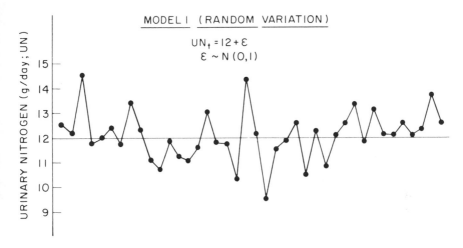

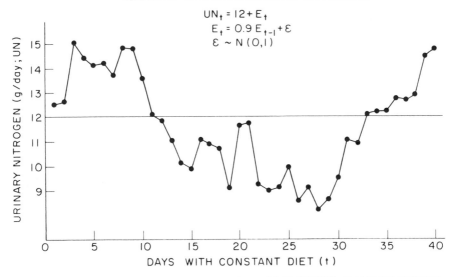

FIG. 7.4.   COMPARISON BETWEEN A RANDOM MODEL (MODEL 1) AND AN AUTORE-GRESSIVE MODEL (MODEL 2) WITH HIGH (r = 0.9) SERIAL CORRELATION
Simulated data, with a mean of 12 and the same sequence of random components (EnN (0.1) for both models have been used to plot these graphs Taken from Rand *et al.* (1979).

from the available literature that the efficiency of N retention is high at very low intakes and falls as N intake approaches requirement levels.

Recently, we have studied this aspect of body protein metabolism by quantifying components of whole body leucine and lysine metabolism in

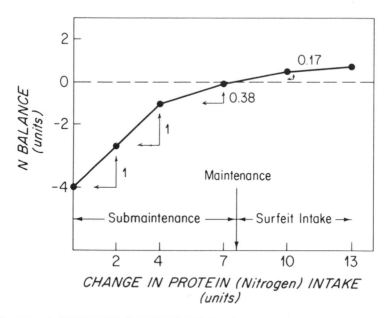

FIG. 7.5. A SCHEMATIC REPRESENTATION OF THE CHANGE IN EFFICIENCY OF NITROGEN RETENTION WITH INCREMENTS IN N INTAKE WITHIN THE SUBMAINTE-NANCE TO SUPRAMAINTENANCE RANGE OF N (PROTEIN) INTAKE

young men receiving protein intakes that approximate "usual" (1.0 g egg protein/kg/day), "requirement" (0.6 g/kg/day) or "minimal" (0.1 g/kg/day) intakes (Motil *et al.*, 1980). The study was carried out in eight healthy young men who received a continuous intravenous infusion of 1-^{13}C-leucine (and ^{15}N-lysine) following an overnight fast (fasted state) or while receiving small meals (fed state) after adaptation to these dietary protein intakes. Details of the ^{13}C and ^{15}N-tracer infusion protocols conducted at the end of each diet period for the different protein intakes have been described (Conway, *et al.*, 1980; Matthews, *et al.*, 1980) and a summary of the results is given in Figure 7.6. From these findings it can be seen that within the surfeit or supra-maintenance range of protein intake, body leucine and amino acid homeostasis is achieved largely by alterations in the rate of leucine oxidation, especially in the fed state. On the other hand, only small changes in the rates of leucine oxidation occur with changes in protein intake below the level for maintenance and the major response of body protein metabolism to changes in protein intake in this deficient range is associated with altered rates of leucine incorporation into proteins (protein synthesis) and entry into the metabolic pool from endogenous sources (protein breakdown). These limited but new data emphasize the fact that

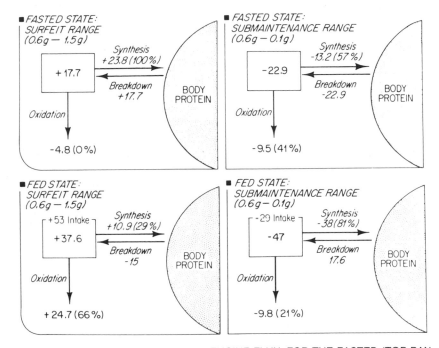

FIG. 7.6. SUMMARY OF CHANGES IN LEUCINE FLUX, FOR THE FASTED (TOP PAN-
ELS) AND FED STATES (BOTTOM PANELS), WITH AN INCREASE IN PROTEIN INTAKE
FROM THE MAINTENANCE REQUIREMENT TO ADEQUATE OR ABUNDANT LEVEL
(SURFEIT RANGE) OR A DECREASE IN INTAKE FROM THE REQUIREMENT TO A
DEFICIT INTAKE *(SUBMAINTENANCE RANGE)*
The figures shown are mean values for changes (H mole/kg/h) in leucine flux and its
component parts (intake, removal from pool via protein synthesis and oxidation, entry
into pool via protein breakdown). Values in parenthesis indicate percentage contribution made by leucine oxidation and removal from pool via protein synthesis in
relation to leucine flux. From Motil *et al.* (1980); based on a study in healthy young men
receiving three levels of protein intake; 0.1, 0.6 and 1.5 g egg protein/kg/day for 5 to
8 days prior to determination of whole body leucine flux and its major components
(oxidation rate of incorporation into body protein and release from body protein via
breakdown).

changes occur in the fate and metabolism of the individual essential amino
acids and in the dynamic status of body protein metabolism with variations
in the level of protein intake. Further they begin to establish a metabolic
basis for alterations in the efficiency of nitrogen retention with changes in
the level of protein intake and underscore the importance of considering the
choice or specific level of protein intake to be used in assessing the nutritional value of a test protein in human subjects.

The precise nature of the N balance response in adult subjects over the
submaintenance to near maintenance range of N intake has not been

determined extensively for different protein sources but studies in experimental animals show that for proteins limiting in lysine the efficiency of dietary N retention is greatly increased at low levels of protein intake (Hegsted, 1973). Experiments in adult human subjects are consistent with this observation (Young, 1975; Hegsted, 1974).

For these metabolic reasons alone, assays involving a single test protein level are not likely to give a precise index of protein quality and investigations designed to determine the N balance response to multiple doses of test protein would appear to offer a more critical approach for assessment of dietary protein quality. Thus, we have conducted studies routinely at several intake levels of test protein. The net effect, of course, is to increase the minimum number of experimental days necessary to obtain an estimate of the quality of a dietary protein under test. In a subsequent section in this review, we will consider a number of important statistical issues that arise in the planning and execution of these studies. However, before doing so, and in reference to the limitations of N balance experiments, we should state the objective of these investigations and raise some problems faced in their interpretation.

OBJECTIVE OF MEASUREMENT OF N BALANCE RESPONSES AND THEIR INTERPRETATION

The purpose of N balance studies for the assessment of dietary protein quality is to determine the minimum intake of a test or food protein required to maintain body protein homeostasis and organ function. Thus, N balance is used as an index of the maintenance or status of body and organ protein metabolism. By inference, the relative amount of test protein that is minimally required to achieve this N balance criterion is taken to be a quantitative and reliable measure of its nutritive value. However, it is important to recognize that the interpretation of N balance results may be difficult (Hegsted, 1976; Oddoye and Margen, 1979). It is pertinent to recall Schoenheimer's classic studies (Schoenheimer, 1942), involving use of stable isotopes to explore the turnover of body constituents, and the now accepted fact that a major proportion of total body protein undergoes continuous synthesis and breakdown in both growing and adult subjects. Thus, the balance between the anabolic and catabolic phases of protein and amino acid metabolism determines cell and organ protein content and, in turn, the efficiency with which dietary nitrogen is retained. Furthermore, it is also recognized that protein synthesis and breakdown are each affected by various factors and their rates regulated through specific control mechanisms (e.g., Schimke, 1970; Munro, 1970; Goldberg and Dice, 1974; Goldberg and St. John, 1976; Ballard, 1977; Walker, 1977) (Figure 7.7). Thus, a given body N balance can, in theory, be achieved within a wide range in the rates of protein synthesis and breakdown and alterations in N balance can be brought about by various combinations of changes in these rates (Figure 7.8). Hence, although N balance estimations can be useful in the evaluation of protein quality, N balance measures do not provide a suffi-

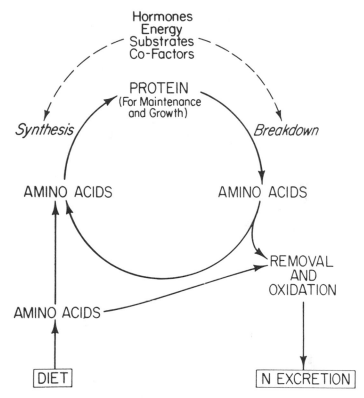

FIG. 7.7.   BODY AND ORGAN PROTEIN CONTENT IS DETERMINED BY THE BALANCE
BETWEEN THE RATES OF PROTEIN SYNTHESIS AND BREAKDOWN AND EACH OF
THESE PHASES OF PROTEIN METABOLISM ARE INFLUENCED BY FACTORS IN-
CLUDING HORMONES, SUBSTRATE (AMINO ACIDS, NITROGEN) AND ENERGY
SUPPLY

ciently secure picture of the status of protein and amino metabolism within
the body and this makes it desirable to supplement N balance measurements of protein quality, wherever it is feasible and worthwhile, with other
measures of protein nutritional status.

In addition to this limitation in the information supplied by N balance
measures, there are additional problems associated with the use of this
approach as a basis for assessing protein quality. For example, the high
retentions of nitrogen in subjects receiving generous intakes of protein
(Oddoye and Margen, 1979) remains a paradox because, according to current concepts, body protein content is essentially constant in adults or
declining only slowly during the advancing years of adult life (Forbes and
Reina, 1970; Young, et al., 1976). However, as discussed by Hegsted (1976),
and illustrated in a recent study by Oddoye and Margen (1979), adults
retain considerably more nitrogen beyond that expected when receiving

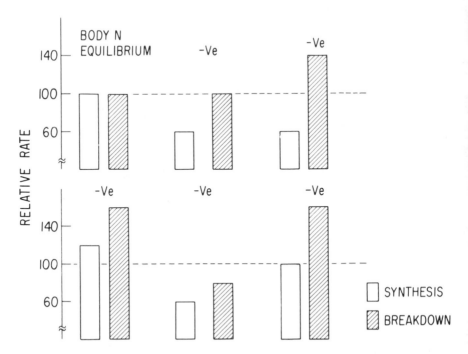

FIG. 7.8.   CHANGES IN RATES OF PROTEIN SYNTHESIS AND BREAKDOWN THAT
RESULT IN NEGATIVE NITROGEN BALANCE: A REDUCED OR NEGATIVE NITROGEN
BALANCE CAN ARISE THROUGH VARIOUS COMBINATIONS OF CHANGES IN WHOLE
BODY PROTEIN SYNTHESIS AND BREAKDOWN (SEE TEXT)

high N intakes and this is not explained adequately on the basis of methodological errors. Although it has been suggested that the physiological adjustment to altered protein intake might take a considerable longer period
of time than commonly assumed (Forbes, 1973) our own findings (Rand *et
al.*, 1979) and those of Oddoye and Margen (1979) fail to support this
suggestion. An alternative explanation was proposed by Costa *et al.* (1968,
1974) who suggest that there is evolution of gaseous nitrogen, arising
during the course of amino acid and nitrogen metabolism. Most reviews of
this topic (e.g. Anonymous, 1974) have concluded that this does not occur
and that there is no known pathway in mammals for the formation of N_2.
However, we are of the opinion that the necessary definitive studies to
validate or refute this point of view remain to be undertaken. Indeed there
is indirect evidence to support the opinion that formation of molecular
nitrogen can occur, possibly through a pathway such as that shown in
Figure 7.9. Thus, nitrite, known to occur in saliva (Tannenbaum *et al.*,
1974), may arise through heterotrophic nitrification by the intestinal micro-

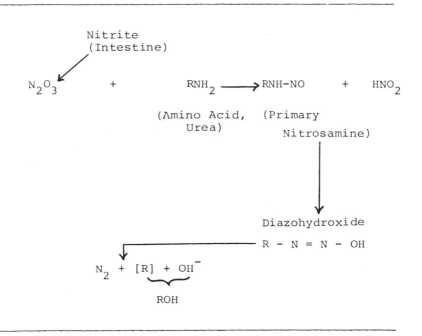

FIG. 7.9.   A POSSIBLE PATHWAY OF FORMATION OF MOLECULAR NITROGEN FROM INGESTED OR ENDOGENOUS N SOURCES

flora or elsewhere in body tissues (Tannenbaum and Young, 1980). Reaction of nitrite with primary amines can, through a series of N-nitroso-reactions, give rise to molecular nitrogen. Because human metabolic experiments have demonstrated a significant endogenous synthesis of nitrate (Tannenbaum, et al., 1978), it will be necessary to carry out further critical studies before it is possible to conclude unequivocally that molecular N or other gaseous or volatile nitrogen compounds cannot be formed in significant quantities during the course of amino acid and nitrogen metabolism in the body. The point that we wish to make is that this problem and the other limitations noted above emphasize the uncertainty surrounding conventional N balance determinations as measures of body and tissue N gain or loss. Our previous long-term metabolic studies in young men (Garza et al., 1977) lead us to the conclusion that it is not sufficient to rely solely on N balance data as a measure of dietary protein adequacy.

From the foregoing, it should be apparent that although there is a strong basis for assessing protein quality with the aid of conventional N balance studies a number of important questions concerning the interpretation of N balance data remain to be answered.

STATISTICAL ANALYSIS

Having considered the rationale for the conduct of nitrogen balance response studies for the evaluation of dietary protein quality it is appropriate to examine the approach in terms of a number of practical and statistical issues. In order to do this and to provide suitable examples, we will make use of data from several experiments conducted in our laboratories, applying standard, statistical techniques (Dunn and Clark, 1974; Afifi and Azen, 1979).

Basic to the discussion is the concept of a nitrogen balance response curve that describes the relationship between nitrogen (protein) intake (I) and nitrogen balance (NB). Whatever its true form, the nitrogen balance response curve can be approximated by a straight line or linear equation for a limited range of total nitrogen intakes. For such a curve, when in the region of submaintenance to near maintenance protein intake, the *slope* of this line can be taken as an index of the efficiency of nitrogen retention while the *intersection* of the N balance response curve with the line of body nitrogen balance equilibrium (zero nitrogen balance) represents the minimum intake necessary to meet requirements (Young *et al.*, 1977; Rand, *et al.*, 1977). Thus, it is important to consider how to estimate this response curve in the individual subject, and then how to aggregate the data for groups of individuals to describe a population response, all with respect to a single test protein source. The final step is that of using these estimates to compare different protein sources.

Given N intake and balance data for an individual studied at several different levels of a test protein, where the measurement of balance is based on a successive number of days long enough to reduce error to an acceptable level and following an adaptation or adjustment period, the nitrogen balance response curve can be estimated by standard least squares regression. An example of this procedure is shown in Table 7.7 and these data reveal that for this, not atypical, subject the nitrogen balance results are quite variable. For this reason the estimate of the response curve cannot be made with great confidence. (With only four data points, statistical significance requires that the ratio of the slope to its standard error be larger than 4.3). Using this approach, Table 7.8 gives the results for each of eight individuals who participated in a nitrogen balance response experiment. In this example, none of the individual slopes statistically differed from zero and, in addition, two individuals showed slopes distinctly different from the re-

TABLE 7.7. REGRESSION OF *NITROGEN BALANCE* (NB) ON N INTAKE (I) FOR SUBJECT CB[1]

I	NB	Predicted	Residual
56	−33	−39	6
71	−36	−28	−8
87	−17	−16	−1
104	0	−3	3

[1] Values are mg N kg^{-1}day^{-1}; NB = −81 + 0.75I; SE (slope) = 0.21; SE (fit) = 17.

TABLE 7.8. ESTIMATED LINEAR RESPONSE CURVES OF *N BALANCE* FOR EIGHT
SUBJECTS RECEIVING SOY ISOLATE AT GRADED LEVELS

| Subj. | Intercept | Slope | Standard Error | |
			Slope	Fit
C.B.	−80.8	.75	.21	16.6
C.C.	−20.0	−.06	.37	13.5
J.F.	−30.8	.27	.08	6.9
F.H.	−48.3	.40	.19	17.0
D.H.	−39.5	.28	.07	6.5
R.F.	−48.1	.29	.15	11.0
R.M.	−37.6	.35	.18	13.0
R.R.	−26.0	.23	.34	26.4

maining subjects. For one of these subjects the slope was about twice that of
the mean and in the other subject the slope appeared to be negative. A
further limitation of this approach is that there is no satisfactory statistical
test to examine the adequacy of the model, in this case a straight line.
However, the standard errors of the fits given in the table suggest that the
model does mimic the actual N balance response in that these errors are
similar to the variabilities observed for daily nitrogen balances discussed
earlier. On this basis, it appears that our experimental design, involving an
initial "protein-free" day and a 10 day diet period with N balances based on
the final 5 days is adequate.

A second point that can be made from these examples is that this statistical approach makes inefficient use of the available data because the nitrogen balance values are derived from a number of separate measurements.
This problem can be explored by analysis of the individual components of
nitrogen balance, with the objective of determining whether additional
useful information can be obtained about the responses of N metabolism to
the various levels of test protein intake. Hence, the regression of urinary
nitrogen on nitrogen intake can be examined and an example of this is given
in Table 7.9. For this particular subject, it appears that the urinary nitrogen
output is not responding linearly to N intake. An analysis for eight subjects,
shown in Table 7.10, reveals that, in general, the regression coefficients
describing the relationship between urinary N output and N intake are
statistically different from zero. In addition, this approach permits a check
of the linearity of the response by comparing the standard errors of the
individual fits with the pooled within standard deviations, as calculated

TABLE 7.9. REGRESSION OF URINARY NITROGEN (U_N) ON NITROGEN INTAKE (I)
FOR SUBJECT CB[1]

I	U_N	Predicted	Residual
56	67	75	−8
71	89	78	11
86	81	81	0
104	81	84	−3

[1] Values are mg N kg^{-1} day^{-1}; $U_N = 64.2 + 0.19I$; SE (slope) = 0.28; SE (fit) = 22.

TABLE 7.10. ESTIMATED RESPONSE CURVES FOR *DAILY URINARY NITROGEN* EXCRETION IN EIGHT SUBJECTS RECEIVING SOY ISOLATE AT GRADED LEVELS

Subj.	Intercept	Slope	Standard Error		
			Slope	Fit	Within
C.B.	64.2	.19	.28	21.9	15.5
C.C.	9.4	.96	.31	22.6	15.4
J.F.	16.0	.70	.05	4.3	6.9
F.H.	29.5	.60	.18	15.6	9.5
D.H.	29.6	.61	.09	8.5	3.9
R.F.	31.9	.70	.14	9.6	8.8
R.M.	22.7	.63	.14	10.1	6.4
R.R.	10.7	.77	.39	30.5	4.5

from the original data for daily urinary nitrogen excretion. In this case only one (C.B.) of the eight subjects shows significant non-linearity.

A similar treatment can be made for fecal nitrogen (Table 7.11) and these findings show only slight variation in fecal N with changes in protein intake. For high digestibility sources this is to be expected. Thus, the value for the slope given in Table 7.11 is much less than its standard error and suggests that for some test proteins it may be legitimate to use an average value for fecal N output during the entire study period. This conclusion is strengthened by the analysis of results for all eight individuals in this particular experiment because, as summarized in Table 7.12, none of the slopes were significantly different from zero. Also, the variabilities of the fits themselves are not different from that which can be predicted from data obtained in long term metabolic experiments at a constant protein intake, as discussed earlier.

TABLE 7.11. REGRESSION OF *FECAL NITROGEN* (F_N) ON NITROGEN INTAKE (I) FOR SUBJECT CB[1]

I	F_N	Predicted	Residual
56	17	15	2
71	13	16	−3
87	18	17	1
104	17	17	0

[1] Values are mg N kg^{-1} day^{-1}; $F_N = 12.8 + 0.04$ I; SE (slope) = 0.07; SE (fit) = 6.

TABLE 7.12. EVALUATION OF *FECAL NITROGEN* EXCRETION IN EIGHT YOUNG MEN RECEIVING GRADED INTAKES OF SOY ISOLATE

Subj.	Intercept	Slope	Standard Error	
			Slope	Fit
C.B.	12.8	.04	.07	5.9
C.C.	5.5	.10	.07	5.2
J.F.	9.7	.03	.08	6.3
F.H.	13.7	.004	.02	1.4
D.H.	4.9	.11	.03	2.9
R.F.	11.3	.01	.03	1.8
R.M.	9.9	.02	.04	3.1
R.R.	10.3	−.002	.07	5.3

It should be evident that combining these two component equations, described in Tables 7.9 and 7.11, results in the original regression values (Table 7.7) and this provides confidence that the linear curve is indeed a good approximation of the N balance response to graded intakes of test protein.

Further examination of these example N balance data suggests a more satisfactory way of incorporating the N balance values into a final analysis and evaluation of the results. Thus, to check for linearity of the urinary nitrogen response we have also calculated the within subject variance for the 5 day balance periods at each level of protein intake. Table 7.13 presents an example of this calculation and shows that the obvious outlier for the four experimental periods is that point with the largest standard deviation. It is, therefore, the balance (or more specifically urinary N) that is least well known for the entire set of diet periods. This suggests application of *weighted regression*, a technique which considers not only the balance of an individual during each specific diet period but also the possible variability during that particular phase of the study. Hence, in Table 7.13 the resulting regression equation better fits those points with small error than it does that point which has a high variance.

TABLE 7.13.   WORKED EXAMPLE OF *WEIGHTED REGRESSION* OF URINARY NITROGEN (U_N) ON NITROGEN INTAKE (I) FOR SUBJECT CB[1]

I	U_N	S_u	Predicted U_N	Residual
56	67	11	70	−3
71	89	27	74	15
87	81	6	79	2
104	81	7	83	−2

[1] Values are mg N/kg/day; $U_N = 53.5 + 0.29$ I; SE (slope) = 0.16; SE (fit) = 5.

A weighted regression of urinary nitrogen excretion can be combined with the regression of fecal nitrogen to provide a *composite nitrogen balance response curve* by simply adding the individual regressions. Results from the use of this type of analysis are shown in Table 7.14 for data obtained with our example subject. As may be noted, the effect of the outlying N balance period has been diminished while the fit to the other points has been improved. For this reason Table 7.15 is given, in which we compare the weighted and unweighted regressions for the entire group of subjects in the

TABLE 7.14.   *COMPOSITE REGRESSION* OF NITROGEN BALANCE (NB) ON NITROGEN INTAKE (I)[1]

I	NB	Predicted	Residual
56	−33	−34	1
71	−36	−24	−12
81	−17	−13	−4
104	0	−2	2

[1] Values are mg N/kg/day; NB = − 71.4 + 0.67 I.

TABLE 7.15. COMPARISON OF SLOPE AND ZERO N BALANCE INTERCEPTS ESTIMATED BY WEIGHTED AND UNWEIGHTED REGRESSION

	Slope		Intercept	
Subj.	Unweighted	Weighted	Unweighted	Weighted
C.B.	.75	.67	108.1	106.5
C.C.	−.06	.23	—	181.4
J.F.	.27	.30	114.1	110.8
F.H.	.40	.32	121.7	133.7
D.H.	.28	.23	142.9	158.4
R.F.	.29	.26	164.4	179.7
R.M.	.35	.24	108.9	117.7
R.R.	.23	.19	111.1	114.2
Mean	.31	.31	124.5	137.8
SD	.22	.15	21.4	31.1
SE	.08	.05	8.1	11.0

experiment and this analysis reveals two important consequences of using a weighted regression; firstly, the slopes are more consistent for the population sample (the between individual standard error being reduced by about half) and secondly, an individual who appeared to behave anomolously in the original regression and would have been considered as an outlier can now be viewed as a subject with an extreme value.

The above considerations apply to the analysis of N balance responses within individual subjects. The next step is to examine estimates of a population's response to a specific test protein source. For this purpose the direct approach is to use estimates of the individual responses in terms of slope (efficiency) and intercept with zero nitrogen balance (requirement) as a sample of the population responses. This suggests use of average values and the standard deviations of the slopes and intercepts that were shown above in Table 7.15.

There are two statistical constraints on this approach. The first is that use of the arithmetic mean as a valid descriptor of a population assumes a single population with no anomolous individuals, or outliers. However, if the population consists of a mixture of mainly normal individuals but with a few individuals who differ fundamentally, then more robust statistical techniques are necessary. The simplest of these is the *trimmed mean*, where a certain percentage of the data (those values high and low) are discarded routinely. However, we cannot recommend any general use of the trimmed mean with so few subjects because it requires the routine elimination of data that is expensive and difficult to obtain and may well be valid. The alternative that is followed by many investigators is to eliminate data for subjects with apparent clinical problems. However, where the investigator's subjective judgment is introduced into the analysis, this may well invalidate the experiment. This is a particular problem which needs attention in relation to human studies, especially including careful examination and retesting of those subjects who display variant responses to determine the reproducibility of any important anomolous values. In addition, investigations in larger population groups will be necessary to determine the actual extent of the problems introduced by so-called outliers.

A second statistical constraint involved in the description of a population response is that of estimation of upper percentiles, in order to determine the level of a test protein or mixture of proteins that would meet the requirement for that protein for a fixed fraction of the population. Statistical methods do exist for doing this, including calculation of tolerance intervals (Rand *et al.*, 1977). However, they require a known distribution and they may result in unreasonably large values when based on small samples. The inherent problem is that of estimating the tail of a distribution when most of the data describes the central portion. Practically, in relation to the assessment of protein quality from N balance response studies, we use, as the best technique available, the estimated mean requirement plus an appropriate number of standard deviations. It must be stated, however, that studies involving relatively small numbers of subjects do not provide sufficient data to establish estimates of the extreme percentiles with confidence.

COMPARISON OF DIFFERENT SOURCES OF PROTEIN BASED ON STUDIES IN SMALL POPULATIONS OF HEALTHY ADULT MEN

In the preceding sections we have examined the variations that exist in N balance experiments. We have also discussed statistical procedures for analyzing the N balance responses observed in multiple-test-level studies designed to assess the nutritive value of food protein sources. Here we will consider briefly the application of the principles developed above in a series of studies in healthy young men conducted to assess the relative quality of five different protein sources (egg, milk, soy protein isolates, strained beef and whole ground wheat) in adult human nutrition. Studies involving beef and wheat have been published in detail (Young *et al.*, 1975) and those concerned with soy have been summarized elsewhere (Young and Scrimshaw, 1979; Scrimshaw and Young, 1979). All experiments followed the same basic design. Four levels of test protein were given in each experiment, except in one involving whole ground wheat alone and this involved three test levels. The experimental diet periods for the wheat and beef study (Young *et al.*, 1975) lasted 15 days at each level; for the other protein studies they lasted 10 days at each intake level. A modified Latin Square was used for allocation of subjects to the various diet periods. Prior to each experimental period the subjects were allowed a free-choice diet providing liberal amounts of protein. All received an initial "protein-free" diet for one day to accelerate adaptation to the lower levels of each protein.

Analysis of the data obtained in these N balance studies followed that outlined in the previous section (see Tables 7.7 and 7.8). This involved calculation of the simple linear regression line to estimate the nitrogen balance response to changes in N intake for each subject. Mean *efficiency* of nitrogen utilization was determined from the average value for the *slopes* of the individual regression lines. The average of the intercepts of separate nitrogen balance regression lines with the line of body nitrogen equilibrium (zero nitrogen balance) estimates the mean nitrogen *requirement* for the test protein source in the study population (Rand *et al.*, 1977).

Table 7.16 summarizes these data, in terms of *slopes* and *intercepts* (with the line of N balance equilibrium) for the example series of studies. These data show the magnitude of the variation in these two related measures of protein quality for groups of subjects receiving various test protein sources. They also provide an indication of the variation among the different experiments. Our experience with respect to the reproducibility of the *actual* values for slope and intercepts among different experiments involving the same or apparently similar protein source has been mixed. As shown in this table the protein quality measures for wheat were quite comparable in two experiments separated by about two years, although the absolute values for soy, without consideration of the value obtained for the reference protein included in a particular experiment, tended to differ between the two series of studies shown here. We have not, however, explored in a prospective manner the reproducibility of an individual's N balance response curve. Furthermore, it does not appear from the published literature that this problem has been examined critically.

TABLE 7.16. SUMMARY OF A SERIES OF EXPERIMENTS IN WHICH PROTEIN SOURCES WERE COMPARED FOR THEIR N BALANCE RESPONSES

Series	Proteins	No. of Subjects	Quality Index Efficiency (slope)	Quality Index Requirement (Intercept)
Egg vs. Soy				
	Egg	8	0.52 ± 0.06[1]	90 ± 11[2]
	Soy[3]	8	0.41 ± 0.03	114 ± 6
Milk vs. Soy				
	Soy[3]	8	0.31 ± 0.05	138 ± 11
	DSM	6	0.34 ± 0.09	145 ± 33
	Soy[4]	5	0.39 ± 0.08	112 ± 9
Wheat vs. Beef				
	Wheat	7	0.27 ± 0.04	147.2 ± 22
	Beef	7	0.50 ± 0.05	83 ± 4
Wheat 1975				
	3 Levels	6	0.21 ± 0.04	135 ± 22

[1] Mean ± SE.
[2] Values are means ± SE; mg N intake/kg for zero N balance.
[3] Supro-620 (Ralston Purina).
[4] Supro-710 (Ralston Purina).

Based on this series of studies, the *comparative* estimates of protein quality for soy and whole wheat in healthy adult men are summarized in Table 7.17. The mean values for relative protein value (RPV) and relative nitrogen requirement (RNR) are given in this table together with their statistical significance. Thus, for well processed soy isolates, their protein

TABLE 7.17. COMPARATIVE ESTIMATES OF THE NUTRITIVE VALUE OF VARIOUS PROTEIN SOURCES

Test Protein	Reference	RPV[1]	RNR[2]
Soy Isolate[3]	Egg	79 (p>0.1)[4]	79 (p>0.05)
Wheat	Beef	54 (p<0.01)	56 (p<0.02)
Soy Isolate[3]	Milk	91 (p>0.1)	105 (p>0.1)
Soy Isolate[5]	Milk	115 (p>0.1)	129 (p>0.1)

[1] RPV (relative protein value) = (slope of N balance response curve with test protein divided by slope of N balance response curve with reference protein) × 100.
[2] RNR (relative nitrogen requirement) = (N intake of the reference protein required for zero N balance divided by the N intake of the test protein required for zero N balance) × 100.
[3] Supro-620 (Ralston Purina).
[4] Significance of difference from reference.
[5] Supro-710 (Ralston Purina).

quality cannot be distinguished from egg or milk, used as reference proteins. On the other hand, whole ground wheat was of significantly lower quality than beef protein as a reference. These results are consistent with preliminary results obtained in pre-school children with the same soy protein isolate (Young, et al., 1979). In this case, the nutritional quality of the soy isolate was found to be close to that of milk. However, these findings differ markedly from those obtained in rat bioassays (Table 7.18) where the soy protein isolates show a significantly lower protein value than casein, unless the former are supplemented with methionine. We have previously discussed these differences in relation to the use of rat bioassay data as a sole basis for formulation of policy concerning protein quality and human foods (Young et al., 1979).

From these findings (Tables 7.16 and 7.17), it is evident that for protein sources with a nutritive value equivalent to that of wheat, or about 50–60% that of a high quality protein standard, studies in a small group of healthy

TABLE 7.18. COMPARISON OF ESTIMATES OF PROTEIN QUALITY OF SOY ISOLATE (SUPRO 620) IN CHILDREN AND ADULTS COMPARED WITH THOSE OBTAINED IN RATS[1]

Species & Group	Protein Quality Measure	SAA Requirement (mg/ g protein)
Human		
Child	NBI (98)[2]	34
Adult	RNR (80)[3]	24
Rat:		
(−Meth)	PER (1.65)[4]	50
(+Meth)	PER (2.5)[4]	50

[1] Summaried from review by Young et al. (1979).
[2,3] Percent compared with milk and egg, respectively.
[4] PER value in parenthesis adjusted to a PER for casein of 2.5. Note: the soy isolate provides 26 mg SAA per g protein.

adults will be able to detect differences. However, where the test protein quality is higher, a larger number of experimental subjects would be required in order to detect a difference in their nutritional values, assuming a difference actually exists and is important from the standpoint of practical nutrition. This point is illustrated in Table 7.19, which provides estimates of the numbers of subjects required to differentiate protein sources with low-, moderate- and high-nutritional value. Thus, to ensure distinguishing between two proteins, one with a nutritional value equivalent to egg or milk and the other 75% of this value, nine subjects would be required, assuming that a 50:50 chance of not detecting a difference between the two proteins (test and reference) was an acceptable level of *beta* error. This reasoning applies equally to RPV and RNR as measures of quality. Hence, to distinguish between proteins of more similar quality this would require experiments involving relatively large numbers of subjects (Table 7.19). This situation contrasts with results based on well standardized studies carried out in genetically homogenous groups of rats, as reflected by the data summarized in Table 7.20. As shown here, the variation observed within a group of rats is about an order of magnitude less than that which we have observed in many human experiments. While this points out a significant limitation in human nitrogen balance studies, conversely, it does suggest

TABLE 7.19. ESTIMATES OF SUBJECT NUMBERS REQUIRED TO DIFFERENTIATE (P = 0.05) PROTEIN SOURCES, BASED ON ACTUAL VARIANCES OBSERVED IN A SERIES OF MIT STUDIES

	For Protein with RPV of:		
Beta error[1]	60	75	90
0.5	4	9	32
0.1	7	23	86
	For N Requirement greater than egg by:		
	15%	30%	50%
0.5	21	7	4
0.1	54	16	7

[1] Probability of finding no difference when the protein values are actually 60, 75 or 90% of reference.

TABLE 7.20. COMPARISON OF VARIATION OBSERVED IN PROTEIN QUALITY MEASURES FOR VARIOUS PROTEINS ASSAYED IN RATS AND ADULT HUMANS

Source	Rat[1] (NPU)	Human (Slope)
Hens Egg	0.94 ± 0.003[2]	0.52 ± 0.06
Soy Isolate (Supro 620)	0.67 ± 0.004	0.41 ± 0.03
Dried Skim Milk	0.89 ± 0.005	0.39 ± 0.08
Soy Isolate (Supro 710)	0.59 ± 0.005	0.34 ± 0.09

[1] Assays conducted by B. Eggum, Copenhagen.
[2] Mean $\pm$ SE; Five rats per group and 5–8 human subjects per group.

that the latter studies may be more informative about the practical signifi-cance of different sources of food protein intended for *direct* consumption by people.

SUMMARY AND CONCLUSIONS

Conventional or "long-term" metabolic balance experiments in humans are often considered to represent the standard or reference method for evaluation of the nutritional quality food protein sources (Bodwell, 1977). However, there has been little detailed discussion and examination of this approach as a reference method for quantifying the nutritive value of the important proteins and protein sources in human diets. Therefore, we have focused on the metabolic rationale for conducting N balance response stud-ies. Because our recent experience has been based on studies in adult subjects, we have examined some of the problems inherent in and/or associ-ated with metabolic N balance investigations for this age group. The extent to which our observations apply to subjects of other ages, particularly pre-school children, will require further study and a detailed analysis of data comparable to those described in the various sections of this review. Nevertheless, it is possible to draw a number of conclusions that may help further exploration of the significance of dietary protein quality in human nutrition.

Specifically, we consider that: 1) Conventional or "long-term" nitrogen balance response studies with human subjects given graded levels of a test protein, represent an essential component of the study of the nutritional quality of food protein sources consumed by humans or intended for use in human diets at levels that are nutritionally significant. 2) These experi-ments should be designed to include an adequate period of adaptation to the test diets and allow collection of data extending over a long enough period for the effect of daily variability in N balance to be acceptable. 3) Statistical analyses should involve the fitting of each individual's data (nitrogen in-take, urinary nitrogen output and fecal nitrogen output) to a linear model, using weighted least squares where appropriate. The individual regression coefficients (the slopes of the response curves) which reflect efficiency of di-etary N utilization, and the intersections of the N balance response curve with zero N balance which represent an estimate of requirement for the test protein, can be used to estimate these two related aspects of the population's response to a test protein. 4) Despite the fact that our studies and those of others reported in the literature show that populations are quite variable with respect to the response of their members to different proteins, food pro-teins of differing quality can be discriminated by N balance methods if ap-propriate samples sizes are used.

Finally, we wish to emphasize that although the N balance response method that we have discussed in detail provides a valuable technique for determining protein quality in humans, we do not consider it to be a sole or only approach that can or should be used in reference to examining various aspects of protein quality (Fig. 7.10). As others have pointed out this is a

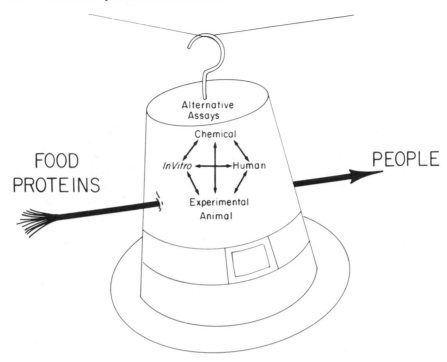

FIG. 7.10.   THE QUANTITATIVE INTERRELATIONSHIPS BETWEEN ALTERNATIVE AS-
SAY PROCEDURES FOR PROTEIN QUALITY EVALUATION MUST BE EXPLORED IN
ORDER TO JUDGE WHAT PARTICULAR ASSAY (S) WE SHOULD *"HANG OUR HAT ON"*
FOR DETERMINATION OF THE CAPACITY OF FOOD PROTEINS TO MEET HUMAN
REQUIREMENTS

difficult, tedious, time-consuming and expensive method to conduct with
success. Shorter-term human assays or those based on single test levels can
provide useful information but these methods also require validation and
critical evaluation for their capacity to predict, in quantitative terms, the
nutritional value of a source of food protein. Indeed we have used a compli-
mentary series of studies involving different specific approaches in order to
establish the nutritive value of soy isolates in human nutrition (Young *et
al.*, 1979). Thus, as summarized in Table 7.21, after initial N balance
response studies, with soy proteins, we have followed these with an assessment of (i) the nutritional significance of the sulfur amino acid content of
soy isolates, (ii) the nutritional value of soy in combination with meat
protein, (iii) the capacity of soy protein alone to support long-term protein
nutritional status when this protein source is consumed at a level predicted
from N balance response studies to be sufficient to meet the requirement of
nearly all healthy subjects in the target population, and (iv) the long-term
tolerance and acceptability in adults on free-choice diets to ingestion of a
significant daily amount of the test proteins. In summary, the findings

TABLE 7.21.   APPROACHES USED FOR EVALUATION OF THE NUTRITIONAL VALUE
OF SOY PROTEIN ISOLATES

1. *N Balance Response Studies*
2. Assessment of the nutritional significance
   of sulfur amino acids in soy protein.
3. Nutritional value in combination with meat
   proteins.
4. Long-term protein nutritional maintenance.
5. Tolerance and acceptability
6. Impact on trace element availability.

obtained with these other approaches confirm and extend the conclusions that were drawn from the more limited data derived with N balance response studies. In this case, where it was of considerable interest and importance to assess the protein quality of a new or non-traditional source of protein for increased use in the U.S. diet, the conventional or long-term N balance response method appears to have been fully vindicated. Now we have begun to explore the interaction, if any, between soy and the trace element constituents of the diet, using stable isotope technology (Young and Janghorbani, 1980), in order to describe more comprehensively the nutritional quality of this vegetable protein food. It is our conclusion that conventional or "long-term" N balance response studies do help to fit securely some of the pieces into the jigsaw puzzle of human nutrition.

ACKNOWLEDGEMENTS

We thank many devoted colleagues, past and present graduate students, and the human volunteers for their significant contribution in our studies of dietary protein quality in human. We are grateful to Dr. B. Eggum for carrying out the rat NPU assays for us.

REFERENCES

AFIFI, A.A., and AZEN, S.P.   1979.   *Statistical Analysis*. A Computer Oriented Approach.   Academic Press. New York.

ANKER, R.M.   1954.   The determination of creatine and creatinine in urine; a correction factor for the determination of twenty-four hour urinary excretion values. J. Lab. Clin. Med. *43*:798.

ANONYMOUS,   1974.   Equality of inspired and expired N_2 in man. Nutr. Rev. *32*:117.

APPLEGARTH, D.A., HARDWICK, D.F., and ROSS, P.M.   1968.   Creatinine excretion in children and the usefulness of creatinine equivalents in amino acid chromatography.   Clin. Chim. Acta *22*:131.

BALLARD, F.J.   1977.   Intracellular protein degradation.   In: *Essays in Biochemistry* (eds. P.N. Campbell and W.N. Aldridge), *13*:1, Academic Press, Inc., New York.

BLEILER, R.E., and SCHEDL, H.P.   1962.   Creatinine excretion: Variability and relationships to diet and body size.   J. Lab. Clin. Med. *59*:945.

BODWELL, C.E.   1977.   Problems associated with the development and application of rapid methods of assessing protein quality.   Nutr. Rept. Intl. *16*:163.

BRESSANI, R., and VITERI, F.   1971.   Metabolic studies in human subjects. In: Proc. SOS/70 Third Intl. Congress Food Sci. Tech. p. 344–357, Institute of Food Technologists, Chicago, Ill.

BRICKER, M., MITCHELL, H.H., and KINSMAN, G.M.   1945.   The protein requirements of adult human subjects in terms of the protein contained in individual foods and food combinations. J. Nutr. *30*:269.

CALLOWAY, D.H., and MARGEN, S.   1971.   Variations in endogenous nitrogen excretion and dietary nitrogen utilization as determinants of human requirements. J. Nutr. *101*:205.

CHATTAWAY, F.W., HULLIN, R.P., and ODDS, F.C.   1969.   The variability of creatinine excretion in normal subjects, mental patients and pregnant women. Clin. Chim. Acta *26*:567.

CHINN, K.S.K.   1967.   Prediction of muscle and remaining tissue protein in man. J. Appl. Physiol. *23*:713.

CLARK, H.E., HOWE, J.M., MAGEE, J.L., and MALZER, J.L.   1972.   Nitrogen balances of adult subjects who consumed four levels of nitrogen from a combination of rice, milk and wheat. J. Nutr. *102*:1647.

CLARK, H.E., YESS, N.J., VERMILLIAN, E.J., GOODWIN, A.F., and MERTZ, E.T.   1963.   Effect of certain factors on nitrogen retention and lysine requirements of adult subjects. iii. Source of supplementary nitrogen. J. Nutr. *79*:131.

CONWAY, J.C., BIER, D.M., MOTIL, K.J., BURKE, J.F., and YOUNG, V.R.   1980.   Whole body lysine flux in young adult men: Effects of reduced total protein and of lysine intake. Am. J. Physiol. (in press)

COSTA, G., KERINS, M.E., KANTOR, F., GRIFFITH, K., and CUMMINGS, W.B.   1974.   Conversion of protein nitrogen into gaseous catabolites by the chick embryo. Proc. Natl. Acad. Sci. *71*:451.

COSTA, G., ULLRICH, L., KANTOR, R., HOLLAND, J.F.   1968.   Production of elemental nitrogen in certain mammals, including man. Nature *218*:546.

CRAMER, K.H., SRAMER, H., and SELANDER, S.   1967.   A comparative analysis between variation in 24-hour urinary creatinine output and 24-hour volume. Clin. Chim. Acta *15*:331.

CURTIS, G., and FOGEL, M.   1970.   Creatinine excretion: diurnal variation and variability of whole and part-day measure. A methodologic issue in psychoendocrine research. Psychosomat. Med. *32*:337.

DUNN, O.J., and CLARK, V.A.   1974.   *Applied Statistics: analysis of variance and regression*, John Willey and Sons, New York.

EDWARDS, O.M., BAYLISS, R.I.S., and MILLER, S.   1969.   Urinary creatinine excretion as an index of the completeness of 24-hour urine collections. Lancet *ii*:1165.

EPSTEIN, S.E., and SCHRIEVER, H.G.   1970.   Creatinine excretion. Lancet *i*:192.

FORBES, G.B.   1973., Another source of error in metabolic balance method. Nutr. Rev. *31*:297.

FORBES, G.B., and REINA, J.C.   1970.   Adult lean body mass declines with age: some longitudinal observations. Metabolism *19*:653.

GARZA, C., SCRIMSHAW, N.S., and YOUNG, V.R.   1977.   Human protein requirements: long-term metabolic nitrogen balance studies in young men to evaluate the 1973 FAO/WHO safe level of egg protein intake. J. Nutr. *107*: 335-352.

GOLDBERG, A.L., and DICE, J.F.   1974.   Intracellular protein degradation in mammalian and bacterial cells. Ann. Rev. Biochem. *43*:835.

GOLDBERG, A.L., and ST. JOHN, A.C.   1976.   Intracellular protein degradation in mammalian and bacterial cells. Part 2. Ann. Rev. Biochem. *45*:747.

GRAYSTONE, J.E.   1968.   Creatinine excretion during growth. In: *Human Growth* (ed. D.B. Cheek) Chpt. 12, p. 182, Lea and Febiger, Philadelphia.

HEGSTED, D.M.   1973.   The amino acid requirements of rats and human beings. In: *Proteins in Human Nutrition* (eds. J.W.G. Porter and B.A. Rolls) Chpt. 18, p. 275-291, Academic Press, New York.

HEGSTED, D.M.   1974.   Assessment of protein quality. In: *Improvement in Protein Nutriture*, National Research Council (Food and Nutrition Board), National Academy of Sciences, p. 64, Washington, D.C.

HEGSTED, D.M.   1976.   Balance studies. J. Nutr. *106*:307.

HOFFMAN, W.S. and McNEIL, G.C.   1949.   The enhancement of the nutritive value of wheat gluten by supplementation with lysine, as determined from nitrogen balance indices in human subjects. J. Nutr. *30*:331.

INOUE, G., FUJITA, Y., KISHI, K., YAMAMOTO, S., and NIIYAMA, Y.   1974.   Nutritive value of egg protein and wheat gluten in young men. Nutr. Reports Intl. *10*:201.

INOUE, G., FUJITA, Y., and NIIYAMA, Y.   1973.   Studies in protein requirements of young men fed egg protein and rice protein with excess and maintenance energy intakes. J. Nutr. *103*:1673.

JACKSON, S.   1966.   Creatinine in urine as an index of urinary excretion rate. Health Phys. *12*:843.

KIES, C., and FOX, H.M.   1970.   Effect of level of total nitrogen intake on second limiting amino acid in corn for humans. J. Nutr. *100*:1275.

KIES, C., and FOX, H.M.   1972.   Protein nutritional value of *Opaque-2* corn grain for human adults. J. Nutr. *102*:757.

KIES, C., WILLIAMS, E., and FOX, H.M.   1965.   Determination of first limiting nitrogenous factor in corn protein for nitrogen retention in human adults. J. Nutr. *86*:350.

KISHI, K., MIYATANI, S., and INOUE, G.   1978.   Requirement and utilization of egg protein by Japanese Young Men with marginal energy intake. J. Nutr. *108*:658.

KREBS, H.A.   1972.   Some aspects of the regulation of fuel supply in omnivorous animals. Adv. Enzyme Reg. *10*:397.

MARTIN, C.J., and ROBISON, R.   1922.   The minimum nitrogen expenditure of man and the biological value of various proteins in human nutrition. Birchen. J. *16*:407.

MATTHEWS, D.E., MOTIL, K.J., ROHRBAUGH, D.K., BURKE, J.F., YOUNG, V.R., and BIER, D.M.   1980.   Measurement of leucine metabolism *in vivo* from primed, continuous infusion of L-[1-^{13}C] leucine in man. Am. J. Physiol. (in press)

MEADOR, C.K., KREISBERG, R.A., FRIDAY, J.P., BOWDOIN, B., COON, P., ARMSTRONG, J., and HAZELRIG, J.B.   1968.   Muscle mass determination by isotope dilution of creatine -^{14}C. Metab. Clin. Exp. *17*:1104.

MOTIL, K.J., BIER, D.M., MATTHEWS, D., BURKE, J.F., MUNRO, H.N., and YOUNG, V.R.   1980.   Whole body leucine and lysine: Response to dietary protein intake in young men. Am. J. Physiol. (In press).

MULDOWNEY, F.P., CROOKS, J., and BLUHM, M.M.   1957.   The relationship of total exchangeable potassium and chloride to lean body mass, red cell mass and creatinine excretion in man. J. Clin. Invest. *36*:1375.

MUNRO, H.N.   1964.   General aspects of the regulation of protein metabolism by diet and by hormones. In: *Mammalian Protein Metabolism* (eds. H.N. Munro and J.B. Allison), Vol. I, p. 381–481, Academic Press, New York.

MUNRO, H.N.   1970.   A general survey of mechanisms regulating protein metabolism in mammals. In: *Mammalian Protein Metabolism*, (ed. H.N. Munro), Vol. IV, p. 3–130, Academic Press, New York.

NICOL, B.M., and PHILLIPS, P.G.   1978.   The utilization of proteins and amino acids based on cassava *(Manihot utilissima)*, rice or sorghum *(Sorghum sativa)* by young Nigerian men of low income. Brit. J. Nutr. *39*:271.

ODDOYE, E.A., and MARGEN, S.   1979.   Nitrogen balance studies in humans: Long-term effect of high nitrogen intake on nitrogen accretion. J. Nutr. *109*:363.

PATERSON, N.   1967.   Relative constancy of 24-hour urine volume and 24-hour creatinine output. Clin. Chim. Acta *18*:57.

RAND, W.M., SCRIMSHAW, N.S., and YOUNG, V.R.   1977.   Determination of protein allowances in human adults from nitrogen balance data. Am. J. Clin. Nutr. *30*:1129.

RAND, W.M., SCRIMSHAW, N.S., and YOUNG, V.R.   1979.   An analysis of temporal patterns in urinary nitrogen excretion of young adults receiving constant diets at two nitrogen intakes for 8 to 11 weeks. Am. J. Clin. Nutr. *32*:1408.

RAND, W.M., YOUNG, V.R., and SCRIMSHAW, N.S.   1976.   Change of urinary nitrogen excretion in response to low protein diets in adults. Am. J. Clin. Nutr. *29*:639.

ROMO, G.S., and LINKSWILER, H.   1969.   Effect of level and pattern of essential amino acids on nitrogen retention of adult men. J. Nutr. *97*:147.

SCRIMSHAW, N.S., HUSSEIN, M.A., MURRAY, E., RAND, W.M., and YOUNG, V.R. 1972. Protein requirements of man: variations in obligatory and fecal nitrogen losses in young men. J. Nutr. *102*:1595.

SCRIMSHAW, N.S., and YOUNG, V.R. 1979. Soy protein in adult human nutrition: A review with new data. In: *Soy Protein and Human Nutrition* (eds. H.L. Wilcke, D.T. Hopkins and D.H. Waggle) p. 121–147, Academic Press, New York.

SCHOENHEIMER, R. 1942. *The Dynamic State of Body*, Harvard University Press, Cambridge, Mass.

SUKHATME, P.V., and MARGEN, S. 1978. Models for protein deficiency. Am. J. Clin. Nutr. *31*:1237.

TANNENBAUM, S.R., FETT, D., YOUNG, V.R., LAND, P.D., and BRUCE, W.R. 1978. Nitrite and nitrate are formed by endogenous synthesis in the human intestine. Science *200*:1487–1489.

TANNENBAUM, S.R., SINSKEY, A.J., WEISMAN, M., and BISHOP, W. 1974. Nitrite in human saliva. Its possible relation to nitrosamine formation. J. Natl. Cancer Inst. *53*:79–84.

TANNENBAUM, S.R., and YOUNG, V.R. 1980. Endogenous nitrite formation in man. J. Environ. Pathol. Toxicol. *3*:357–368.

THOMAS, M.R., ASHBY, J., SPEED, S.M., and O'REAR, L.M. 1979. Minimum nitrogen requirement from glandless cottonseed protein for nitrogen balance in college women. J. Nutr. *109*:397.

WALKER, R.P. 1977. The regulation of enzyme synthesis in animal cells. In: *Essays in Biochemistry* , (eds. P.N. Campbell and W.N. Aldridge), *13*:39– Academic Press, New York.

WATERLOW, J.C. 1968. Observations on the mechanism of adaptation to low-protein intakes. Lancet *2*:1091.

YOUNG, V.R., and JANGORBANI, M. 1980. Soy proteins in human diets in relation to bioavailability of trace elements: A brief overview. Cereal Chem. *58* (1): 12–18.

YOUNG, V.R., FAJARDO, L., MURRAY, E., RAND, W.M., and SCRIMSHAW, N.S. 1975. Protein requirements of man: comparative nitrogen balance response within the submaintenance to maintenance range of intakes of wheat and beef proteins. J. Nutr. *105*:534.

YOUNG, V.R., RAND, W.M., and SCRIMSHAW, N.S. 1977. Measuring protein quality in humans: a review and proposed method. Cereal Chem. *54*:929.

YOUNG, V.R., and SCRIMSHAW, N.S. 1978. Nutritional evaluation of proteins and protein requirements. In: *Protein Resources and Technology: Status and Research Needs*. (eds. M. Milner, N.S. Scrimshaw and D.I.C. Wang), Chapt. 10 p. 136–173, AVI Publ. Co., Inc., Westport, Conn.

YOUNG, V.R., SCRIMSHAW, N.S., TORUM, B., and VITERI, F. 1979. Soybean in human nutrition: An Overview. J. Am. Oil Chem. Soc. *56*:110.

YOUNG, V.R., TAYLOR, Y.S.M., RAND, W.M., and SCRIMSHAW, N.S. 1973. Protein requirements of man: Efficiency of egg protein utilization at maintenance and submaintenance levels in young men. J. Nutr. *103*:1164.

YOUNG, V.R., WINTERER, J.C., MUNRO, H.N., SCRIMSHAW, N.S. 1976. Muscle and whole body protein metabolism in aging, with reference to man. In: *Special Reviews, Of Experimental Aging Research* (eds. M.F. Elias, B.E. Eleftherious, and P.K. Elias) p. 217–252. EAR Inc., Bar Harbor, Maine.

ZEZULKA, A.Y. and CALLOWAY, D.H. 1976. Nitrogen retention in men fed isolated soybean protein supplemented with L-methionine, D-methionine, N-acetyl-L-methionine or inorganic sulfate. J. Nutr. *106*:1286.

DISCUSSION

DR. MARGEN: I think that this morning's program is particularly important and I think this first paper is one that is of extreme importance because it addresses something that really disturbed me last night. I think it's a thing that I would really like to mention because I think that we nutritionists, contrary to so many other biologists, tend to think that we are dealing with figures which seem to be engraven in stone and have some very great mystical significance. We forget one of the principle factors in biology which is really the study of variability. We very frequently become disturbed when we see large variations and variability but this is what the problems of biology and nutrition in the real world are and this is what we have to deal with. I think that our attempts to avoid dealing with this has really led us down many strange pathways and into many errors in thinking. So I think that we're really quite fortunate to have Dr. Young set us on the correct path. He has pointed out some of the problems and why we are in many of the difficulties that we are in. Yesterday, we covered many factors. I think Dr. Harper did an excellent job last night in this entire matter of protein quality but I think that we should point out that we're not as far along as we think. For instance, if we were to have this meeting for the next two and a half days on the problem of protein quantity and protein requirements, we could probably have just as lively a time as we are having on protein quality.

In many ways, there is difficulty in addressing this question of protein quality when we have still not really solved the protein quantity question, but since we are not to get involved with this I will only make some remarks and then pass on. We're going to assume that the data presented is sufficiently reliable that we are at least in some general ballpark as far as quantitative requirements are concerned. Now the same difficulty has to be true as far as quality is concerned. The thing that was not addressed last night and which we are addressing today, is the question of the accuracy and the reliability of all of the various measures used to estimate protein quality and, particularly, the question of variability. And, as I said, the variability is a characteristic of biological systems. It varies between individuals. It varies within the same individual over time, probably in a stationary stochastic manner. It also varies secularly and contains random errors. These are due to indigenous functions of the organism and measurement errors.

There is also the variability of the "source". The "source" may be changed as a result of many factors and the interaction of the source and its variability and the host and its variability leads to even greater variability. Finally, we have the variations and variability due to measurement errors.

Now all of these are generally additive. How do we handle these problems or how do we avoid them? I think this is the main purpose of our discussion this morning and it will be one of the problems that the task forces will be addressing in great detail this afternoon.

I think that this first paper was an absolutely beautiful presentation of an extremely complex subject. It's always a pleasure to hear the members of the MIT group discuss and present material.

On the whole, I agree with virtually everything Dr. Young has said. There will be a few minor points along the way as to some of the interpretations of certain data but I think the important thing which was emphasized both by Dr. Rand and Dr. Young, and also by Dr. Scrimshaw and is brought forth so beautifully, is the problem that we have in dealing with balance studies in terms of their inter- and intra-individual variability and the problems that these cause in interpreting the data when we are dealing with such small numbers of individuals. And I think that one of the questions which we are going to have to answer and will be asked of the task forces, is just how much variability or variation can one accept. We can talk at great length regarding the statistical significance and we can talk at great length as to where to set limits, but at what point do these statistical differences become of biological or even potentially regulatory significance. This is especially important. We must consider total diets, the matter of feeding of people, and their benefits to health.

The other things which I think are extremely important that Dr. Young has pointed out are the serious differences which occur between animal assays and human assays. These differences are not due only to variability and number of subjects. I am convinced, no matter how many individuals you are going to study, you are going to get these differences. These are species differences, in part, and they are not necessarily methodological differences. We are constantly caught in this bind, as you know, of the matter of looking for animal models to help us in explaining and making leaps of inference to the human condition. And when we find such models, we are very often criticized for using them because we very often come across phenomenon such as this. We appear to have excellent animal models but then when the findings are extrapolated to people, we find that the models do not necessarily serve our purposes.

These are all questions which are going to be troubling us but I think that what we have learned is that utilizing the human experimental guinea pig, if we want to call him that, or the human experimental rat, we can come up with answers albeit they'd still be rather crude though statistically significant estimates of differences in terms of protein quality. We have a large degree of variability which means that either we have to deal with large numbers or we have to accept the limitations of dealing with small numbers and realize the amount of variability and the accuracy we are going to deal with. We then have to ask how do we apply this data and when do we use this data? Because as Dr. Young pointed out, it's not only the matter of the dollar expense, but it's the matter of the expense in terms of the amount of time that investigators have to spend in carrying out these types of experiments. I don't know about the MIT group, but after having done this for a fair number of years, I think, at least I for one, am kind of getting a little tired of going through these types of studies repeatedly. Certainly, to just do them on a routine basis is something that I would not look forward to. Again, as Dr. Young said, there are more ways then one to skin a cat and maybe we can come up along the way with some shorter method which will not be as complex and will be of as much value to us in terms of human experimentation.

DR. HARPER: I would like to ask Dr. Young one question. It has to do with the high variability of the individuals who were consuming 70 milligrams of nitrogen per kilogram of body weight. This is close to the average requirement. One would expect the maximum of variability with some who come into positive and some who fail to come into positive balance at that point. Differences in adaptability might also be at the maximum at that point. Should one, therefore, use a uniform time in such studies for all individuals? Is there not a problem of different individuals coming into the steady state at different times?

DR. YOUNG: The example that Dr. Rand presented did indicate in that particular example that the variation for that particular subject within that particular experiment was greatest at 70 milligrams than at the other test levels. However, I do not believe data show this is a consistent observation and so, perhaps that example in reference to your thinking, is a little misleading now.

The question concerning the length of period is a fundamental one and a very good one. I think on the basis of the observations that we've made, we build in an adaptation period to take into account variability with respect to rates of adaptation on the premise that this is the component of the response that we're attempting to minimize in terms of variability among individuals. But the point's well taken, Alf, and that's about as well as I can do in that context.

DR. RAND: Just a very brief comment. One of the reasons that we weight the data that we have is to, in a sense, equalize the variability and we weight by one over the variance, which allows us to deal with different variability at different levels.

DR. MARGEN: Dr. Rand has essentially made the comment that I wanted to make. We've done a fair number of analyses on the nature of the response curve prior to reaching equilibrium. You have the same type of variability but in fact greater at least initially, in terms of standardization than you do once the individual is "standardized".

But the interesting thing, and I think that this is confirmed in almost all of our results, is that the coefficient of variation remains approximately the same irrespective of the amount of protein that you feed. So, in terms of absolute numbers, the more protein you have, the greater the variability is on a day-to-day basis. Why this occurs, as we've said, we are still uncertain but this is a very constant phenomenon for a given individual. Obviously, there are some individuals who have greater variability than others, but once an individual has established his or her pattern, then that pattern tends to remain for very long periods of time and you just do not see any type of diminution in this with time.

8

A Short-Term Procedure to Evaluate Protein Quality in Young and Adult Human Subjects

R. Bressani, B. Torún, L.G. Elías, D.A. Navarrete and E. Vargas

The short-term method for evaluating protein quality to be discussed is being developed in consideration of the need to decrease the cost of assays in human subjects. The data are interpreted in the same way as the data derived from use of the conventional long-term procedure. Thus, the estimates of protein quality obtained should be comparable to estimates obtained with the conventional procedure. Besides cost and its implications in relation to protein quality assays, there are other desirable advantages to short-term assays which are well recognized (e.g., more rapid, more convenient for subjects). It is also recognized, however, that there are disadvantages, one of which is the danger that because of the short duration of the assay, it may over estimate protein quality when compared to estimates obtained from longer-term assays.

The method as performed up to the present time is based on nitrogen balance data obtained from relatively small changes in protein intake, taking advantage of the straight line relationship between nitrogen intake and nitrogen retention, which in theory should exist at low levels of protein intake. It is, therefore, a multiple-point assay which includes as one of the points the excretion of nitrogen in feces and urine when nitrogen intake is zero. In this discussion, the theoretical basis of the method is first described. Results obtained with adults and children are then discussed and application of the method to evaluate supplementary protein effects is considered. Lastly, some observations about improving the procedures used for protein quality evaluation are presented.

BASIS FOR THE SHORT-TERM ASSAY

The concept of biological value of protein is fundamentally a function of the relationship between nitrogen retention and nitrogen intake, a relationship which without assumptions can be used to evaluate the quality of dietary proteins. The high correlation between nitrogen intake (NI)[1] and nitrogen retention (NR) was first established in 1937 (Melnick and Cowgill) and later confirmed by various workers (Allison and Anderson, 1945; Allison et al., 1946) who stated that the relationship is linear in the region of negative nitrogen retention. This linearity extends over into the positive side of nitrogen retention, but becomes curvilinear well on the positive side of nitrogen retention. There are two ways to express the quality of a protein, one is to relate nitrogen intake (NI) to nitrogen retention (NR) and the second, to relate nitrogen absorbed (NA) to NR. The first would be equivalent to net protein utilization (NPU) and the second to biological value (BV). The empirical equation for the linear portion of the relationship between N retention and N intake is shown in Figure 8.1, equation 1, and for N retention and absorbed N in equation 2 of the same Figure. The constant b in both equations is equivalent to biological value. The difference in these equations is that the NEo when using NI is equal to the sum of urine and fecal N obtained when a protein-free diet is fed, while in the expression with NA, NEo is equal to total endogenous urine N. This point is evident from actual experimental results shown in Figure 8.1, for casein tested in dogs (Bressani et al., 1978). For one example, the intercept is 94.4 at NI equal zero and 54.1 when NA is equal to zero. These values are similar to experimental data of 96 (NEo = NFo + NUo) and 58 (NUo), respectively. These equations suggest, therefore, the importance of the nitrogen retention value which is obtained when NI or NA is zero when estimating protein quality by this assay.

Preliminary Observations with Dogs

The first studies conducted to explore the possibility of reducing the experimental time for protein quality evaluation using a multiple point assay, were conducted with semi-adult dogs fed various protein sources

---

[1] The terminology used in this manuscript is defined as follows:

NI   = Nitrogen intake
NF   = Fecal nitrogen at any nitrogen intake, except zero.
NFo = Fecal nitrogen at zero nitrogen intake, also known as endogenous fecal nitrogen
UN   = Urine nitrogen at any nitrogen intake, except zero
UNo = Urine nitrogen at zero nitrogen intake, also known as endogenous urine nitrogen
NA   = Nitrogen absorbed = NI−NF
NR   = Nitrogen retention = NI−NF−UN, numerically equal to NB or nitrogen balance. Nitrogen retention represents the numerical difference between total nitrogen excretion and nitrogen intake, and, in this sense, is equal to nitrogen balance. The latter term, however, is often used to describe the methodology in general or the periods under which nitrogen retention is measured.
NEo = Total endogenous nitrogen excretion when NI is zero. Is equal to the sum of NFo + UNo.
NI_{ne} = Nitrogen intake for nitrogen equilibrium, which is the same as NR = zero for the relationship between NI to NR
NA_{ne} = Nitrogen absorbed for nitrogen equilibrium, which is the same as NR = zero for the relationship between NA to NR.

$$NR = b \, (NI) - NEo$$
$$NI - NF - NU = b \, (NI) - NEo$$
For NI = 0
$$-NF - NU = b \, (0) - NEo$$
$$NF + NU = NEo$$

$$NR = b \, (NA) - NEo$$
$$NR = b \, (NI - NF) - NEo$$
$$NI - NF - NU = b \, (NI - NF) - NEo$$
For NA = 0   NI = NF
$$NF - NF - NU = b \, (NF - NF) - NEo$$
$$+ NU = + NEo$$

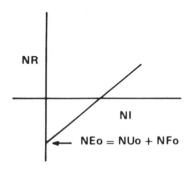

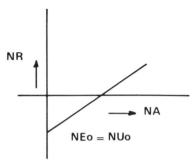

ACTUAL RESULTS
CASEIN–DOGS

$$NR = -94.4 + 0.78 \, (NI)$$
NEo (Calculated): 94.4
NEo (Experimental): 96
(Urine + feces)

$$NR = -54.1 + 0.78 \, (NI)$$
NEo (Calculated): 54.1
NEo (Experimental): 58
(Urine)

Incap 80–328

FIG. 8.1.   BASIS FOR THE SHORT-TERM ASSAY. Equation 1: $NR = b(NI) - NE_o$; Equation 2: $NR = b(NA) - NE_o$.
(See footnote 1 on page 99 for interpretation of letters.)

(Bressani et al., 1978). In these studies protein intake was from about 0.5 to 2.0 g protein/kg/day with adequate intake of calories, water and other essential nutrients. Protein intake changes were made daily in a series of nitrogen balance studies and every four days in another; these changes, however, were always small in magnitude, from 0.15 to 0.25 g/kg/day. The changes were made either in an ascending or a descending order. Some representative results relating nitrogen intake to nitrogen retention for a protein blend (TRL) are shown in Figure 8.2. The data show the relationship to be linear between balances done daily and those done every four days. The two lines are parallel and the coefficients of regression are statistically the same. Not all results, however, were as clear cut as those shown. Figure 8.3 summarizes the data obtained with casein. This graph shows 3 lines, corresponding to the relationship between N intake (NI) and N retention

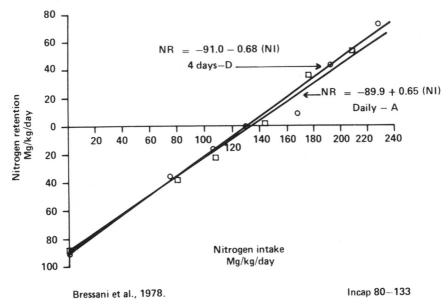

Bressani et al., 1978.   Incap 80–133

FIG. 8.2. RELATIONSHIP BETWEEN NI AND NR IN YOUNG ADULT DOGS FED A VEGETABLE PROTEIN MIXTURE (TRL)
(See footnote 1 on page 99 for interpretation of letters.)

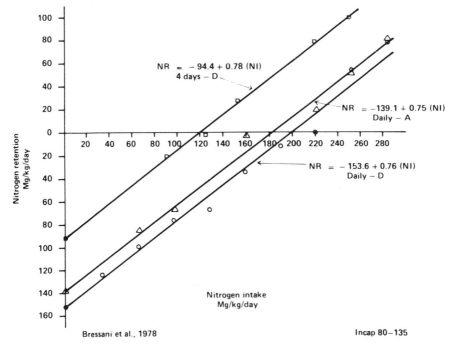

Bressani et al., 1978   Incap 80–135

FIG. 8.3. RELATIONSHIP BETWEEN NI AND NR IN YOUNG ADULT DOGS FED CASEIN
(See footnote 1 on page 99 for interpretation of letters.)

(NR), from daily balances in an ascending and descending order, and from 4-day balances in a descending order. The regression coefficients of the lines are also shown and they are not statistically different. Therefore, the lines are essentially parallel. The coefficient of regression is equivalent to net protein utilization (NPU). Therefore, even if the lines do not fall close to each other, as was the case for the line derived from the 4-day balance data (Fig. 8.3), the three approaches give the same protein quality value as estimated by the regression coefficient. However, two aspects must be indicated. Firstly the value of NR, at an N intake of 0, is different for all three cases, particularly for the 4-day balance periods. Secondly, the N intake point for N equilibrium (NR = 0) is essentially the same for the 2 values obtained from the daily N balance data, but both of these values are significantly different from the results obtained with the 4-day N balance data. These results were interpreted to mean that the experimental animals were possibly in a different protein status when each study was carried out (i.e., animals showing lower NEo values in the equation were more protein depleted than animals showing the higher NEo values).

To elaborate further on this problem, the results from studies on 3 protein sources are shown in Table 8.1. The coefficients of regression show casein to be highest in quality and TRL to be superior to VM9. Of the 7 regression equations, 5 indicate a value for total nitrogen excretion (NEo) at zero intake between 88.6 and 94.4 mg, while two indicate values of 139 and 153.6 mg. The 2 high values were obtained when the animals used for the study had been on relatively high protein intakes before being placed under the experimental conditions of the assay, while the 5 lower values were obtained on the same animals which had been used in previous assays at low protein intakes. The question may be raised as to which are, in fact, true values. No answer can be given, since this value which represents the total nitrogen excretion from feces and urine when NI is zero may vary in accordance to experimental conditions before placing the subjects on zero N intake. This may be seen in Table 8.2. These results were also obtained with semi-adult dogs fed nitrogen free diets (NFD). In the first two cases, before feeding the NFD, the level of dietary N fed was decreased gradually from

TABLE 8.1.   RELATIONSHIP BETWEEN NITROGEN INTAKE (NI) AND NITROGEN RETAINED (NR) WITH VALUES OBTAINED DAILY AND EACH 4 DAYS WHEN 3 PROTEIN SOURCES WERE FED TO SEMI-ADULT DOGS

Test protein	Type of balance	Regression of NI to NR [NR = NEo + b (NI)]	r
Casein	4 days (D)[1]	NR = −94.4 + 0.78 NI	0.99
Casein	Daily (D)	NR = −153.6 + 0.76 NI	0.99
Casein	Daily (A)	NR = −139.1 + 0.75 NI	0.99
TRL	4 days (D)	NR = −90.9 + 0.68 NI	0.99
	Daily (A)	NR = −89.9 + 0.65 NI	0.99
VM9	4 days (D)	NR = −88.9 + 0.51 NI	0.99
	Daily (D)	NR = −88.6 + 0.52 NI	0.99

[1] D = Descending; A = Ascending (see text).
See footnote on page 99 for interpretation of letters.

TABLE 8.2. ENDOGENOUS NITROGEN EXCRETION IN SEMI-ADULT DOGS OB-
TAINED UNDER DIFFERENT CONDITIONS[1]

N intake prior to feeding NFD[1] (mg/kg/day)	Days on NFD[1]	Total N loss in urine (mg/kg)	Endogenous N		Total (NEo)
			Fecal	Urine	
			(mg/kg/day)		
(625) 61[2]	15	779	20	66	86
(625) 68[2]	15	780	30	63	93
571	14	1398	31	96	127
600	16	1277	34	81	115
683	12	1375	29	94	123

[1] NFD = Nitrogen-free diet.
[2] From 625, slowly decreased to 61 or 68 mg N/kg/day.

625 mg to 61 or 68 mg in 24 days. In the other cases, the transfer from a protein to a non-protein diet was immediate. These treatments affected the endogenous N excretion (mainly in urine) and resulted in different total endogenous nitrogen losses (Bressani et al., 1972). These results suggest, therefore, the need to standardize the experimental subjects as much as possible before using them in protein quality assays. These results also indicate that a prolonged time on low protein intakes may deplete the animals or experimental subjects in a manner which might not be detected during an assay and thus introduce more variability in the results.

DEVELOPMENT OF THE SHORT-TERM ASSAY IN YOUNG ADULT HUMANS

Description of Method

Following the extensive observations made with experimental animals, the short-term assay was tested in adult human subjects according to the scheme shown in Table 8.3. The subjects are first screened on the basis of age, energy and protein intake, health and physiological state (particularly in terms of digestive performance). All subjects are then standardized for energy and protein intake with a good quality diet for a period of 6 days. Nitrogen balances are determined during the last two days. Two protein feeding sequences follow, maintaining energy intake constant. In the descending sequence the subjects are fed 0.7, 0.5, 0.3 (or 0.6, 0.4, 0.2) and 0 grams protein/kg body weight/day for two days each, collecting daily urine and fecal outputs and making two day pools for nitrogen analysis. For the ascending protein feeding sequence, after the 6-day standardization period, the subjects are fed a nitrogen-free diet (NFD) for 3 days, with quantitative collections of urine and feces during the last two days. Protein feeding follows at levels of 0.3, 0.5, and 0.7 (or 0.2, 0.4 and 0.6) g protein/kg body weight/day, for two days each as indicated above. The nitrogen intake from the NFD made of a variety of foods has ranged from 9 to 24 mg N/kg/day in the various studies carried out so far (Bressani et al., 1979; Navarrete et al., 1977; Navarrete et al., 1979).

TABLE 8.3.   DESCRIPTION OF THE SHORT-TERM ASSAY IN ADULTS

	Total days	Days on nitrogen balance	Protein intake (g/kg/day)	Energy intake (kcal/kg/day)
Standardizing period	4	–	0.6/0.7	45–50
Descending study period	8			
nitrogen balance 1		2	0.6/0.7	45–50
nitrogen balance 2		2	0.4/0.5	45–50
nitrogen balance 3		2	0.2/0.3	45–50
nitrogen balance 4		2	0	45–50
Ascending study period	9			
nitrogen balance 1		2[1]	0	45–50
nitrogen balance 2		2	0.2/0.3	45–50
nitrogen balance 3		2	0.4/0.5	45–50
nitrogen balance 4		2	0.6/0.7	45–50

[1] 3 days on the NFD with a nitrogen balance during the last two days.

Summary of Results

Up to the present time, around 12 protein sources have been tested in adults; results are summarized in Table 8.4. The quality of the proteins, as indicated by the regression coefficients ranged from a low value of 0.54 for common black beans, a protein source known to be of low protein digestibility and markedly deficient in sulfur amino acids, to a value of 1.03 for milk.

TABLE 8.4.   RELATIONSHIP BETWEEN NITROGEN INTAKE (NI) AND NITROGEN RETENTION (NR). SUMMARY OF RESULTS WITH YOUNG ADULTS

Protein source		Short-term assay			Long-term assay		
		NR = NEo + b (NI)			NR = NEo + b (NI)		
		NEo	b	r	NEo	b	r
Egg	(A)[1]	−55.2	0.86	0.88	−57.6	0.70	0.89
Milk	(A)	−78.8	0.91	0.84	−72.3	0.82	0.92
	(D)	−73.6	0.98	0.92	−70.4	0.77	0.91
	(A)	−81.3	1.03	0.84			
	(D)	−70.6	1.00	0.82			
Casein	(D)	−60.3	0.64	0.90			
Beef	(D)	−74.4	0.87	0.95			
TVP	(D)	−65.7	0.68	0.92			
TVP/Beef (50/50)[2]	(D)	−79.9	0.87	0.90			
Soy isolate	(D)	−72.2	0.84	0.90			
H.P. cookies	(D)	−75.4	0.90	0.91			
Common black beans	(D)	−62.7	0.54	0.75			
Maize:beans (50/50)[2]	(D)	−86.8	0.89	0.89			
Rice:beans (50/40)[2]	(A)	−71.6	0.75	0.87			
Rice:beans (60/40)[2]	(D)	−74.6	0.95	0.92			
Milk	(D)	−74.6	0.95	0.92			

[1] (A) = ascending; (D) = descending (see text).
[2] Protein distribution.

The second lowest value observed was 0.64 for casein, a protein also deficient in sulfur amino acids. It is of interest to point out the reproducibility of the method which is evident from the values obtained with milk in 3 consecutive years which ranged from 0.91 to 1.03. The method is also capable of measuring improvement in quality as judged by the results from beans alone (0.54) and from a 70:30 maize:bean mixture (0.89) or as indicated by the value (0.75) for a 60:40 rice:bean mixture compared to the value (0.95) for a 55:35:10 rice:bean:milk mixture to be discussed later (Vargas & Bressani, 1980). The intercept for zero nitrogen intake (the value NEo) ranged from 55.2 for egg protein to 86.8 for the maize:beans mixture, with an average value of 72.2.

Only a limited number of proteins have been assayed using the longer-term conventional assay. Results are also shown in Table 8.4. The regression coefficients tend to be smaller as compared to values obtained with the short-term approach. However, the NEo intercept values at zero nitrogen intake are statistically similar to those from the short-term assay. The difference in the regression coefficients between long- and short-term feeding, at a constant level of intake, is due to adaptation as shown by the extensive studies carried out with milk protein (Bressani et al., 1979). However, the physiological state of the experimental subject may be of importance. Furthermore, the short-term assay measures NR when the individual is not in a steady state of N catabolism in contrast to the long-term assay (Bressani et al., 1979). The values for the short-term assay are similar to values reported from other laboratories on the same proteins as indicated previously (Bressani et al., 1979a).

DEVELOPMENT OF THE SHORT-TERM ASSAY IN CHILDREN

Description of Method

The application of the short-term assay to studies with children has not been extensive. Up to the present time, however, the results which have been obtained are encouraging. Table 8.5 describes the method as first

TABLE 8.5. SHORT-TERM METHOD AS APPLIED TO EVALUATE PROTEIN QUALITY IN CHILDREN

Period		Protein intake level (g/kg/day)	Days
Standardization of subjects		1.2–1.4	8
NFD[1]	1	0	2–3
Balance	2	0.3 (0.5)[2]	2
	3	0.6 (0.8)	2
	4	0.9 (1.1)	2
	5	1.2 (1.4)	2

[1] NFD = Nitrogen-free diet.
[2] Intake levels of 0.3, 0.6, 0.9 and 1.2 were used for higher quality proteins; 0.5, 0.8, 1.1 and 1.4 for poorer quality proteins (see text).

applied to children. For a period of 8 days all subjects, usually 10 per experimental trial, were fed 1.2–1.4 g of protein/kg/day for 8 days as a standardization period. This was followed by feeding a nitrogen free diet (NFD) for 2 or 3 days, followed in turn by the experimental diets which provided increasing (ascending) levels of protein intake every two days. The levels used were 0.3, 0.6 and 0.9 and 1.2 g/kg/day for the better quality proteins and levels of 0.5, 0.8, 1.1 and 1.4 for the other proteins. Energy intake was fixed at 100 kcal/kg/day and all diets were supplemented with vitamins, minerals and electrolytes.

Summary of Results

Table 8.6 summarizes the values obtained by the short-term assay with children and compares these values to those obtained with children by the conventional (long-term) technique. The coefficients of regression are consistently higher for the short-term assay, as well as the value for the intercept (NEo) when NI is zero. These differences are also evident for N intake required for N equilibrium, that is, the short-term assay gives higher values. On the other hand, if one judges the validity of the method on the basis of the theoretical intercept when feeding a nitrogen free diet, the short-term method would appear to evaluate the quality of the protein better, because it is closer to the experimental obligatory nitrogen losses. This is suggested by the ratio NEo_{EXP} to NEo_{TNE}, which should be equal to 1. However, if the values obtained are expressed within each method, relative to milk, the differences between the two methods become smaller. This observation, however, will need further confirmation before it is accepted. Additional results are shown in Table 8.7. Two aspects are of interest. Firstly, the short-term assay is reproducible as judged by the values obtained in two years. Secondly, the value for the intercept at zero N intake is relatively close to the theoretical value of 1 (Torún and Bressani, 1978).

TABLE 8.6. PROTEIN QUALITY OF VARIOUS PROTEINS TESTED IN CHILDREN BY THE SHORT-TERM AND CONVENTIONAL NITROGEN BALANCE INDEX METHODS

Protein source	Assay	Regression NI to NR NR = NEo + b (NI)	NI for N Equil. (mg/kg/day)	$\dfrac{NEo_{EXP}[1]}{NEo_{TNE}}$	Relative quality, %
Milk	C_D	Y = −45 + 0.72 X	62	0.86	100
Milk	S_A	Y = −65 + 0.89 X	73	1.25	100
S-710	C_D	Y = −35 + 0.59 X	59	0.67	82
S-710	S_A	Y = −57 + 0.78 X	73	1.09	88
VM9	C_D	Y = −35 + 0.35 X	100	0.67	50
VM9	S_A	Y = −59 + 0.65 X	91	1.13	73
S-220	C_D	Y = −40 + 0.53 X	75	0.77	74
S-220	S_A	Y = −50 + 0.78 X	64	0.96	88

[1] NEo_{EXP} = The value from the linear regression equation; NEo_{TNE} = The total nitrogen excretion in feces and urine upon feeding a nitrogen-free diet.

TABLE 8.7.   PROTEIN QUALITY OF MILK PROTEINS TESTED IN CHILDREN BY THE SHORT-TERM N BALANCE INDEX ASSAY[1]

Protein source	NFD^{3} (days)	Number of children	Year of test	Regression of NI to NR $NR = NEo + b\,(NI)$	NI for N Equil. (mg/kg/day)	$\dfrac{NEo_{EXP}^{2}}{NEo_{TNE}}$
Milk	3	10	1978	$y = -65 + 0.89\,X$	73	1.25
Milk	2	7	1979	$y = -58 + 0.91\,X$	64	1.11

[1] Ascending feeding of protein (see text).
[2] NFD = Nitrogen-free diet.
[3] $NEo = (NFo + NUo) = (19 + 33) = 52$ mg/kg/day (in theory) and $\dfrac{NEo_{EXP}}{NEo_{TNE}} = 1.0$.

APPLICATION OF THE METHOD TO ADULTS

The short-time assay technique is now being used in our laboratories to measure the protein quality of diets based on cereal grain and beans.

An example is shown in Table 8.8 in which data on the effect of adding maize to bean protein is given. The results indicate an improvement in protein quality when beans are consumed with maize in a 30:70 weight ratio, as it has been demonstrated in other studies. Using milk as a reference (100%), protein quality values for beans ranged from 60 to 75% and from 88 to 98% for the beans:maize mixture depending on whether the comparisons were made by use of the coefficients of regression, or the values of nitrogen intake required for zero nitrogen retention.

TABLE 8.8.   PROTEIN QUALITY OF COMMON BLACK BEANS CONSUMED ALONE AND WITH MAIZE USING NI DATA[1]

Protein source	n	Regression of NI to NR NR = NEo + b (NI)			NI_{ne} (mg/kg/day)	% of milk	
		NEo	b	r		b	NI_{ne}
Black bean	6	−62.7	0.55	0.75	114	60.4	75.4
Black bean: maize, 30:70	6	−86.9	0.89	0.89	98	97.8	87.7
Milk	10	−78.8	0.91	0.84	86	100.0	100.0

[1] NI = Nitrogen intake; NR = Nitrogen retention; n = number of subjects; NI_{ne} = Nitrogen intake for nitrogen equilibrium; from Navarrete et al. (1979).

In these studies a high variability was observed in the subjects fed the bean diet. This variability was significantly reduced when the beans were consumed with maize. This was attributed to differences in digestibility between subjects when all the protein was derived from beans with some being able to digest the protein better than others. The data in Table 8.9 indicate that if the digestibility problem is eliminated by working with NA information, the estimated difference between protein sources becomes smaller (Navarrete and Bressani, 1979).

Additional studies with a bean:rice mixture have been conducted. In this case, however, the objective was to find out if a mixture of rice:bean (60:40

TABLE 8.9.   PROTEIN QUALITY OF COMMON BLACK BEANS CONSUMED ALONE AND WITH MAIZE USING NA DATA[1]

Protein Source	n	NR = NEo + b (NA)			NA_{ne} (mg/kg/day)	% of milk	
		NEo	b	r		b	NA_{ne}
Black bean	6	−57.4	0.81	0.82	70.9	89.0	84.5
Black bean: maize, 30:70	6	−63.8	0.95	0.92	67.1	104.0	89.3
Milk	10	−54.5	0.91	0.93	59.9	100.0	100.0

[1] NA = Nitrogen absorbed; NR = nitrogen retention; NA_{ne} = nitrogen absorbed when nitrogen retention = zero; n = number of subjects.

on a protein basis) could be improved by adding supplementary milk protein and whether this improvement could be detected by the short-term assay. Some representative results are shown in Table 8.10.

The coefficients of regression indicate that addition of 10% of the protein intake as milk improved the quality from a value of 0.75 to a value of 0.95, which is similar to the value for whole milk of 0.91. When these values as well as those of N equilibrium (NR = 0) are expressed relative to milk, they indicate that the mixture of beans, rice and milk is of a very high quality (Vargas and Bressani, 1980).

TABLE 8.10.   RELATIONSHIP OF NITROGEN INTAKE (NI) TO NITROGEN RETENTION (NR) FOR RICE:BEANS, RICE:BEANS: MILK, AND MILK DIETS[1]

Protein source	NR = NEo + b (NI)			NI_{ne} (mg/kg/day)	% of milk	
	NEo	b	r		b	NI_{ne}
Rice: bean	−71.6	0.75	0.87	95.6	82.4	90.6
Rice: bean: milk	−74.6	0.95	0.92	78.4	104.4	110.4
Milk	−78.8	0.91	0.84	86.6	100	100

[1] NI_{ne} = Nitrogen intake for nitrogen equilibrium; see footnote 1 for other definitions.

Similar studies have been carried out with Textured Vegetable Protein (TVP) and meat and the results are summarized in Table 8.11. The coefficients of regression for meat and the 50:50 mixture were the same, while that for TVP alone was significantly lower. When the comparison is made on the basis of NI for NR equal zero (NI_{ne}) less meat nitrogen is required, followed by the mixture and then by the TVP product. These differences are also similar when either the regression coefficients or the NI_{ne} values are expressed as percentages of milk (Table 8.11) (Navarrete et al., 1979).

TABLE 8.11.   RELATIONSHIP BETWEEN NITROGEN INTAKE (NI) TO NITROGEN RETENTION (NR) FOR DIETS OF MEAT AND TVP[1]

Protein source	n	NR = NEo + b (NI)			NI_{ne} (mg/kg/day)	% of milk	
		NEo	b	r		b	NI_{ne}
TVP	5	−65.7	0.69	0.92	95.2	75.8	91.0
Meat	5	−74.4	0.88	0.95	84.5	96.7	102.5
TVP/Meat	5	−79.9	0.88	0.90	90.8	96.7	95.4
Milk	10	−78.8	0.91	0.84	86.6	100	100

[1] TVP = Textured Vegetable Proteins; NI_{ne} = Nitrogen intake for zero nitrogen balance; also see definitions given in footnote 1 (text).

SOME OBSERVATIONS OF INTEREST

From the various studies carried out during the last 3 years with adult human subjects several observations have been made which may be of interest to point out at this time.

Protein Digestibility

Is has often been assumed that protein digestibility is constant as protein intake varies from a value of 0.2 to 0.6 g/kg/day. The data available, however, indicate that, in general (since there are some exceptions) the apparent protein digestibility increases as protein intake increases. A representative example is shown in Table 8.12 for the studies with TVP, meat and milk (Navarrete et al., 1979). An increase in apparent digestibility with

TABLE 8.12. EFFECT OF PROTEIN INTAKE LEVEL ON APPARENT PROTEIN DIGEST-IBILITY[1]

Protein level (g/kg/day)	Apparent protein digestibility, %			
	TVP	Meat	TVP:Meat	Milk
0.2	67.0	60.1	71.5	62.7
0.4	73.0	77.0	71.3	76.1
0.6	77.6	83.8	80.7	77.0
0.7	—	—	—	79.0

[1] Navarrete et al. (1979).

increased protein intake is observed in every case, therefore, the application of a constant digestibility factor for all levels of protein intake is not recommended. The same type of relationship also applies for the long-term assay in adults as well as in children. This is shown in Figure 8.4 for two protein sources, for both apparent and true digestibility (Bressani, 1977). The values of apparent digestibility became stable at protein intake levels of around 1.5 g/kg/day for children. For adult humans the apparent protein

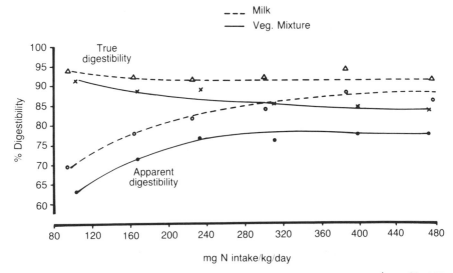

Incap 80 – 136

FIG. 8.4. EFFECT OF PROTEIN INTAKE ON PROTEIN DIGESTIBILITY

digestibility becomes stable between 0.4 and 0.5 g protein/kg/day. The lower digestibility at low levels of protein intake may be responsible for the curvature which has been reported when NI is related to NR at low levels of protein intake. This can be seen from the equations shown in Table 8.13.

TABLE 8.13. RELATIONSHIP BETWEEN PROTEIN DIGESTIBILITY AND NITROGEN RETENTION[1]

$$
\begin{aligned}
1.\ & NR = NA - UN \\
2.\ & NA = (NI)\,(D) \\
3.\ & NR = (NI)(D) = UN \\
4.\ & D = \dfrac{NA - UN}{NI}
\end{aligned}
$$

[1] NR = Nitrogen retention, calculated on the basis of absorbed N only; NR = Nitrogen retention; UN = Urinary nitrogen; NI = nitrogen intake; D = Digestibility (%).

At a fixed level of low nitrogen intake, differences in digestibility for the same protein will affect NR. Since at low levels of protein intake, values for digestibility are low, for the same amount of urine nitrogen less will be retained at low levels of protein intake than at higher levels where digestibility is higher. But low digestibility values are probably not enough to explain the curvature often observed. A higher digestibility at a low fixed level of protein intake implies greater absorption and increased loss of N in urine. However, a higher retention compensates for this since the balance studies in the short-term method proceed from a NFD. In most studies carried out, it has been observed that for proteins with a high protein digestibility the regression equations relating NR to NI and NR to NA are parallel, as shown by the results presented in Figure 8.5 for a soybean protein isolate (Navarrete et al., 1979). However, for proteins with a low digestibility, such as common beans (Navarrete and Bressani, 1979), the two lines converge towards the zero NI or NA value as indicated in Figure 8.6. In the first case, the protein quality value given by the coefficient of regression is the same for both expressions. In the second case, however, the true quality of the food is given by the regression coefficient of the relationship between NI and NR and the coefficient of the relationship between NA and NR represents an overestimation of the quality of the protein, since urinary nitrogen (which in part is dependent upon the amino acid pattern of the protein source) is the same for both expressions.

Total Nitrogen Excretion at Zero N Intake

One of the arguments against the use of the nitrogen excretion at zero N intake in the calculation of the quality of the protein by regression has been that the value derived from the regression equation is not the same as the one obtained by direct feeding of a NFD. Therefore, the question is raised as to which is really the true value for purposes of the assay. The results in dogs discussed previously suggested a different value which was obtained depending on the condition used to obtain the data used. If the approach to feeding zero N intake is slow, that is with small decreases, the final value

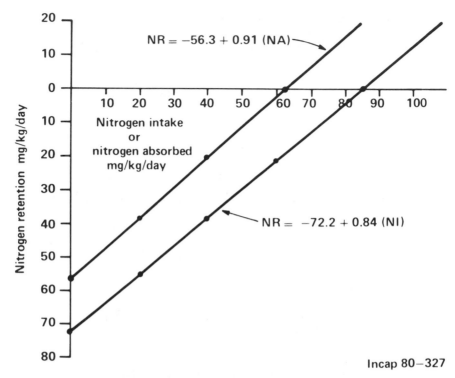

FIG. 8.5.   RELATIONSHIP BETWEEN NI AND NR, AND NA AND NR, FOR A SOYBEAN PROTEIN ISOLATE (SUPRO 620) FOR A PROTEIN OF HIGH APPARENT DIGESTIBILITY (See footnote 1 on page 99 for interpretation of letters.)

would be lower than if the excretion is measured following an immediate change from a high N to a zero N intake. Values obtained by use of the latter approach are probably more representative of the true value, particularly in the context of the methodology employed. Furthermore, the endogenous losses as determined by the conventional method require feeding a nitrogen-free diet for 10 to 14 days, while in the short-term approach a NFD is fed for only 3 days.

From the studies carried out with the short-term assay, the values for total N excretion at NI = 0 calculated from the regression coefficients range from 60.0 to 86.6 with an average of 72.5, as shown in Table 8.14. Values from actual measurements range from 61.8 to 94.0 with an average of 72.5; these are not different from the calculated values. Of the 72.5 mg representing total actual excretion at zero nitrogen intake, 69% represents urine nitrogen and 31% fecal nitrogen. From the studies reported by Scrimshaw et al. (1972) it was estimated that on the third day on a NFD, 19 subjects were excreting close to 50 mg/kg/day in urine, a value which is close to the value found in the studies conducted by the short-term method. This

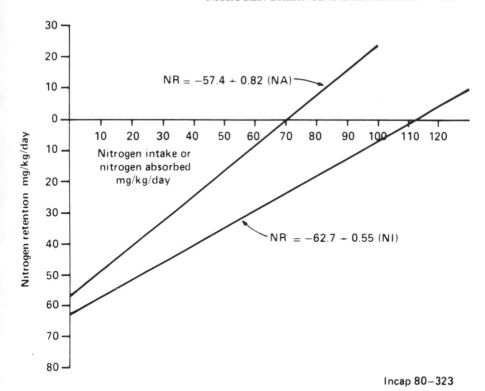

Incap 80–323

FIG. 8.6.   RELATIONSHIP BETWEEN NI AND NR, AND NA AND NR, FOR COMMON BEANS FOR A PROTEIN OF LOW APPARENT DIGESTIBILITY

point is illustrated in Figure 8.7. This figure was constructed by plotting the urinary nitrogen excretion on a NFD by young human adults as reported by Scrimshaw et al. (1972). At the end of 3 days, total N excretion in urine was, on average, 52 mg/kg/day. The second part of the graph includes the line given by the regression equation reported by Navarrete et al. (1979) for meat protein using the short-term assay. It may be seen that at zero nitrogen absorbed, the intercept point or NR was 55 mg/kg/day which is close to the above value and within the standard deviation reported. This observation lends more support to the use of the short-term method for the estimation of protein quality.

TABLE 8.14.   TOTAL NITROGEN EXCRETION AT ZERO NITROGEN INTAKE[1]

Nitrogen loss at NI = 0	Range (mg/kg/day) NEo	Mean	SD	CV
Actual	61.8–94.0	72.5	10.1	13.9
Calculated	60.0–86.9	72.5	7.2	9.9

[1] Based on average values (groups of subjects) from 14 studies; NI = nitrogen intake; SD = Standard deviation; CV = Coefficient of Variation (%).

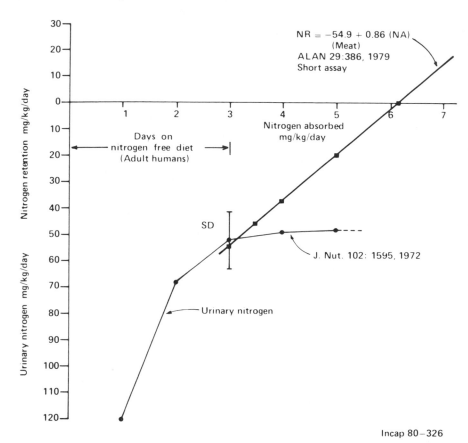

FIG. 8.7. URINE NITROGEN EXCRETION WITH RESPECT TO TIME AND RELATION-
SHIP BETWEEN NA AND NR OF HUMAN ADULTS

Effect of Eliminating Various Points in the Calculation

From the experimental information obtained in the rice:bean studies
(Vargas and Bressani, 1980) the regression coefficient and the N equilibrium value for NI = 0 were calculated using all the data, omitting either the
data for the 0.2 g intake level or the NFD period, or by using only the values
from the NFD and 0.6 g periods. The results are shown in Table 8.15. They
indicate little differences in the regression coefficients, when omitting from
the calculation, data from the 0.2 g or the 0.2 and 0.4 g intake levels, except
when the NFD data is not taken into consideration. In this case, the
regression coefficients decrease significantly which is interpreted to mean
that protein quality decreases. In the lower part of Table 8.15, nitrogen
intake values for nitrogen equilibrium have been calculated from the regression equation. All values under each column are not statistically differ-

TABLE 8.15.   AVERAGE REGRESSION COEFFICIENTS AND NITROGEN INTAKES RE-
QUIRED FOR NITROGEN EQUILIBRIUM CALCULATED WITH OR WITHOUT OMISSION
OF SELECTED DATA[1]

	60:40 Rice:bean	55:35:10 Rice:bean:milk	Milk
b with all information	0.75	0.95	1.06
b without 0.2 intake	0.79	0.99	1.11
b with NFD and 0.6	0.78	0.95	1.06
b without NFD	0.56	0.65	0.83
NI_{ne} with all information	95.6	78.4	74.3
NI_{ne} with NFD + 0.6	101.1	87.0	80.9
NI_{ne} without NFD	100.3	75.6	71.1

[1] "b" = regression coefficient from equation: $NR = NEo + b (NI)$; see previous Tables; NI_{ne}
= nitrogen intake required for zero nitrogen balance; NFD = data from nitrogen-free
diet period.

ent. Even when the NR values from the NFD are not used in the calculation
of the equation, nitrogen intake for zero nitrogen balance is similar to the
other values. The significance of this finding contradicts the significance of
the lower coefficients of regression obtained when NR data from zero NI is
not used in its calculation, since a lower coefficient indicates a lower protein
quality and implies a greater need of protein for nitrogen equilibrium. In
fact, when NFD results are not used, all proteins are overestimated in
quality, more so for better quality proteins, such as milk, than for the poorer
quality protein foods (as judged by the data shown in Table 8.15 and other
unpublished results). Similar conclusions are reached when the values for a
test protein are expressed as a percentage of a reference protein such as
milk. Although the data in Table 8.15 refer to rice, beans and/or milk
(Vargas and Bressani, 1980), similar findings are also obtained with the
other protein sources tested. Therefore, it is concluded that it is necessary to
utilize NB data from zero N intake periods. Otherwise proteins are overestimated when NI values for NR = 0 are used to evaluate protein quality.

CONCLUSIONS

As previously indicated, the protein quality assay results presented in
this Chapter suggest that the short-term approach to evaluate protein
quality yields data which is essentially the same as that obtained by the
conventional long-term method. Furthermore, the method of calculation
and the interpretation are equivalent. It is recognized, however, that the
short-term assay measures protein quality under non-steady state conditions. This is in contrast to the long-term assay in which a steady state
seems to exist due to the fact that the balance measured at each fixed level of
intake is preceded by an adaptation period at the same level.

The short-term method can be used to discriminate between proteins of
different quality and to measure the effects of protein complementation.
Furthermore, the method is reproducible since assays on the same protein
conducted at different times yield essentially the same values.

The data obtained with both children and young human adults by the short-term assay, suggest that the method is of value as a technique to evaluate protein quality, even though there is a small tendency to obtain higher values than when the longer-term approach is used. The short-term method offers obvious advantages and is reproducible and sensitive enough to detect small changes in protein quality. Unpublished data indicate that the results obtained correlate highly with estimates of protein quality obtained with rats. The method should be subjected to testing in other laboratories and deserves further development.

REFERENCES

ALLISON, J.B. and ANDERSON, J.A.   1945.   The relation between absorbed nitrogen, nitrogen balance and biological value of proteins in adult dogs. J. Nutr. *29*:413–420.

ALLISON, J.B., SEELEY, R.D., BROWN, J.H. and ANDERSON, J.A.   1946. The evaluation of proteins in hypoproteinemic dogs. J. Nutr. 31:237–247.

BRESSANI, R.   1977.   Human assays and applications. In: Evaluation of Proteins for Humans. (C.E. Bodwell, ed.). Chapter 5. The AVI Publishing Company, Inc. Westport, Connecticut, p. 81–118.

BRESSANI, R., ELÍAS, L.G., OLIVARES, J. y NAVARRETE, D.A.   1978. Comparación del índice de balance de nitrógeno de 3 proteínas calculado de períodos de balance de nitrógeno diarios o de 4 días. Arch. Latinoamer. Nutr. 28:318–336.

BRESSANI, R., GÓMEZ BRENES, R.A. y ELÍAS, L.G.   1972.   Nitrógeno urinario de perros adultos alimentados con una dieta sin nitrogeno y con diversas ingestas de calorías. Arch. Latinoamer. Nutr. 22:451–466.

BRESSANI, R., NAVARRETE, D.A., de DAQUI, V.A.L., ELÍAS, L.G., OLIVARES, J. and LACHANCE, P.A.   1979.   Protein quality of spray-dried whole milk and of casein in young adult humans using a short-term nitrogen balance index assay. J. Food Sci. 44:1136–1149.

BRESSANI, R., NAVARRETE, D.A., ELÍAS, L.G. and BRAHAM, J.E.   1979a. A critical summary of a short-term nitrogen balance index to measure protein quality in adult human subjects. In: Soy Protein and Human Nutrition. (H.L. Wilcke, D.T. Hopkins, D.H. Waggle, eds.) Proc. Keystone Conference on Soy Protein and Human Nutrition. Keystone, Colorado, Academic Press. p. 313–324.

MELNICK, D. and COWGILL, G.R.   1937.   The protein minima for nitrogen equilibrium with different proteins. J. Nutr. 13:401–424.

NAVARRETE, D., ELIAS, L.G. and BRESSANI, R.   1979.   Protein digestibility and protein quality of common beans, alone and with maize, in young adult humans. INCAP Annual Report 1979, p. 12.

NAVARRETE, D.A., de DAQUI, V.A.R., ELÍAS, L.G., LACHANCE, P.A. and BRESSANI, R.   1977.   The nutritive value of egg protein as determined by the nitrogen balance index (NBI). Nutr. Rep. Intern. 16:695–704.

NAVARRETE, D.A., ELÍAS, L.G., BRAHAM, J.E. and BRESSANI, R.   1979. The evaluation of the protein quality of soybean products by short-term bioassays in adult human subjects. Arch. Latinoamer. Nutr. 29:386–401.

TORÚN, B. and BRESSANI, R.   1978.   Uso de un método corto basado en el índice de balance de nitrógeno (IBN corto) para evaluar la calidad nutricional de proteínas en niños de edad preescolar. Informe Anual INCAP 1978, p. 105–110.

SCRIMSHAW, N.S., HUSSEIN, M.O., MURRAY, E., RAND, W.M. and YOUNG, V.R.   1972.   Protein requirements of man: variations in obligatory urinary and fecal nitrogen losses in young men. J. Nutr. 102:1595–1604.

VARGAS, E. and BRESSANI, R.   1980.   Protein quality of rice/bean diets with or without protein and energy supplements to estimate protein requirements in young adult humans. Report to the United Nations University Program INCAP.

DISCUSSION

DR. MARGEN: It's rather an interesting phenomenon that Dr. Bressani was describing for us but it's not at all unique in either biological or chemical determinations. Chemists are well acquainted with dealing with this phenomenon of a nonequilibrium situation which essentially, is what Dr. Bressani is telling us he is doing. As long as there is a very specific time period in the reaction, and the reaction is reproducible, there is no necessity to carry the experiment through to final equilibrium. I believe that you are dealing with and that you are assuming that the responses of a given individual under nonequilibrium conditions, if time is kept constant, are sufficiently reproducible and the variability is sufficiently low to make these experiments worthwhile.

This is a very interesting area and one which should be discussed. Further investigation will be required and, in particular, a great deal of very careful statistical analyses must be done.

9

Biochemical Parameters and Protein Quality

G. Richard Jansen

Dietary protein supplies the amino acids necessary to sustain life. Amino acids are precursors for vitamins or growth factors such as niacin and carnitine and neurotransmitters such as norepinephrine and serotonin. However, quantitatively the most important function of amino acids is their role as the building blocks for tissue protein synthesis and maintenance. Protein quality is an attribute of protein related to the efficiency with which protein is utilized for these purposes. Proteins of high quality are utilized with the highest efficiency and are therefore required in the smallest amounts.

In very general terms, ingested protein is digested, absorbed as amino acids, transported in plasma as free amino acids, taken up by cells as amino acids and eventually synthesized into tissue protein by assembly on polyribosomes. Munro (1970) has summarized the important role of free amino acid pools in regulating protein metabolism. In the biochemical approach to protein quality evaluation, inferences about the efficiency of utilization of dietary protein are deduced from measurements made on amino acids and/or their metabolites in blood, urine or the tissues.

Over the years, a wide variety of biochemical methods have been proposed for use in assessing dietary protein quality. These include amino acid levels in blood, muscle, urine and feces, metabolite levels in plasma and urine, and enzyme levels in liver, kidney and plasma. These methods have been exhaustively reviewed recently by Bodwell (1975, 1977). Emphasis in this review will be placed on clinical methods that have been demonstrated to be useful, or potentially useful, in assessing the efficiency of protein utilization directly in human subjects. These are 1) urinary nitrogen or urea, 2) blood urea and 3) plasma free amino acids.

URINARY NITROGEN AND PROTEIN QUALITY

It has been known for the better part of a century that urinary nitrogen excretion, most of which is urea, varies directly with protein intake (Folin, 1905). The increased excretion of urinary nitrogen associated with amino acid deficiency, i.e. reduced protein quality, also is an old observation going back at least thirty years ago to the observations by Rose et al. (1950) with valine and methionine deficient diets.

More recent experiments with both swine and human subjects confirm these early observations. Brown and Cline (1974) evaluated the effect on urea excretion of adding lysine to a basal corn diet fed to swine. The results are shown in Table 9.1. Urea excretion declined with the addition of the limiting amino acid (lysine), under both pair feeding and ad libitum intakes. Addition of the second limiting amino acid (tryptophan), together with lysine, reduced urea excretion even further.

TABLE 9.1.   EFFECT OF LYSINE ADDITION TO CORN DIETS ON UREA EXCRETION IN SWINE[1]

Diet	Urinary Urea N(g/day)	
	Pair Fed	Ad Libitum Fed
Basal Corn Diet	3.16 ± 0.34[2]	3.79 ± 0.45
Basal Corn Diet + 0.02% L-Lysine	2.38 ± 0.13	3.16 ± 0.37
Basal Corn Diet + 0.24% L-Lysine	1.70 ± 0.21	1.99 ± 0.23
Basal Corn Diet + 0.36% L-Lysine	1.73 ± 0.27	2.38 ± 0.15

[1] Source: Data from Brown and Cline (1974).
[2] Mean ± standard error.

Scrimshaw et al. (1973) evaluated the effect on urea excretion of lysine fortification of wheat gluten fed to young men at two protein and two energy levels. As illustrated by the data in Table 9.2, lysine addition reduced urea excretion at both protein levels and energy intakes, but the effect of protein quality was overshadowed by the effect of protein level.

Taylor et al. (1974) analyzed the results of a number of their own earlier nitrogen balance experiments in young men. They concluded that serum urea nitrogen (SUN) was highly correlated with urinary urea nitrogen expressed as g/day (r=0.87) or as a percentage of total urinary nitrogen (r=0.93). The efficiency of nitrogen utilization, as indicated by NPU (Net

TABLE 9.2.   EFFECT ON UREA EXCRETION OF LYSINE FORTIFICATION OF WHEAT GLUTEN IN YOUNG ADULTS[1]

Level of Protein g/Kg/day	Lysine	Urinary Urea (g/day)	
		Restricted Energy	Adequate Energy
0.27	−	2.93 ± 0.58[2]	2.21 ± 0.44
	+	2.76 ± 0.99	2.04 ± 0.59
0.73	−	7.53 ± 0.91	6.97 ± 0.80
	+	7.31 ± 0.98	6.43 ± 0.92

[1] Source: Data of Scrimshaw et al (1973).
[2] Mean ± standard deviation.

Protein Utilization Values), was highly inversely correlated with SUN levels (r = −0.89). Bodwell (1977) further analyzed these data of Taylor et al. (1974) and concluded that urinary urea nitrogen excretion and NPU both were highly correlated with protein intake, making the relationship between urea excretion and NPU difficult to interpret.

To clarify the effect of protein quality independent of changes in protein intake, Bodwell and coworkers (1979) investigated the utility of using urinary nitrogen as a measure of protein quality with protein fed at the low level of 0.4g/Kg body wt./day. Six protein sources widely varying in quality ranging from egg to wheat gluten were fed to adult men for 11 days. Total urinary nitrogen and urea nitrogen were measured. A portion of their data is summarized in Table 9.3. The NPU's for all sources except wheat gluten did not differ significantly. The NPU for gluten was clearly lower, but not significantly so because of large variability and the more limited number of subjects fed wheat gluten than fed egg. The statistical analysis of the individual subject data showed that NPU was highly inversely correlated with total urinary nitrogen (r = −0.89) and urinary urea nitrogen (r = −0.89).

TABLE 9.3.   RELATIONSHIP BETWEEN URINARY NITROGEN AND NET PROTEIN UTILIZATION[1]

Protein Source	Nitrogen Intake mg N/Kg BW/day	Urinary Nitrogen Excretion mg N/Kg BW/day	Net Protein Utilization (NPU)
Spray Dried Whole Egg	63.0 ± 1.0[2]	52.3 ± 4.2	73 ± 6
Cottage Cheese	63.4 ± 0.8	57.7 ± 4.7	71 ± 7
Tuna	63.1 ± 0.3	53.7 ± 6.5	68 ± 10
Peanut Flour	61.9 ± 0.4	53.8 ± 2.6	68 ± 5
Soy Isolate	64.9 ± 0.9	58.9 ± 2.8	66 ± 5
Wheat Gluten	64.9 ± 0.9	70.7 ± 5.9	48 ± 10

[1] Source: Data of Bodwell et al (1979).
[2] Mean ± standard error.

These results confirm that urinary nitrogen can be used as an indirect indication of protein quality, but even under closely controlled conditions, quantitative relationships are somewhat difficult to discern because of small differences and large variability. Additional complications are observations that urinary nitrogen excretion is influenced significantly by urine flow (Simmons and Korte 1972) as well as physical activity (Marable et al. 1979). It is apparent that more work is required before urinary nitrogen or urea can be used as reliable quantitative indicators of protein quality in man.

BLOOD UREA LEVELS AND PROTEIN QUALITY

A strong inverse correlation between biological value and blood urea levels in growing rats and pigs was observed by Munchow and Bergner (1967). The relationship was further explored and confirmed by Eggum

(1970, 1973). This investigator carried out nitrogen balance and blood urea measurements in young growing rats using forty-two protein sources ranging in protein quality from egg (BV [Biological Values] = 95.5) to fodder beet tops (BV=21.7). The diets were isonitrogenous and isocaloric as consumed, and food was removed from the cages 4–5 hours before blood samples were taken. As shown by the data in Figure 9.1, BV was highly inversely correlated with blood urea levels (r=−0.95). In addition, using casein supplemented with methionine as the protein source, Eggum (1973) showed that blood urea levels were highly positively correlated with protein level (r=0.95). A gradual reduction in plasma urea levels associated with increasing levels of lysine added to a basal corn diet fed to young growing pigs was reported by Brown and Cline (1974). The inverse relationship between biological value and blood urea levels in rats and swine would appear to be reasonably well established.

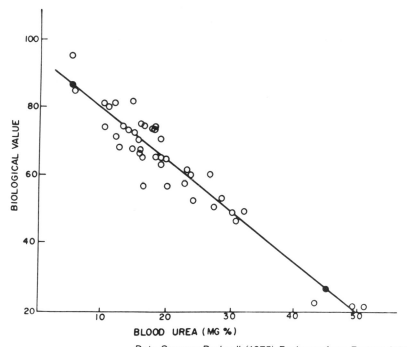

Data Source: Bodwell (1975) Redrawn from Eggum (1970)

FIG. 9.1. RELATIONSHIP BETWEEN PROTEIN BIOLOGICAL VALUE AND BLOOD UREA LEVELS IN RATS

The relationship between serum urea levels and the efficiency of protein utilization in human subjects was evaluated by Taylor and colleagues (1974) based on a series of nitrogen balance experiments previously carried out in young men. In these experiments, diet periods were usually 15 days in duration. In Figure 9.2 their plot of net protein utilization (NPU) as a function of serum urea level is shown. These data, in which NPU is inversely correlated with serum urea (r = −0.89), apparently confirm the relationship between protein quality and blood urea noted by Eggum (1970). However, the data on which this figure is based were obtained in nitrogen balance experiments in which protein intake varied from 0.2 to 0.7 grams/kg body weight/day. As was previously pointed out (Bodwell 1977), in these experiments as NPU decreased, serum urea is increased, but so did protein intake. As Eggum showed previously, blood urea is highly positively correlated with protein intake. Therefore it was not completely clear what the protein quality-serum urea relationship in humans would be when dietary protein intake remained constant.

To clarify the relationship between protein quality and plasma urea in human subjects, Bodwell et al. (1979) fed six protein sources to adult men at a constant protein intake of 0.4 grams/kg body weight/day. Calorie intakes were adjusted to maintain body weight and ranged from 33 to 43 kcal/kg body wt./day. Plasma urea nitrogen was determined in samples taken after an overnight fast and NPU and BV were determined by the usual nitrogen balance techniques. These workers found that, even with protein fed at

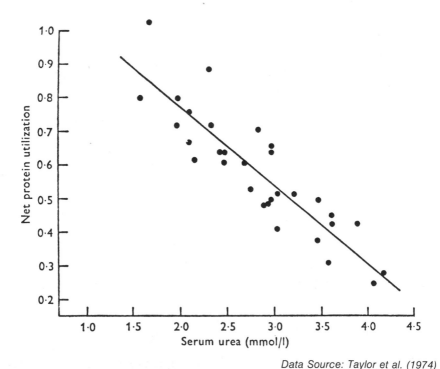

Data Source: Taylor et al. (1974)

FIG. 9.2. RELATIONSHIP BETWEEN NET PROTEIN UTILIZATION (NPU) AND SERUM UREA CONCENTRATION IN YOUNG MEN

isonitrogenous levels, plasma urea nitrogen was highly negative correlated with NPU ($r=-0.84$) in good confirmation of the previous animal work reported by Eggum (1970). However, the scatter of NPU values in Figures 9.1 and 9.2 corresponding to similar urea values suggest that the use of serum urea values to predict protein quality in human subjects would likely be imprecise.

Additional caution in the use of this method to predict quality is suggested by the study of Bolourchi et al. (1968) in which the influence of consumption of wheat flour on blood urea concentrations was evaluated in human subjects. In this study 12 young men were fed for 20 days their usual diet consisting of a mixture of protein sources of which approximately half were of animal origin. This was followed by a 50 day experimental period in which over 90% of the protein was supplied by wheat products, primarily white flour. The other 10% was supplied by fruits and vegetables. In the control phase the diets provided 12.2 grams of nitrogen per day per subject while during the experimental period nitrogen intake was 11.8g/day/person. At the completion of the control phase the blood urea nitrogen level (BUN) was 16.9 mg N/dl. After 50 days on the experimental diet BUN decreased to 6.4 ± 2.5 mg/dl. A later study showed that the greater part of this reduction occurred within 48 hours of starting to eat the high bread diet, and furthermore the BUN level increased back up to the control level within 48 hours of reinstituting the control diet. These results are difficult to interpret in the light of the inverse relationship between protein quality and blood urea levels reported by Eggum (1970), Taylor et al. (1974) and Bodwell et al. (1979).

In some of the studies analyzed by Taylor et al. (1974) energy was fed at several levels, but the influence of the protein/energy ratio was not isolated. In the course of an investigation into amino acid metabolism during parenteral nutrition in human subjects, Chen et al. (1974) examined the influence of the calorie/nitrogen ratio on BUN levels. Their data, shown in Figure 9.3, show that as expected, there is a highly significant inverse correlation between the calorie/nitrogen ratio and BUN levels ($r=-0.90$, $p<0.001$).

As an alternative to the use of fasting blood urea levels, Bodwell et al. (1978) explored the use of post-prandial changes in plasma urea nitrogen to predict protein quality. The subjects, consisting of 6 young men, prior to the test consumed similar diets which supplied 0.7 or 1.0 g protein/kg/day depending on the study. The test proteins were consumed at an intake of 0.45−0.50g/kg body weight after a 12 hour fast.

The results of one of their experiments are summarized in Table 9.4. As was the case for fasting urea levels, a strong inverse correlation between the postprandial increase in plasma urea levels and protein quality was noted ($r=0.87$). However a number of discrepancies are apparent, as pointed out by the authors. Based on the results of these experiments Bodwell and colleagues concluded that it would be difficult to standardize this assay to make it useful in predicting the protein value of a wide variety of protein sources. The authors suggested some limited utility to predict the value of similar protein sources.

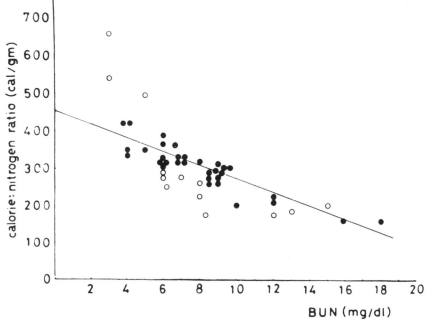

Data Source: Chen et al. (1974)

FIG. 9.3.   RELATIONSHIP IN THREE CHILDREN BETWEEN BLOOD UREA NITROGEN LEVELS AND THE CALORIE: NITROGEN RATIO OF THE INFUSATE
Data (open circles) obtained during first 3 days, first 3 post operative days or days on which blood transfusions were obtained were excluded (r= −0.90).

TABLE 9.4.   POSTPRANDIAL CHANGES IN PLASMA UREA AS RELATED TO PROTEIN QUALITY[1]

| Protein Source | Plasma Urea Nitrogen (mg/dl) | | | RPV[2,3] |
	Fasting	3 ½ Hours	Δ(3 ½ Hrs)	
Egg White-A	11.35 ± 0.96[4]	12.76 ± 1.07	1.41 ± 0.14	—
Egg White-B	11.00 ± 0.59	13.43 ± 0.45	2.43 ± 0.22	123
Tuna	10.75 ± 0.52	14.17 ± 0.45	3.43 ± 0.24	87
Whole Egg	10.49 ± 0.87	14.09 ± 0.80	3.60 ± 0.21	117
Peanut Flour	10.43 ± 0.29	15.26 ± 0.41	4.84 ± 0.36	55
Soy Isolate B	11.18 ± 0.53	16.22 ± 0.72	5.04 ± 0.55	59
Cheese C	10.20 ± 0.61	15.77 ± 0.49	5.57 ± 0.22	80
Wheat Gluten	10.73 ± 0.53	16.96 ± 0.59	6.23 ± 0.02	26

[1] Source: Bodwell et al 1978.
[2] Based on rat assay; RPV = Relative Protein Value.
[3] Correlation between Δ 3½ hours postprandial and RPV= −0.87 (p<.01).
[4] Mean ± standard deviation.

In spite of the apparent difficulties in assessing changes in either fasting or postprandial plasma urea levels as related to protein quality, Fomon et al. (1979) have found them useful in demonstrating the improvement in the efficiency of nitrogen utilization associated with the addition of methionine to a soy formula fed to infants. In this study a number of clinical parameters were compared in infants fed for 110 days a soy formula with and without addition of methionine. The formula, which was fed ad libitum with volume of intake measured, supplied protein at 9% of calories. Growth and nitrogen retention did not vary significantly as a function of methionine intake. However serum albumin and gain per unit energy consumed were significantly higher and fasting serum urea nitrogen levels were significantly lower in the methionine fortified group. The urea nitrogen data are shown in Table 9.5. In this study, serum urea was a more sensitive indicator of protein quality than either growth or nitrogen retention. The authors concluded that methionine fortification improved the protein quality of the soy formula for infants. However, McLaughlan (1979) recently reported that when soy protein isolate was fed to weanling rats with protein at 7.4% of calories, addition of methionine resulted in the development of fatty livers due to a deficiency of threonine. These results suggest that the desirability of fortifying soy formulae with methionine is not completely established.

TABLE 9.5.   EFFECT OF METHIONINE ADDITION TO A SOY FORMULA ON SERUM UREA NITROGEN IN INFANTS[1]

| Age (days) | Urea Nitrogen (mg/dl) | |
	Unsupplemented	Methionine Added
28	11.5 ± 1.3[2]	9.5 ± 1.7
56	10.1 ± 1.3	8.6 ± 1.4
84	10.1 ± 1.7	8.1 ± 1.3
112	9.4 ± 1.6	7.5 ± 1.6

[1] Data Source: Fomon et al (1979).
[2] Mean ± standard deviation.

PLASMA AMINO ACIDS

Nutritional Significance of Plasma Amino Acids

The important role played by free amino acid pools in regulating protein metabolism has been appreciated for a number of years (Munro 1970). Not surprisingly, over the last twenty years the nutritional significance of plasma free amino acid levels has been extensively evaluated and many excellent reviews are available (Longenecker 1961, 1963; Leathem, 1968; Berry 1970; den Hartog and Pol 1972; Young and Scrimshaw 1972; Eggum 1973, McLaughlan 1974, Longenecker and Lo 1974 and Bodwell 1975, 1977).

The nutritional significance and potential utility in human subjects of changes in plasma free amino acids, reside in three major areas: 1) determination of protein nutritional status, 2) determination of requirements for essential amino acids and 3) the evaluation of protein quality. The first two areas will be considered briefly, but major attention will be directed toward the usefulness of post-prandial changes in plasma free amino acids in assessing the quality of dietary protein directly in human subjects.

Plasma Amino Acids and Protein Nutritional Status

The changes in the free amino acid patterns in the plasma of children suffering from protein-calorie malnutrition have been reviewed by Young and Scrimshaw (1972). A simplified field assessment to diagnose protein-calorie malnutrition involving plasma free amino acid level was proposed by Whitehead and Dean (1964). These authors suggested that the ratio of plasma levels for glycine, serine, glutamine and taurine (N) to plasma levels of leucine, isoleucine, valine and methionine (E) may be useful in diagnosing kwashiorkor. Normal values for the N/E ratio were found to be <2.0 with the ratio in children suffering kwashiorkor exceeding this value. Subsequently, it was observed that the N/E ratio in marasmic children was in the normal range (McLaren et al. 1965) suggesting the use of this ratio to diagnose differentially kwashiorkor and marasmus. Although Young and Scrimshaw (1972) considered the N/E ratio to be generally useful for this purpose, they suggested caution in interpreting the results, primarily because of the influence of infections and recent protein intake on free amino acid levels.

Plasma Amino Acids and Amino Acid Requirements

It was observed twenty years ago by Morrison and coworkers (1961) that, in the rat, when graded levels of lysine were fed and lysine was first limiting, plasma lysine levels remained low until the lysine requirement for growth was met, at which point plasma lysine levels increased rapidly. This phenomenon has been confirmed by others and has been used to establish lysine or lysine and tryptophan requirements for growth of the rat (Stockland et al. 1970), Lewis et al. 1976a), lysine, tryptophan, isoleucine, leucine and histidine requirements for growth of pigs (Mitchell et al. 1968, Lewis et al. 1976b) and the lysine and tryptophan requirements for lactation in sows (Lewis and Speer 1974).

Studies in which this technique was used to determine amino acid requirements of adult human subjects have been carried out primarily by Young, Scrimshaw and their colleagues. Application to the determination of the tryptophan requirement of young men was made by Young et al. (1971). A total of 14 young men were fed an amino acid diet made up of 17 L-amino acids patterned after egg and glycine. Blood was drawn before breakfast and three hours later with plasma tryptophan determined by fluorometry and nitrogen balance by the usual procedures.

The responses of both fasting and post-prandial tryptophan levels as related to tryptophan intake are shown in Figure 9.4. It is apparent that food consumption decreased plasma tryptophan. By either the fasting or post-prandial values, a tryptophan requirement of 3 mg/kg body weight/day was predicted. This compares favorably to the estimate of 2 to 2.6 mg/kg body weight/day derived from the nitrogen balance data. This technique has been further applied by this group of investigators for the estimation of the valine and lysine requirement of young men (Young et al. 1972), the

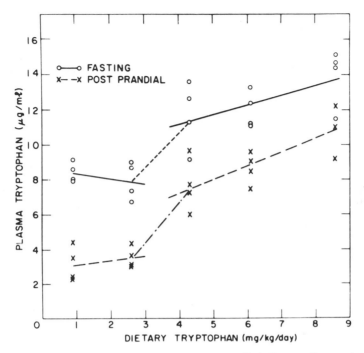

Data Source: Young et al. (1971)

FIG. 9.4. FASTING AND POST PRANDIAL PLASMA TRYPTOPHAN CONCENTRATIONS OF SUBJECTS FED GRADED LEVELS OF TRYPTOPHAN

tryptophan requirement of elderly men and women (Tontisirin et al. 1973a), the threonine requirement of young men and elderly women (Tontisirin et al. 1974) and the tryptophan requirement of children with Down's Syndrome (Tontisirin et al. 1973b).

Postprandial Amino Acids and Protein Quality

The earliest observations in which postprandial changes in free amino acid patterns were related to the pattern of amino acids in the dietary protein were reviewed by Almquist (1956). This author suggested that post absorptive changes in plasma free amino acid would eventually be useful in assessing the efficiency of utilization of dietary protein, i.e. protein quality. However Longenecker and Hause (1959) were the first to develop a practical procedure for evaluating protein quality using plasma amino acid changes. This now classical paper of Longenecker and Hause (1959) was based on observations on relatively few dogs, although it was followed up by important confirmatory studies in dogs (Longenecker, 1961) and in human subjects (Longenecker and Hause 1961; Longenecker, 1963). Nevertheless there is no doubt it was a potent stimulus to work on plasma amino acids as related to protein quality and amino acid requirements, and can be considered a landmark in this field.

The premise on which the work of Longenecker and Hause is based is that postprandial plasma amino acid changes are influenced both by the entry of amino acid into plasma from the intestines and the removal of amino acids from plasma in proportion to amino acid requirements. They proposed a plasma amino acid (PAA) ratio which was calculated for each essential amino acid as follows:

$$\text{PAA Ratio}_i = \frac{B_i - A_i}{\text{Amino Acid Requirement}_i} \times 100$$

Where B equals the average concentration in mg/100 ml of amino acid i for five hours after a meal, and A equals the concentration of that amino acid just prior to the meal in mg/100 ml while the amino acid requirement for that amino acid is expressed in grams per 16 grams nitrogen.

The changes in the plasma free amino acids following feeding wheat gluten with or without supplemental lysine to dogs are shown in Figure 9.5 for lysine, threonine and tryptophan. In these experiments a meal of casein, sucrose and corn oil was fed at a calorie intake of 1000 kcal and protein at

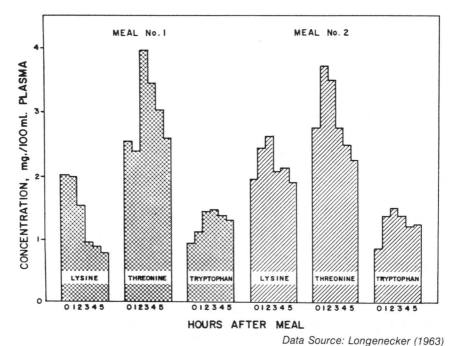

Data Source: Longenecker (1963)

FIG. 9.5. POST PRANDIAL AMINO ACID CHANGES AFTER THE INGESTION OF WHEAT GLUTEN WITHOUT (MEAL NO. 1) OR WITH (MEAL NO. 2) L-LYSINE

26% of calories. Protein intakes ranged from 2 to 6 g/kg body weight. A reduction of the plasma lysine level in the dog fed wheat gluten was very apparent. Supplemental lysine effectively prevented this decline. In Table 9.6 the plasma amino acid ratios for wheat gluten with or without lysine are related to chemical scores. Lysine was clearly identified as the first limiting amino acid, and the order of the other amino acids was close enough to the order calculated from chemical scores as to be very encouraging and to merit future work.

TABLE 9.6.   PLASMA AMINO ACID RATIOS (PAAR) AND CHEMICAL SCORES (CS) FOR WHEAT GLUTEN[1]

Amino Acid	Wheat Gluten PAAR[1]	CS	Gluten + Lysine PAAR[1]	CS
Lysine	$-10.8(1)^2$	27	$3.9(3)^2$	49
Methionine	2.9(2)	31	2.2(1)	31
Arginine	3.2(3)	56	2.6(2)	56
Valine	15.9(5)	68	16.7(6)	68
Threonine	13.8(4)	70	7.3(4)	70
Isoleucine	17.3(6)	71	10.8(5)	71
Tryptophan	36.4(9)	82	45.5(10)	82
Leucine	23.8(8)	89	23.9(9)	89
Histidine	42.5(10)	95	21.0(7)	95
Phenylalanine + Tyrosine	17.9(7)	102	21.1(8)	102

[1] Data Source: Longenecker (1963).
[2] Order of limiting amino acids in parentheses.

In order to make their procedure more applicable to human subjects, these authors developed a shortened procedure in which the post prandial amino acid concentrations were determined on a single pooled blood sample, rather than individual hourly samples (Longenecker and Hause 1961). The changes in PAA ratios in a human subject fed a high protein bread with and without supplemental lysine are presented in Figure 9.6 (Longenecker 1963). In this experiment, protein was consumed at 0.83 g/kg body weight. As was the case for the earlier work with dogs, the PAA data suggested that lysine was clearly first limiting.

Longenecker and Lo (1974) later demonstrated that the PAA ratio technique could reveal the reduction in availability of lysine and methionine resulting from heat treatment of a wheat gluten-egg-milk mixture and soybean meal, respectively. Other investigators have confirmed the utility of using the postprandial changes in free amino acid levels to assess the reduction in digestibility and/or amino acid availability associated with heat treatment of lactalbumen (Vaughan et al. 1977) or non fat dry milk (Ljungqvist et al. 1979). The utility of the PAA ratio method to predict the limiting amino acids of a variety of protein sources directly in human subjects was confirmed by McLaughlan et al. (1963) and Yearick and Nadeau (1967).

An important modification of the PAA ratio method was proposed by McLaughlan (1964). This investigator proposed a PAA score which was

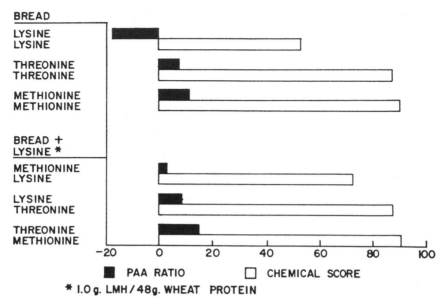

FIG. 9.6. COMPARISON OF PAA RATIOS AND CHEMICAL SCORES FOR THE MOST LIMITING AMINO ACIDS OF BREAD FED TO A HUMAN SUBJECT WITH OR WITHOUT SUPPLEMENTAL L-LYSINE

calculated for each amino acid by dividing the free amino acid concentration in the plasma of ad libitum fed rats by the concentration of that amino acid in the plasma of rats starved for 12 hours. Lysine, threonine, methionine and tryptophan concentrations were determined by microbiological methods which reduced the amount of blood necessary for analysis.

The PAA score method was applied to the determination of the limiting amino acid in twenty-one animal and plant protein sources fed to rats (McLaughlan et al. 1967). In this study threonine and isoleucine were included in the analysis. A summary of the results of this experiment is shown in Table 9.7. It is apparent that the PAA score procedure and rat growth studies in all cases predicted the same amino acid to be first limiting.

The plasma amino acid studies reviewed are important because they have resulted in a new clinical method and have demonstrated a conceptual and experimental basis for its general utility. However essential amino acid needs for growth are known to be more intense than for maintenance, and in this context, plasma amino acid studies in human infants would appear to be of more relevance than studies carried out in adults.

Graham, MacLean and Placko, in a series of studies, have explored the utility of using post-prandial changes in amino acids to assess the limiting amino acids in proteins limiting in methionine or lysine (Graham et al. 1976, MacLean et al. 1977). The results of a preliminary experiment explor-

TABLE 9.7. LIMITING AMINO ACIDS PREDICTED BY PAA SCORES AND RAT GROWTH[1]

Protein Source	Limiting Amino Acid by	
	PAA Score	Rat Growth
Whole wheat	Lysine	Lysine
Oatmeal	Lysine	Lysine
Rye	Lysine	Lysine
Corn	Lysine	Lysine
Rice	Lysine + Threonine	Lysine + Threonine
Millet	Lysine	Lysine
Soy flour	Methionine	Methionine
Chickpeas	Methionine	Methionine
Lentils	Methionine	Methionine
Lima beans	Methionine	Methionine
Navy beans	Methionine	Methionine
Cottonseed flour 1	Lysine and Threonine	Lysine and Threonine
Cottonseed flour 2	Lysine	Lysine
Peanut flour 1	Threonine	−
Peanut flour 2	Lysine, threonine Methionine	Lysine, threonine + Methionine
Sesame	Lysine	Lysine
Egg powder	Lysine	Lysine
Milk powder	Methionine	Methionine
Fish-potato	Tryptophan	−
Corn-fish	Tryptophan	Tryptophan
Corn-blood	Isoleucine	Isoleucine

[1] Data Source: McLaughlan et al (1967).

ing the utility of the plasma amino acid method for predicting limiting amino acids in proteins fed to human infants were described by Graham and Placko (1973). The protein sources, fed at 6.4 to 8.0% of calories to convalescing infants 7–23 months of age, were soy isolate, an oat-soy mixture, white flour, flour plus lysine, and whey-soy-milk and wheat-soy mixtures. Their data, shown in Figure 9.7, show that free methionine concentrations declined postprandially when methionine was calculated to be limiting, but not when other amino acids were limiting. It should be noted that these observations (which are in good confirmation of the earlier work in rats, dogs, and human adults) were made under conditions where protein supplied only 6.4 to 8.0% of calories. These data are important because they suggest that plasma amino acid levels in infants respond to changes in dietary amino acid levels even when protein is fed at or even below the level needed for optimum growth.

Graham et al. (1976) later reported the results of a study in which plasma amino acid changes in convalescing infants fed soy protein or soy protein fortified with DL-methionine (12 to 20 mg/day) were evaluated. In this study the protein sources were fed at 6.4 to 6.7% of calories for 3 days prior to the test meal. The results of this experiment are presented in Table 9.8. In the original paper essential amino acid concentrations were presented as ratios and PAA scores were not calculated. In Table 9.8, actual concentrations were computed from the ratios and then the PAA scores were calculated, as proposed by McLaughlan (1964).

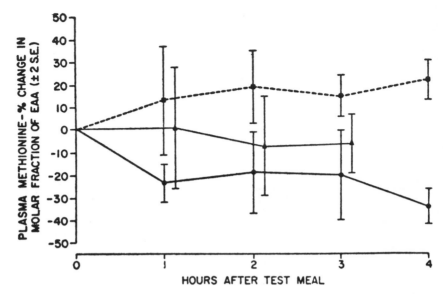

Data Source: Graham and Placko (1973)

FIG. 9.7.   POST PRANDIAL CHANGES IN PLASMA FREE METHIONINE AFTER MEALS LIMITING IN METHIONINE (CLOSED CIRCLES, SOLID LINE) LIMITING IN ANOTHER AMINO ACID (CLOSED SQUARES, BROKEN LINE) OR NOT LIMITING IN ANY AMINO ACID (TRIANGLES, SOLID LINE). VERTICAL BARS ± 2 SEM

The results are consistent with the earlier animal studies in showing methionine to be clearly the limiting amino acid in soy protein. When DL-methionine was added to the soy, methionine was no longer first limiting (the PAA score for methionine increased from 96 to 178). It is known that D-methionine is not utilized by infants, increases plasma methionine and is excreted in the urine (Stegink et al. 1971). The D-methionine most likely increased fasting as well as postprandial amino acid levels, but it may still have increased the PAA score somewhat more than would have been the case had L-methionine alone been used.

Maclean et al. (1977) also evaluated the use of postprandial changes in amino acids as an indicator of lysine adequacy in infants fed wheat protein. In one experiment whole wheat or white flour in a purified diet was fed at levels to provide protein at 6.4% of calories. In another experiment whole wheat or white flour was fed to provide protein at 8% of calories along with casein providing another 4% of calories for a total protein intake of 12% of calories.

The results are presented in Table 9.9. Consistent with the earlier studies, lysine was clearly first limiting in wheat, as judged by a PAA score of 78 compared to a score of 101 for total essential amino acids and a score of 104 for threonine. Making up the lysine deficiency by feeding one third of the

TABLE 9.8.   POSTPRANDIAL AMINO ACID CONCENTRATIONS RELATED TO METHIONINE ADEQUACY[1]

Protein Source[2]	Concentration (umoles/100 ml)				
	Total Amino Acids	Essential Amino Acids	Lysine	½ Cystine	Methionine
Soy protein					
Fasting	255.0	69.4	12.4	1.73	1.80
3 hrs. post prandial	293.0	78.2	13.7	2.19	1.72
PAA Score[2]	115	113	110	127	96
Soy protein + DL-methionine					
Fasting	225.3	54.8	8.8	2.19	1.92
3 hrs post prandial	270.3	66.8	11.5	2.41	3.41
PAA score[3]	120	122	131	110	178

[1] Data Source: Derived from data published by Graham et al (1976).
[2] The protein sources (isolated soy protein with and without DL-methionine addition) were fed at 6.4 to 6.7% of calories.
[3] PAA Score $= \dfrac{\text{3 Hrs Postprandial concentration}}{\text{Fasting Concentration}} \times 100$

TABLE 9.9.   POSTPRANDIAL AMINO ACID CONCENTRATIONS AS RELATED TO LYSINE ADEQUACY[1]

Protein Source	Total Amino Acids	Concentrations (umoles/100ml) Essential Amino Acids	Lysine	Threonine
Wheat[2]				
Fasting	258.1 ± 35.0[3]	74.1 ± 11.3	13.0 ± 2.0	10.7 ± 2.9
3 hrs. Post-prandial	287.3 ± 54.1	74.8 ± 12.0	10.2 ± 2.8	11.1 ± 3.6
PAA Score[4]	111	101	78	104
Wheat/casein[5]				
Fasting	207.8 ± 28.1	63.9 ± 10.7	8.7 ± 2.2	9.7 ± 4.5
3 hrs. Post-prandial	242.9 ± 32.9	78.6 ± 12.0	10.7 ± 2.1	11.5 ± 5.6
PAA Score[4]	117	123	123	119

[1] Data Source: MacLean et al (1977).
[2] Whole Wheat or white flour supplying protein at 6.4% of energy.
[3] Mean ± standard deviation.
[4] PAA score = $\dfrac{\text{3 Hrs. Postprandial Concentration}}{\text{Fasting Concentration}} \times 100$
[5] Whole wheat or white flour supplying protein at 8% of calories and casein supplying protein at 4% of calories.

protein in the form of casein and two thirds as wheat increased the PAA score for wheat to 123 compared to a PAA score of 123 for total essential amino acids and 119 for threonine. Interpretation of this experiment is complicated by the fact that almost twice as much protein was fed in the casein supplemented group, than when wheat alone was fed. Nevertheless, if lysine had remained limiting, based on the earlier literature, the increased protein level would have caused a decrease rather than an increase in postprandial plasma lysine (Longenecker 1963). Similar results to these with wheat were obtained with rice protein, although the differences were lessened due to a lowered digestibility and a less intense lysine deficiency (MacLean et al. 1978). These workers also have used postprandial plasma amino acid changes to evaluate the utilization of dietary protein in children consuming a diet with uneven distribution of protein relative to energy (MacLean et al. 1976). In this work all the protein was present in one daily feeding with dietary energy supplied evenly in five feedings per day. Based on this work, the authors concluded that essential amino acids were available for protein synthesis for up to 17 hours following a single protein feeding.

Over the years many factors have been shown to influence plasma free amino acid concentrations in normal subjects including age, sex, exercise, carbohydrate level, protein level and time of day (Swendseid et al. 1967, Weller et al. 1969, Armstrong and Stave 1973, Milsom et al. 1979). The important role of intestinal secretions in modulating the dietary amino acid pattern in the intestine during digestion is recognized (Nasset and Ju 1969). The fasting pattern of amino acid is by no means an ideal reference pattern (McLaughlan et al. 1967) and indeed, there is no theoretical reason why it should be. Nevertheless it is reasonable to conclude that the PAA ratio and PAA score methods of evaluating protein quality directly in human subjects (including infants) are useful in predicting or at least confirming the first limiting amino acid in those protein sources that have a clear cut first limiting amino acid. These methods also appear in a semi-quantitative sense, to have merit in providing an indication of the extent of the deficiency.

IMPORTANCE OF PROTEIN QUALITY IN HUMAN NUTRITION

Before considering the value of biochemical methods of assessing the efficiency of protein utilization, it would be well to review briefly the importance of protein quality in human nutrition. This topic is discussed more extensively elsewhere (Jansen 1979; also, see Chapters 1–6).

In considering all the nutrition or nutrition related issues in the United States, dietary protein quality would appear to be one of lesser importance. However several trends in U.S. society suggest that protein quality may become of increasing significance even in the United States (Jansen 1978). There is growing interest in vegetarianism in the United States. The Report on Dietary Goals issued by the McGovern Committee, while controversial, has clearly stimulated much public and government interest in reducing animal protein consumption and increasing consumption from plant sources. Perhaps the most significant reason why plant proteins will likely become of increasing significance in the United States combines both economic and technological considerations. Generally speaking, plant proteins are cheaper than animal proteins, and due to advances in food technology, industry is increasing its ability to use cheaper ingredients in the formulation of food products with good acceptability.

Nevertheless, as discussed in more detail elsewhere (Jansen 1979; also, see Chapter 1), protein quality is a much more important consideration in developing countries where cereals are the major suppliers of both protein and calories. Animal studies have shown that improving the protein quality of a cereal diet improves growth even when food energy is restricted (Jansen and Verburg 1977). In studies with young infants, MacLean and Graham (1979) have confirmed that improving the protein quality of a cereal diet increases the efficiency of energy utilization. Jansen and Monte (1977) demonstrated that increasing the protein quality of a cereal diet fed to rats during pregnancy and lactation increased the growth of the offspring, even when food energy was restricted. Comparable studies on protein quality-energy interactions have, for obvious reasons, not been carried out with lactating women. However, if one calculates the level of essential amino acids, especially lysine, secreted in breast milk, it would appear that, for a mother consuming a poor quality cereal diet while secreting high quality breast milk, dietary protein quality is, indeed, an important consideration.

PRACTICAL VALUE OF BIOCHEMICAL PARAMETERS

Protein quality is more important for growth than for maintenance, primarily because of the higher need for essential amino acids especially lysine. Growth is an important consideration not only during early life, but also during lactation and during recovery after sickness or surgery. The best way to assess protein quality needs for growth in humans, is to carry out studies in young growing infants. For ethical and economic reasons, only a limited number of such studies can be carried out. When studies are carried out with infants or young children, it would appear desirable to make biochemical assessments of quality such as plasma urea (Fomon et al. 1979) or post prandial free amino acids (Graham et al. 1976, MacLean et al. 1977) along with nitrogen balance and growth measurements. These techniques have provided useful confirmatory evidence on the importance of protein quality directly in human infants. Such data are, however, supportive, rather than definitive in nature.

In view of the difficulties and cost of studies in infants, other methods are used to predict protein quality in the context of human needs. These include 1) rat growth assays, 2) nitrogen balance studies in adults, 3) biochemical studies in adults and 4) chemical methods.

As discussed elsewhere (Jansen 1978) rat growth methods are good predictors of protein quality for humans. The two major problems with these methods are the more rapid growth of the rat than the infant (thus heightening the need for essential amino acids) and differences in digestibility between rats and human infants. The former problem can be minimized by using a method such as net protein ratio (NPR) that deemphasizes growth rather than methods such as the protein efficiency ratio (PER) or relative protein value (RPV) that emphasize growth to a greater extent. However digestibility is a more serious problem and results in a more significant limitation of the predictive value of rat growth methods.

Nitrogen balance studies carried out in human adults provide important confirmatory data on protein quality directly in human subjects. They are expensive and time consuming, although a short-term nitrogen balance method being developed by Bressani and colleagues (1979; also see Chapter 8) appears to have considerable promise. These types of studies have the most importance in evaluating protein quality for human adults. For example, it is likely that formulated liquid diets being used by obese adults in weight reduction programs can be successfully evaluated in nitrogen balance studies in adults. It is not clear, however, that such studies have revealed much about protein quality for growth of infants that wasn't already known from amino acid composition and rat growth studies.

Biochemical studies in adults also have similar limitations. Plasma and urinary urea levels are in general correlated with protein quality, but considerable variation is apparent. When carried out under carefully controlled conditions, determination of plasma or urinary urea levels can provide useful confirmatory but not definitive data. As is the case for nitrogen balance studies in adults, it is difficult to document many, if any, cases where urea levels have demonstrated differences in protein quality that were not known on the basis of amino acid composition and rat growth data. Post prandial plasma amino acid levels have provided confirmatory data directly in human subjects on the importance of protein quality and the concept of limiting amino acids. However, changes in plasma free amino acids have shown the most utility in revealing differences in amino acid availability directly in humans and in providing an additional independent method for estimating amino acid requirements.

Chemical methods that combine the determination of amino acid composition with an estimation of digestibility have been demonstrated to have considerable predictive value (Satterlee 1978). The chief difficulty with these methods is the problem of adequately assessing the availability of amino acids. It is not at all clear that digestibility studies carried out in human adults can predict the availability of amino acids for young infants. Also it has been demonstrated that digestibility studies in rats do not always predict protein digestibility for infants. Nevertheless, for routine regulatory purposes, it would appear that chemical methods have reached the point where they are or will soon become the methods of choice. However, a chemical assay can only be used as the primary regulatory method if provision is made for follow up animal and human studies in those cases where use of either new ingredients or a new process would indicate that they are needed.

SUMMARY

Biochemical methods useful in assessing protein quality have been considered. Although a variety of methods are available, including amino acid, urea, and enzyme levels in various body fluids or tissues, those most useful to assess quality directly in human subjects are plasma and urinary levels of urea, and postprandial changes in free amino acid levels.

Fasting plasma and urinary urea levels are highly inversely correlated with net protein utilization as is the postprandial plasma urea level. These methods provide significant insights into protein quality differences directly in human subjects. However, enough variability is observed in these general relationships as to make these techniques less than totally satisfactory predictive tools. Postprandial changes in plasma free amino acid have been related to protein quality in both infants and adults. Free amino acid levels have been particularly useful in confirming directly in human subjects the concept of limiting amino acids, in yielding information on amino acid availability, and in providing a useful independent method for assessing amino acid requirements.

The biochemical methods considered provide useful confirmatory data in assessing the efficiency of amino acid utilization. None, however, can be considered definitive methods for the evaluation of protein quality directly in human subjects.

BIBLIOGRAPHY

ALMQUIST, H.J. 1956. "The requirements for amino acids." *In* Amino Acid Handbook. R.J. Block and K.W. Weiss (Eds.). C.C. Thomas, Springfield, Illinois.

ARMSTRONG, M.D. and STONE, U. 1973. A study of plasma free amino acid levels. III. Variations During Growth and Aging. Metabolism 22:571.

BERRY, H.K. 1970. "Plasma amino acids." *In* Newer Methods of Nutritional Biochemistry, Vol. IV, A.A. Albanese (Ed.). Academic Press, New York.

BODWELL, C.E. 1975. "Biochemical parameters as indexes of protein nutritional value." *In* Protein Nutritional Quality of Foods and Feeds, Part 1. Assay Methods — Biological, Biochemical, and Chemical, M. Friedman (Ed.). Marcel Dekker, New York.

BODWELL, C.E. 1977. "Biochemical indices in humans." *In* Evaluation of Proteins for Humans, C. E. Bodwell (Ed.). Avi Publishing Co., Westport, Connecticut.

BODWELL, C.E., WOMACK, M., SCHUSTER, E.M., and BROOKS, B. 1978. Biochemical indices in humans of protein nutritive value. II. Postprandial plasma urea nitrogen. Nutr. Reports. Int. 18:579.

BODWELL, C.E., KYLE, E.M., SCHUSTER, E.M., VAUGHAN, D.A., WOMACK, M., AHRENS, R.A., and HACKLER, L.R. 1979. Biochemical indices in humans of protein nutritive value. III. Fasting plasma urea nitrogen and urinary metabolites at a low protein intake level.   Nutr. Reports 19:703.

BOLOURCHI, S., FEURIG, J.S., and MICKELSEN, O.   1968.   Wheat flour, blood urea concentrations and urea metabolism in adult human subjects. Am. J. Clin. Nutr. 21:836.

BRESSANI, R., NAVARRETE, D.A., ELIAS, L.G., and BRAHAM, J.E.   1979. "A critical summary of a short-term nitrogen balance index to measure protein quality in adult human subjects." In Soy Protein and Human Nutrition, H.L. Wilcke, D.T. Hopkins, and D.H. Waggle (Eds.).   Academic Press, New York.

BROWN, J.A. and CLINE, T.R.   1974.   Urea excretion in the Pig: An indicator of protein quality and amino acid requirements.   J. Nutr. 104:542.

CHEN, W.J., OHASHI, E., and KASEI, M.   1974.   Amino acid metabolism in parenteral nutrition with special reference to the calorie: Nitrogen ratio and the blood urea nitrogen level. Metabolism 23:1117.

DEN HARTOG, C. and POL, G.   1972.   "Biological evaluation of protein quality methods based on measurements of plasma-free amino acids." In Protein and Amino Acid Function, E.J. Bigwood (Ed.).   Pergamon Press, New York.

EGGUM, B.O.   1970.   Blood urea measurement as a technique for assessing protein quality.   Brit. J. Nutrition 24:983.

EGGUM, B.O.   1973.   "The levels of blood amino acids and blood urea as indicators of protein quality." In Proteins in Human Nutrition, J.W.G. Porter and B.A. Rolls (Ed.).   Academic Press, New York.

FOLIN, O.   1905.   Laws governing the chemical composition of urine. Amer. J. Physiol. 13:66.

FOMON, S.J., ZIEGLER, F.F., FILER, L.J., NELSON, S.E., and EDWARDS, B.B.   1979.   Methionine fortification of a soy protein formula fed to infants. Am. J. Clin. Nutr. 32:2460.

GRAHAM, G.G. and PLACKO, R.P.   1973.   Postprandial plasma free methionine as an indicator of dietary methionine adequacy in the human infant. J. Nutr. 103:1347.

GRAHAM, G.G., MacLEAN, JR., W.C., and PLACKO, R.P.   1976.   Plasma amino acids of infants consuming soybean proteins with or without added methionine.   J. Nutr. 106:1307.

JANSEN, G.R.   1978.   Biological evaluation of protein quality. Food Technology 32:52.

JANSEN, G.R.   1979.   "The importance of protein quality in human nutrition. In Soy Protein and Human Nutrition, H. L. Wilcke, D. T. Hopkins, and D.H. Waggle (Eds.).   Academic Press, New York.

JANSEN, G.R. and MONTE, W.C.   1977.   Amino acid fortification of bread fed at varying levels during gestation and lactation in rats.   J. Nutr. 107:300.

JANSEN, G.R. and VERBURG, D.T.   1977.   Amino acid fortification of wheat diets fed at varying levels of energy intake to rats.   J. Nutr. 107:289.

LEATHEM, J.H.   1968.   Protein nutrition and free amino acid patterns.

Rutgers University Press, New Brunswick, New Jersey.

LEWIS, A.J. and SPEER, V.C.   1974.   Plasma amino acid response curves in lactating sows.   J. Animal Sci. 38:785.

LEWIS, A.J., PEO, JR., E.R., CUNNINGHAM, P.J., and MOSER, B.D.   1977a. Determination of the optimum dietary proportions of lysine and tryptophan for growing rats based on growth, food intake, and plasma metabolites.   J. Nutr. 107:1361.

LEWIS, A.J., PEO, JR., E.R., CUNNINGHAM, P.J., and MOSER, B.D. 1977B.   Determination of the optimum dietary proportions of lysine and tryptophan for growing pigs based on growth, food intake, and plasma metabolites.   J. Nutr. 107:1369.

LJUNGQVIST, B.G., BLOMSTRAND, E., HELLSTROM, A., LINDELL, I., OLSSON, M.E., and O. SVANGERG, V.S.   1979.   Title Res. Exp. Med. (Berl.) 174:209.

LONGENECKER, J.B.   1961.   Relationship between plasma amino acids and composition of ingested protein. III. Effect of dietary protein in plasma amino acid and clinical chemistry of dogs. In Progress in Meeting Protein Needs of Infants and Preschool Children.   Publ. 843, National Research Council — National Academy of Sciences, Washington, D.C.

LONGENECKER, J.B.   1963.   "Utilization of dietary proteins." In Newer Methods of Nutritional Biochemistry, A.A. Albanese (Ed.).   Academic Press, New York.

LONGENECKER, J.B. and HAUSE, N.L.   1959.   Relationship between plasma amino acids and composition of the ingested protein. Arch. Biochem. Biophys. 84:46.

LONGENECKER, J.B. and HAUSE, N.L.   1961.   Relationship between plasma amino acids and composition of the ingested protein. II. A shortened procedure to determine plasma amino acid (PAA) ratios.   Am. J. Clin. Nutr. 9:356.

LONGENECKER, J.B. and LO, G.S.   1974.   "Protein digestibility and amino acid availability - assessed by concentration changes of plasma amino acids." In Nutrients in Processed Foods Proteins, P.L. White and D.C. Fletcher (Eds.).   Publishing Sciences Group, Inc., Acton, Massachusetts.

MACLEAN, JR., W.C., PLACKO, R.P., and GRAHAM, G.G.   1976.   Plasma free amino acids of children consuming a diet with uneven distribution of protein relative to energy.   J. Nutr. 106:241.

MACLEAN, JR., W.C., PLACKO, R.P., and GRAHAM, G.G.   1977.   Postprandial plasma lysine as an indicator of dietary lysine adequacy in infants. J. Nutr. 107:567.

MACLEAN, JR., W.C., PLACKO, R.P., and GRAHAM, G.G.   1979.   Postprandial plasma free amino acid changes in preschool children consuming exclusively rice protein.   J. Nutr. 109:1285.

MACLEAN, JR., W.C. and GRAHAM, G.G.   1979.   The effect of level of protein intake in isoenergetic diets on energy utilization.   Am. J. Clin. Nutr. 32:1381.

MARABLE, N.L., HICKSON, JR., J.F., KORSLAND, M.K., HERBERT, W.G., DESJARDINS, R.F., and THYE, F.W. 1979. Urinary nitrogen excretion as influenced by a muscle building program and protein intake variation. Nutr. Reports Int. 19:795.

McLAREN, D.S., KAMEL, W.W., and AYYOUB, N. 1965. Plasma amino acids and the detection of protein-calorie malnutrition. Am. J. Clin. Nutr. 17:152.

McLAUGHLAN, J.M., NOEL, F.J., MORRISON, A.B., and CAMPBELL, J.P. 1963. Blood amino acid studies. IV. Some factors affecting plasma amino acid levels in human subjects. Can. J. Biochem. Physiol. 41:191.

McLAUGHLAN, J.M. 1964. Blood amino acid studies. V. Determination of the limiting amino acid in diets. Can. J. Biochem. 42:1353.

McLAUGHLAN, J.M., VENKAT RAO, S., NOEL, F.J., and MORRISON, A.B. 1967. Blood amino acid studies. VI. Use of plasma amino acid score for predicting limiting amino acid(s) in dietary proteins. Can. J. Biochem. 45:31.

McLAUGHLAN, J.M. 1974. "Nutritional significance of alterations in plasma amino acids and serum proteins." In Improvement of Protein Nutriture. National Academy of Sciences, Washington, D.C.

McLAUGHLAN, J.M. 1979. Fatty liver in rats fed soya protein isolate with added methionine. Nutr. Reports Int. 19:27.

MILSOM, J.P., MORGAN, M.Y., and SHERLOCK, S. 1979. Factors affecting plasma amino acid concentrations in control subjects. Metabolism 28:313.

MITCHELL, JR., J.R., BECKER, D.E., JENSEN, A.H., HARMON, B.G., and NORTON, H.W. 1968. Determination of amino acid needs of the young pig by nitrogen balance and plasma-free amino acids. J. Animal Sci. 27:1327.

MORRISON, A.B., MIDDLETON, E.J., and McLAUGHLAN, J.M. 1961. Blood amino acid studies. II. Effects of dietary lysine concentration, sex, and growth rate on plasma free lysine and threonine levels in the rat. Can. J. Biochem. Physiol. 39:1675.

MUNCHOW, H. and BERGNER, H. 1967. Untersuchungen zur proteinbervertung von futtermitteln. 2. Mitteilung. Arch. Tierernahrung 17:141.

MUNRO, H.N. 1970. "Free amino acid pools and their regulation." In Mammalian Protein Metabolism, Vol. III. H.N. Munro (Ed.). Academic Press, New York.

NASSET, E.S. and JU, J.S. 1969. Amino acids and glucose in human blood plasma after beef and non-protein meals. Proc. Soc. Exp. Biol. Med. 132:1077.

ROSE, W.C., JOHNSON, J.E., and HAINES, W.J. 1950. The amino acid requirements of man. I. The role of valine and methionine. J. Biol. Chem. 182:541.

SCRIMSHAW, N.S., TAYLOR, Y.S.M., and YOUNG, V.R. 1973. Lysine supplementation of wheat gluten at adequate and restricted energy intakes in young men. Am. J. Clin. Nutr. 26:965.

SIMMONS, W.K. and KORTE, R.   1972.   The excretion of urea in relation to protein intake and diuresis. Arch Latinoamer. Nutr. 22:33.

STEGINK, L.D., SCHMITT, J.L., MEYER, P.D., and KAIN, P.H.   1971.   Effects of diets fortified with DL-methionine on urinary and plasma methionine levels in young infants. J. Pediatrics 79:648.

STOCKLAND, W.L., MEADE, R.J., and MELLIERE, A.L.   1970.   Lysine requirement of the growing rat: Plasma-free lysine as a response criterion. J. Nutr. 100:925.

SWENDSEID, M.E., TUTTLE, S.G., DRENICK, E.J., JOVEN, C.B., and MASSEY, F.J.   1967.   Plasma amino acid response to glucose administration in various nutritive states.   Am. J. Clin. Nutr. 20:243.

TAYLOR, Y.S.M., SCRIMSHAW, N.S., and YOUNG, V.R.   1974.   The relationship between serum urea levels and dietary utilization in young men. Br. J. Nutr. 32:407.

TONTISIRIN, K., YOUNG, V.R., and SCRIMSHAW, N.S.   1973a.   Plasma tryptophan response curve and tryptophan requirements of elderly people. J. Nutr. 103:1220.

TONTISIRIN, K., YOUNG, V.R., and SCRIMSHAW, N.S.   1973b.   Plasma tryptophan response curve and tryptophan requirements of children with Down's syndrome.   Am. J. Clin. Nutr. 25:976.

TONTISIRIN, K., YOUNG, V.R., RAND, W.M., and SCRIMSHAW, N.S. 1974.   Plasma threonine response curve and threonine requirements of young men and elderly women.   J. Nutr. 104:495.

VAUGHAN, D.A., WOMACK, M., and McCLAIN, P.E.   1977.   Plasma free amino acid levels in human subjects after meals containing lactalbumin, heated lactalbumin, or no protein.   Am. J. Clin. Nutr. 30:1709.

WELLER, L.A., MARGEN, S., and CALLOWAY, D.H.   1969.   Variations in fasting and postprandial amino acids of men fed adequate or protein-free diets.   Am. J. Clin. Nutr. 22:1577.

WHITEHEAD, R.G. and DEAN, R.F.A.   1964.   Serum amino acids in kwashiorkor. II. An abbreviated method of estimation.   Am. J. Clin. Nutr. 14:320.

YEARICK, E.S. and NADEAU, R.G.   1967.   Serum amino acid responses to isocaloric test meals.   Am. J. Clin. Nutr. 20:338.

YOUNG, V.R., HUSSEIN, M.A., MURRAY, E., and SCRIMSHAW, N.S. 1971.   Plasma tryptophan response curve and its relation to tryptophan requirements in young adult men.   J. Nutr. 101:45.

YOUNG, V. and SCRIMSHAW, N.S.   1972.   "The nutritional significance of plasma and urinary amino acids." In Protein and Amino Acid Function, E. J. Bigwood (Ed.).   Pergamon Press, New York.

YOUNG, V.R., TONTISIRIN, K., OZALP, I., LAKSHMANAN, F., and SCRIMSHAW, N.S.   1972.   Plasma amino acid response curve and amino acid requirements in young men: Valine and lysine.   J. Nutr. 102:1159.

DISCUSSION

DR. MARGEN: We have been involved with this whole matter of plasma amino acid concentrations. In an attempt to either evaluate protein quality or just to see what the matter of variability in replication is, about 8 or 10 years ago, we did a series of experiments with Dr. Felix Held and Dr. Sigmund Nasset. It took us about 8 years to publish that paper because we could never make heads or tails out of it. In the experiment, adult males, under ideally controlled metabolic ward situations on constant diets for relatively long periods of time, were fed the same meal. The whole spectrum of amino acids was measured over a period of six hours and each individual varied all over the map in terms of each of the amino acids measured. When this was repeated, there was poor replication for any given individual. Now, what this meant to us was that we had just better get out of this and let other people work at this who might be able to spend more time and have more energy. So when I see these types of very interesting results, I shudder and wonder what went wrong with us.

The other thing is the matter, again, of blood urea nitrogen. I think we'll have some comments on that. Certainly, we have to really consider a number of things in these measurements. As you pointed out the differences, particularly in adults, are extremely small. The methodology itself, in terms of measuring the blood amino nitrogen, especially the chemical methods, have a built-in potential error in terms of artifactual results which you can sometimes see.

The urease methods, even in the best hands, have a coefficient of variation, I believe, of some place between 5% to 10% as do most enzyme methods do. So again, these small differences which you see have to be interpreted with great caution. They have to be interpreted both in terms of the altered physiological state and the nonconstant state in which the individual might be. These may effect urea concentrations, the biological individual variation and the methodological errors. All of these add up to give a variability which I think very often exceeds the variability which occurs under the most ideal type of circumstances.

Again, I think we have here the difficulties in terms of methodology and variability which constantly plague us as biologists. Anyway, with those comments, my own prejudice shows again. I believe that this remains a very difficult area and as Dr. Jansen has so beautifully pointed out, one which has not been worked out. I would certainly like to hear from Dr. Eggum on this because he's probably had more experience than anyone else. Would you like to make a comment?

DR. EGGUM: We have been using the blood urea measurement for determining protein quality and, as was mentioned here by Dr. Jansen, we have found very good agreement between protein quality and blood urea levels. But if you need to detect small differences, I don't think this is good enough as was mentioned by you. But it's very simple and very fast. You mentioned that the analysis procedure could be difficult. We use the old

Conway method and we seem to be happy with that. But there are many new results now on pigs and there is very good agreement between nitrogen balance and blood urea values obtained with pigs as well as with rats. I think this is one of the very best indirect methods for estimating protein quality.

DR. TORUN: My comments are along the same line as Dr. Margen's. First, as far as the serum urea levels are concerned, if you withdraw from all these very nice graphs the extreme values of NPU's, you end up with such a scatter of points that I don't think you would find any significant correlation between the NPU and the serum urea values except in very isolated cases.

The other comment has to do with the ratio of pre-postprandial plasma amino acids. One of the problems that we have had with other types of studies has to do with the dynamics of gastric emptying. I find it very difficult to interpret changes in post-prandial responses after giving a specific diet. Maybe in the case of the rats, who supposedly are not so subject to stress and other factors as we humans are, the dynamics of gastric emptying might be simpler and this could explain why, in some cases, you can get a better result when you look at pre- versus post-prandial responses in the rat than in the human.

DR. MARGEN: One further comment on your statement which is rather interesting to me. You know, data is frequently presented of, let's say, wheat with a high lysine concentration, and then wheat with lysine added. The assumption here is that the lysine which is added is going to be handled and absorbed exactly as lysine contained in the wheat protein itself. I think the data comparing such experiments have to be interpreted with caution. The two conditions are not the same. In a very neat experiment, Dr. Nancy Canolty, working with Dr. Sigmund Nassett, demonstrated, by using tagged free methionine fed simultaneously with non-labeled protein methionine, that the peaking of the methionine from the two different sources followed entirely different patterns. In other words, the material that was added in the free state peaked in the serum very early and came down and then the material which was in the protein reached a second peak later on. We have to be careful in comparing fortified or free amino acids compared with that incorporated in protein.

DR. YOUNG: There has been recent interest with respect to the biochemical approach to the evaluation of protein nutritional status in the measurement of transport proteins in plasma with a rapid turnover such as thyroxine-binding pre-albumin, and retinol binding protein. Is there an effort that you're aware of, or would it be worthwhile, to explore the measurement of these proteins as a biochemical approach to the evaluation of protein quality?

DR. MARGEN: I guess, as you suggested that the pre-albumins or other rapidly synthesized liver proteins are more sensitive indicators of protein nutritional status in young children, but I don't think anybody that I know of is using it as a means of looking at protein quality through rapid changes in their levels. One would think that the free amino acids might be more sensitive indicators and I think what all the commentators have said are essentially what I was trying to say, namely, that these techniques may have utility, but they're not definitive in what they can do as far as protein quality measurement is concerned. As far as the post-prandial changes are concerned, of course, it has been well known for many years, particularly because of the work of Nassett and his colleagues, that the intestinal secretions can modulate the amino acid pattern to a significant extent. However, I think the studies that have been carried out and are reported, do show that the plasma changes may reflect the dietary protein changes. While the time course of absorption is quite likely to be different for free amino acids and protein bound amino acids, it does not mean that one is more or less effectively utilized. I think what Alf Harper said many years ago—"what the intestine giveth, the intestine taketh away" is an important link in the chain which we have not emphasized enough.

DR. CHENG: I would like to have either Dr. Jansen or Dr. Bressani answer this question. A few years ago when I was in Chile, I did a feeding study with infants of an average age of six months. We were using isocaloric diets with different levels of protein. We used most of the biochemical parameters to test their response to the different levels of protein intake, such as plasma amino acids, urinary urea, and serum urea. Most of the parameters did correspond to the different levels of protein intake. However, with some of our subjects, we could not correlate their highest, or let me say their maximum growth rate, with their nitrogen retention. As we know, when using infants to determine protein quality, it is the maximum growth rate that should be taken as the criteria. We were very confused about how to interpret our data. I know that Dr. Bressani has done much work in the area of infant nutrition so perhaps he could help us solve this puzzle.

DR. BRESSANI: Maybe Dr. Torun should answer the questions related to small children. In studies carried out with animals, such as dogs and children, sometimes one observes situations where nitrogen balance does not correlate with weight gain. We have observed, when testing vegetable protein foods in children, that after reaching a plateau in nitrogen retention, higher intakes of nitrogen result in unexpected high nitrogen retention values, at least higher than those at the plateau. We have never been able to explain such results.

DR. TORUN: This lack of correlation between growth rate and nitrogen retention is something that is very commonly found and I think the answer was partially given by Dr. Young during his presentation. The same thing that is seen in adults when they eat the high protein diet is seen in children and we cannot explain the fate of the excess nitrogen that apparently is retained by the body. In terms of growing infants with the type of experiment that you described, and I'm assuming that you gave the children adequate amounts of protein, I would expect that nitrogen retentions were much higher than at their requirement levels. I think that nitrogen retention is not a sensitive indicator of protein quality under those conditions.

DR. EDWARDS: I want to insert a word of caution as we look at short-term indices of evaluation for protein quality in comparison with long-range experiments.

We did a study at A & T State University on the extent to which wheat based diets would maintain 12 college men in balance over a 77-day experiment. Briefly, for two weeks we fed the men on a basal diet containing wheat, mainly as white bread, with potatoes, fruit and orange juice. And then, in a Latin square design, they were fed either the wheat diet alone, wheat plus pinto beans, wheat plus rice, or wheat plus peanut butter. We found that even though the men were in nitrogen balance during the 60 days on the various wheat-based diets, there was a drop in plasma albumen with a corresponding increase in the alpha and beta globulins. And even though we looked at blood and urinary changes, the first one that was most prominent was a drop, a significant drop in the level of plasma alpha amino butyric acid. We hypothesized that in the case of methionine and threonine of which alpha amino butyric acid is the metabolite, there is an adaptation mechanism, that there is a conditioning response that is occurring which manifests itself over a period of time.

Looking at the urinary alpha amino butyric acid levels, we did not see this drop until about 60 to 77 days when the level was significantly lower. We're of the opinion that possibly one might look at both plasma and urinary alpha amino butyric acid in long-term experiments. Now, the other thing I want to mention is that though these men were in nitrogen balance, we found that there were significant decreases in the retention of selenium and iron at the end of the experiment. So I think that as we look at the selection of an index for protein quality, there needs to be an assessment of the long-range kind of experiments, particularly in the case of proteins of poor quality.

DR. JANSEN: Just a philosophical comment on using animals for predicting protein quality in humans. I think that there certainly are some differences, but in my opinion, in the main, rat growth can be used as a good predictor if one realizes that the human grows slower and basically has a lower need for essential amino acids and a higher proportion of maintenance to growth than does the rat, and also that there is a lower sulfur amino acid requirement for hair and that digestibility is not as good. With those limitations in mind, rat growth predicts protein quality for the human fairly well. I would like to support what Dr. Harper said last night and that is that relative NPR is a good protein assay from a regulatory standpoint. It tends to emphasize maintenance a little more and tends to deemphasize growth more than either the RPV or the PER and I think it has merit from a regulatory standpoint.

DR. MARGEN: I certainly agree with both of the comments but I think these comments point out a very important difference in the question posed. That is the matter of the question of the relative ranking of protein quality versus the question of what are the long-range effects and possible adaptation to any type of protein, both quantitatively and qualitatively in humans. This whole matter is certainly one which has not yet been settled and is probably only going to be settled from long-term controlled studies in different populations. I think that the adaptation times may be so long that we really cannot do them in laboratories. We will probably have to study people in laboratories as close to field conditions as possible and utilizing the proper subjects as Dr. Scrimshaw has already organized, or devise methods for field study. We must examine people in different parts of the world, eating different types of foods and diets, and then begin comparing the data from such peoples.

10

Energy-Protein Relationships

Doris Howes Calloway

The conveners of this meeting asked me to address the following topics: What are the quantitative relationships between protein utilization and nutritional and calorie intake level in different population groups? What is the extent of variation in energy intakes in the U.S. population? What are the implications in relation to defining dietary protein needs (quantitatively and qualitatively) in the U.S. diet? In relation to estimating nutritional value in general and for specific purposes such as nutritional labeling? This is a rather fulsome assignment for the time allowed and this review cannot be comprehensive.

The older literature demonstrates that at any given level of dietary protein, addition of energy improves nitrogen (N) balance of adults until a plateau is reached reflecting adequacy of protein intake (Calloway and Spector, 1954). The curve relating N balance to energy intake rises steeply to about 1500–1800 kcal/day and the rate of improvement is less with calories added beyond this point (Fig. 10.1). Analysis of these data indicated that an adult diet providing 7 to 8% of energy from protein is advantageous, in that it does not impose significant additional demands for urinary water (for excretion of urea) at low energy intake levels and maximum benefit of dietary N is obtained at increasing energy levels. Studies carried out since this early review have made quantitative refinements but not altered the concept.

N BALANCE AND ENERGY INTAKE OF ADULTS

Experiments in which high quality proteins have been fed to men at levels near the maintenance range of both N and energy have not yielded entirely consistent values for the gain in N balance with added energy (Table 10.1). Garza et al. (1976) fed whole egg protein at the FAO/WHO Safe Level of 0.57g/kg body weight and varied energy intake 4–20% between test periods; N retention ranged from 1.74 to 4.14 mg N/kg/kcal for the three male subjects. Calloway (1975) determined individual energy and egg-

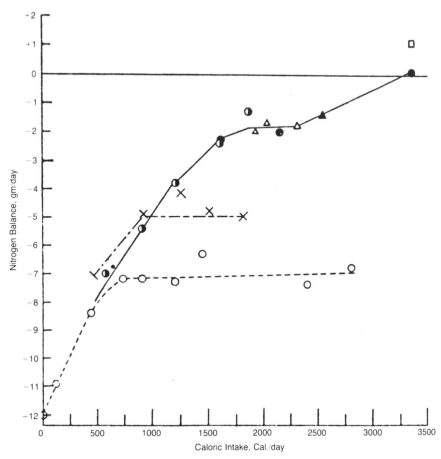

Source: Calloway and Spector (1954)

FIG. 10.1.   NITROGEN BALANCE AT VARIOUS LEVELS OF CALORIC INTAKE
○ Protein-free or Nitrogen intake, Gm/day: 1.0−1.9, × 2.4−5.0, ◐ 5.4−7.7, ● 8.1−9.7, △ 10.4−11.7, ▲ 12.4, □ 15.4.

white protein requirements of six men and then gave constant protein with energy 0.85, 1.0 and 1.15 times requirement. Slopes relating N balance to energy intake were significant in four cases and indicated retention of 1.40−2.02 mg N/kg/kcal. In the third study cited in Table 10.1 (Calloway and Chenoweth, 1973), milk protein was fed to men at the FAO/WHO Safe Level and energy intake was adjusted each third day if necessary to achieve essentially constant body weight. Energy intakes were shifted a maximum of 15−30% during the study. Correlations between intake and balance were significant for four of the men, with values of 1.19−2.59 mg N/kg/kcal. The average of all 14 values tabulated is 1.94 mg N/kg/kcal with a coefficient of variation of 52%.

TABLE 10.1.   RELATIONSHIP BETWEEN NITROGEN BALANCE AND ENERGY INTAKE
N Balance (mg/kg) = b + m Energy Intake (kcal/kg)

Protein Fed	b	m
Whole egg[1]	−161	3.45
91 mg N/kg	−85	1.74
	−206	4.14
Egg White[2]	−23	0.49[3]
$\overline{X}$ 90 mg N/kg	−47	1.40
	−87	2.02
	−64	1.85
	−70	1.70
	−56	1.34[3]
Milk[4]	−4	0.48[3]
94 mg N/kg	−89	2.59
	−78	2.36
	−99	2.41
	−46	1.19

[1] Garza et al. (1976).
[2] Calloway (1975).
[3] $.05 < p < .10$; other values $p < .05$.
[4] Calloway and Chenoweth (1973).

Scrimshaw et al. (1973) evaluated the efficacy of lysine supplementation to wheat gluten at two levels of energy in each of two experiments, in one of which protein intake was 0.27 g/kg and in the other, 0.73 g/kg. Individual balances are not provided in the published report but the average figures for improvement in N balance were 0.91 and 0.92 mg N/kg per additional kcal/kg for wheat gluten at both protein levels. (These values differ from those in the published report which were calculated as ΔN balance mg/d ÷ ΔEnergy kcal/d, not adjusted to body weight). With lysine added the figure was 0.82 mg/kg/kcal at the lower protein intake level and 1.23 mg at the higher protein level.

Anderson et al. (1969) compared N balances of men given casein and amino acids patterned on casein at two levels of energy intake, 44 and 50 kcal/kg. N intake was held constant at 6.3 g/d (about 86 mg/kg and approximately the same as the studies cited above). Reading values from graphs provided suggest that N balance improved more with intact protein than with amino acids, values being about 1.2 mg and 0.83 mg/kg/kcal with casein and amino acids respectively.

The N balances obtained by Anderson et al. (1969) appeared to be more positive with casein than with the amino acids although not significantly different. If the N balance response to energy were linear across the range of energy intakes, a higher intake would be required to promote N balance with amino acids than with casein, a conclusion reached earlier by Rose (1949). This result could also be viewed as an indication of poorer utilization of amino acid N. Because Anderson et al. held total N intake constant and body weight of subjects varied it is possible, although perhaps not quite legitimate, to construct an equation linking N balance and N intake at two levels of energy intake. The least squares fit of the data expressing N balance (mg/kg) as a function of N intake (mg/kg) are: casein, lower energy

-41 + 0.55 NI, higher energy -31 + 0.52 NI; and amino acids, lower energy -26 + 0.37 NI, higher energy, -65 + 0.87 NI. The slope ratio indicates that the NPU (net protein utilization) of casein was not affected by energy intake whereas that of amino acids was improved with increased energy.

Another experimental design that has been used to study the effect of energy intake on protein utilization provides uniform daily intakes of protein and energy to all subjects irrespective of individual needs, with different treatment levels of both protein and energy. In their study, Nageswara Rao et al. (1975) included two protein levels, 40 and 60 g/d, from mixed foods and several energy intakes differing by 300 kcal/d across a range of 1800 to 3000 kcal/d. There was substantial variation in response among the 2–4 subjects per protein-calorie treatment group. N balance improved by averages of 0.32 to 4.64 mg/kg/kcal between incremental energy periods. The protein sources were not held constant among treatments so 40- and 60-g protein periods cannot be used to compute NPU.

The only comparable study of women indicates that their N balance response to added energy is not different from men (Leverton et al. 1951). N balance improved by 2.0 mg/kg/kcal with energy increased from 1800 to 2400 kcal/d at 6.9 g/d N intake, and by 1.3 mg/kg/kcal with 10.0 g/d N intake. Leverton and co-workers introduced a third variable in this study— animal protein was divided among two or three daily meals. N balance was not affected by meal-distribution at the level of 10.0 g N/d but at the lower level of N intake, balance was better with the three-meal distribution.

Experimental Variables Affecting N Balance in Relation to Energy Intake of Adults

Another study in which men were given additional carbohydrate with mixed foods diets yielded interesting results. Munro and Wikramanyake (1954) found N retention to be improved by 2.9 mg/kg/kcal when the additional carbohydrate was given with meals (calories increased about 25%) and by 2.1 mg when it was given separately. These findings and previously published evidence led the authors to conclude that "protein utilization is affected in two ways by the other energy-yielding nutrients in the diet. First, protein utilization is favorably influenced by the presence of some carbohydrate in the protein-containing meals. Close proximity in time of eating carbohydrate and protein is necessary in order that this interaction may take place, and fat cannot be used in place of carbohydrate. Secondly, carbohydrate and fat act interchangeably as energy sources in sparing protein; in order to exert this sparing action they do not need to be taken along with dietary protein."

Richardson et al. (1979) have now tested the effect of carbohydrate on N balance of men at isoenergetic intakes and confirm the previous conclusion. Diets containing the FAO/WHO Safe Level of milk protein supported better N balance when energy was derived from a 2:1 ratio of carbohydrate to fat than when energy was derived equally from the two sources. Utilizing the range of energy intakes among subjects within treatment groups, the investigators derived regression equations relating N balance to energy intake (Fig. 10.2). N balance increased by 2.6 mg/kg/kcal with the higher carbohydrate diet and by 1.3 mg/kg/kcal with the lower level. However, the significance of the regression depends heavily on the difference in N balances at low energy intake (38–40 kcal/kg with subjects experiencing weight loss). Calloway (1975) has also reported that added energy has more impact when the level is raised from low to adequate than from adequate to luxus levels.

Their data led Richardson et al. (1979) to conclude that the FAO/WHO Safe Level of protein intake is inadequate to support N balance in young men. However, at energy intakes adequate to maintain body weight in their

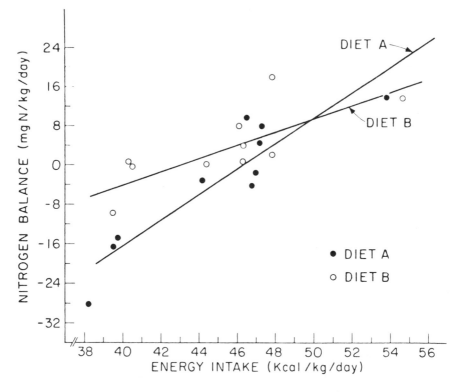

FIG. 10.2.   RELATIONSHIP BETWEEN N BALANCE (MG N/KG BODY WEIGHT PER 24 HR) AND ENERGY INTAKE (KCAL/KG BODY WEIGHT PER 24 HR) DURING DIET PERIODS A (O) AND B (●)
Diet A provides carbohydrate and fat in a ratio of 1:1. Diet B provides carbohydrate and fat in a ratio of 2:1. The regression equations are: Diet A: N balance = 2.576 × energy intake −119.49, s 5.586; s_b = 0.393; r = 0.918. Diet B: N balance = 1.297 × energy intake −55.832; s = 5.767; s_b = 0.435; r = 0.733. Statistical difference between regression slopes = P<0.05.

subjects, N balances appeared to be randomly distributed about the equilibrium point and the group average balance was 3 ± 8 mg N/kg with the higher carbohydrate diet. This finding accords entirely with those of Calloway and Chenoweth (1973) that the FAO/WHO Safe Level of milk protein meets average protein needs (not the mean plus two standard deviations as intended) and that at this N intake level, energy intake determines the sign and magnitude of N balance.

It is, unfortunately, the case that energy balance as judged by the criterion of essentially constant body weight (a just-not-noticeable-difference in statistical terms) is an inadequate descriptor of energy requirement for our present purposes. Calloway and colleagues (Calloway, 1975; Calloway and Chenoweth, 1973; Calloway and Margen 1971) have shown that N balance is more responsive to changes in energy intake than is body weight and, of course, body composition may change even if weight remains constant. Now the Berkeley group have demonstrated that N balance differs at equivalent energy balance states with energy intake level altered according to different work requirements (Butterfield-Hodgdon and Calloway 1977). Subjects of these studies performed a standard amount of exercise (one hour of walking at 3 mph up a 10% grade) or had activity increased by an amount equal to 15% of standard energy requirement (about 420 kcal/d). Added cycle work was of the same intensity as the treadmill work so that there was not a training effect. Energy intake was first increased by 15% without added work (i.e. luxus intake). Then equivalent work was added (i.e. energy balance again achieved at higher activity), and finally another 420 kcal/d was added to the diet with increased work continued (luxus energy at higher work level). (Subsequent studies, as yet unreported, utilized different activity schedules to induce different energy balance and work states).

Average crude N balance of six men was 0.12 g/d under standard conditions and 0.66 g/d when 15% energy was added without changing work level, or about 1.3 mg/kg/kcal added per kg/day. When a balancing work requirement was added, N balance was 0.42 g/d. With luxus energy intake at the higher work level, N balance was 1.29 g/d, or 12 mg/kg/kcal added. Thus, under conditions of energy balance or excess, physical activity promotes N retention, indicating an apparent improvement in NPU. Marable et al. (1979) have also shown in college men that urinary nitrogen excretion is less with training exercise than under control conditions with diets providing generous protein and energy intakes. Variation in exercise protocols may be one of the factors accounting for between-study variation in estimates of protein utilization.

Berkeley studies of pregnant women have led to reexamination of the phenomenon of body weight change in relation to N balance, the purpose being to evaluate the composition of tissue deposited. N balance proved to be positively associated with weight gain in both pregnant (r=0.72) and non-pregnant women (r=0.86) (Calloway et al. 1980). For 34 balances in non-pregnant women fed 0.8 g egg white protein/kg, some gaining and others losing weight, N balance (mg/d) = 189 + 16.2 weight change (g/d). The intercept is very close to the error term for precision of N balances performed in this laboratory (Calloway 1975). The slope predicts that tissue gained or lost contains about 10% of protein; for pregnant women, the estimate is about 12% of protein.

Two studies reported in sufficient detail have been used to construct equations relating weight change and N balance of men given the FAO/WHO Safe Level of protein (0.57 g/kg). For data of Richardson et al. (1979), N balance (mg/d) = 40 + 9.6 weight change (g/d) (r=.74) and data of Calloway (1975) yield N balance = 113 + 7.8 weight change (r=.82). Under these conditions, gain or loss of weight in men appears to represent tissue containing 5% of protein.

NET PROTEIN UTILIZATION BY ADULTS IN RELATION TO ENERGY INTAKE

To determine an effect of energy intake on NPU by the accepted slope-ratio method it is necessary to include two or more levels of protein in the range below requirement and two or more levels of energy in the near-maintenance range. Inoue et al. (1973) have used this design to measure NPU of rice and egg proteins. For each protein source, two groups of men were studied, one having an adequate level of energy intake (about 45 kcal/kg) and the other an energy surplus (about 57 kcal/kg). Within each group 2 or 3 subjects received a given level of N within a range of about 50 to 120 mg/kg. Linear regression equations relating N intake to N balance (mg/kg/d) were, for adequate and excess energy groups, respectively: egg protein −37 + 0.41 NI and −26 + 0.36 NI (Note: this value was reported in the original paper as 0.538; recalculation from the published balances for week 3 gives a figure of 0.358); and rice protein −32 + 0.27 NI and −38 + 0.47 NI. (See Table 10.2). Thus, the NPU of egg protein was unrelated to energy intake level and that of rice protein appeared to be about 75% higher with excessive than with adequate energy intake.

NPU can be estimated from N balance in the study of lysine-supplementation of wheat gluten by Scrimshaw et al. (1973) cited previously, if it is assumed that obligatory N loss remains constant over the range of energy levels fed. The investigators have made similar calculations from their studies of methionine-supplementation of chick peas. The data, in Table 10.2, demonstrate the expected decline in NPU with increased protein level and increase in NPU due to supplementation with amino acids. The fall in NPU due to reduction of energy intake amounted to 24 to 37% of the values observed at adequate energy levels.

PROTEIN-ENERGY RELATIONSHIPS IN CHILDREN

The same test designs have been used to evaluate protein-energy relationships in children as in adults, but generally with reduction in duration of the higher risk treatments. Most of the data have been generated from infants and young children with poor dietary and health histories and in varying stages of recovery from malnutrition and associated diseases. For

TABLE 10.2. PROTEIN QUALITY ASSESSED IN MEN UNDER CONDITIONS OF MARGINAL, ADEQUATE AND EXCESSIVE ENERGY INTAKE

Protein Source	N Intake mg/kg	Energy Intake kcal/kg	Protein Quality
			slope ratio
Egg[1]	50–100	46 ± 2	0.41
	44–74	58 ± 2	0.36[2]
Rice[1]	50–121	45 ± 2	0.27
	51–96	57 ± 2	0.47
			Calculated NPU
Wheat Gluten[3]	39 ± 1	41 ± 2	44
	39 ± 1	53 ± 3	70
	118 ± 6	38 ± 5	21
	118 ± 6	48 ± 7	29
Wheat Gluten[3] + Lysine	39 ± 1	41 ± 2	55
	39 ± 1	53 ± 3	84
	118 ± 6	38 ± 5	26
	118 ± 6	48 ± 7	36
Chick Pea[4]	43	"Restricted"	49 ± 14
		"Adequate"	72 ± 9
Chick Pea[4] + Methionine	43	"Restricted"	61 ± 16
		"Adequate"	80 ± 7

[1] Inoue et al. (1973).
[2] This value was reported in the original paper as 0.538; recalculation from the published balances for week 3 gives a figure of 0.358.
[3] Scrimshaw et al. (1973).
[4] Young (1972).

at least the younger children, balance studies involve some days of physical restraint, repeated at several time intervals; this cannot help but induce stress, especially when coupled with maternal separation. Thus, variation within and between studies may be large. An early study is an example:

Holemans and co-workers (1955) studied 29 cases of uncertain age but ranging from months to 8–9 years, weighing 4–16 kg and having either typical symptoms of Kwashiorkor or chronic malnutrition but growing slowly. Most had uncorrected hookworm, 19 had malaria without fever, and all had some degree of anemia. Four cases are presented in which two levels of energy were fed while N intake was kept nearly constant for a given subject. N retentions were improved by 5.7, 5.3, 1.8 and 1.1 mg/kg/kcal.

A large number of studies have been conducted by Graham and his colleagues, in Peru. (Subjects are partially recovered, malnourished infants). In a typical study (Graham et al. 1969) cottonseed flours were fed at two levels of energy intake, 100 and 125 kcal/kg, with N intake constant at 320 mg/kg. Commercial milk-based formula served as the control with infants receiving alternating test and control treatments. With milk formula, N retention improved by 1.16 mg/kg/kcal added (Table 10.3). N retention from cottonseed flour was affected by its gossypol content and ranged from 0.84 mg/kg/kcal with a degossypolized product to a low of 0.16 mg/kg/kcal for products with 0.84–0.92% of gossypol.

Iyengar et al. (1979) gave 4-year-old children two levels of energy with diets providing 1.0 to 2.0 g protein/kg (Table 10.4). As varying proportions

TABLE 10.3.   EFFECT OF ENERGY INTAKE ON NITROGEN RETENTION BY MALNOUR-
ISHED INFANTS AGED 4–38 MONTHS FED 320MG N/KG FROM DIFFERENT SOURCES
OF PROTEIN[1]

Protein Source	Crude Nitrogen Balance mg/kg		N retained mg/kg/kcal added
	100 kcal/kg	125 kcal/kg	
Cow's milk ("Similac")	65(9)	94(10)	1.16
Cottonseed flour Gossypol			
<.02%	80(2)	101(1)	0.84
.13–.17%	58(3)	73(2)	0.60
.84–.92%	42(2)	58(4)	0.16

[1] Recalculated from Graham, et al. (1969). Numbers in parentheses are number of balances included in the mean.

TABLE 10.4.   EFFECT OF ENERGY INTAKE ON NITROGEN RETENTION BY FOUR
CHILDREN AGED 3.5–4.5 YEARS AT DIFFERENT LEVELS OF PROTEIN INTAKE[1]

Protein Intake g/kg	Rice	Diet % N from Legumes	Milk	Crude Nitrogen Balance mg/kg 80 kcal/kg	100 kcal/kg	N retained mg/kg/kcal added
1.0	74	11	10	−30.8	4.2	1.75
1.3	59	18	15	26.3	35.9	0.48
1.6	45	27	22	44.9	67.7	1.14
2.0	44	33	18	56.8	75.8	0.95

[1] Recalculated from Iyengar et al. (1979).

of the protein were derived from rice, legumes and milk, the data cannot be
used to compute a slope ratio. However, increasing energy intake from 80 to
100 kcal/kg improved N retention by 1.8 mg/kg/kcal at 1.0 g protein/kg, by
0.5 mg at 1.3 g/kg, 1.1 mg at 1.6 g/kg and 1.0 mg at 2.0 g protein/kg.

NET PROTEIN UTILIZATION BY INFANTS IN RELATION
TO ENERGY INTAKE

Another study by Graham et al. (1964) does allow calculation of slope
ratios from two levels of milk N intake, 320 and 480 mg/kg, at five levels of
energy intake, 75–175 kcal/kg. All of the infants were recovering from
malnutrition and the population was separated into groups according to the
presence or absence of low serum albumin levels (group HA). At the lowest
energy intake (75 kcal/kg) there were not sufficient normoproteinemic
(group C) subjects to compute a ratio; the slope for HA infants was 0.90 and
the intercept (which should equal obligatory N output, expected to be about
100 mg/kg in infants given adequate calories) was −193 (Fig. 10.3). The
slope was greater for HA than C subjects at 100 kcal/kg (0.46 and 0.26,
respectively) but correlation between N balance and N intake was so low at
the 125–130 kcal/kg level for the HA group as to question the data (n=6,
r=0.32). Slope-ratios were clearly improved in group C infants at 150 and
175 kcal/kg. The maximum value was 0.66.

A recent study by Morasso et al. (1979) essentially confirms in malnourished infants the conclusions derived from data for adults presented in Fig.

REGRESSION OF NITROGEN BALANCE AND INTAKE OF
MALNOURISHED INFANTS AT DIFFERENT ENERGY INTAKES

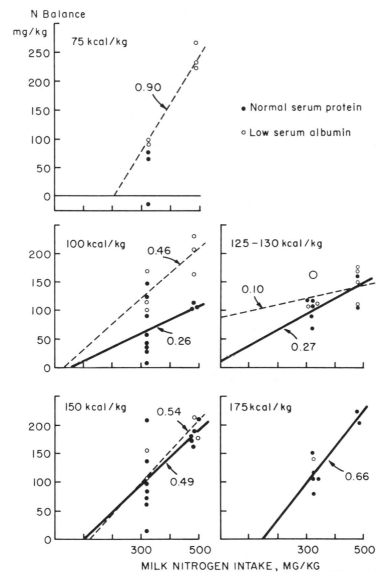

Calculated from Graham et al. (1964)

FIG. 10.3.   REGRESSION OF NITROGEN BALANCE AND INTAKE OF MALNOURISHED
INFANTS AT DIFFERENT ENERGY INTAKES

10.1. In their very young (1½–17 months) undernourished (weight deficit for age > 25%) infants, addition of energy beyond 110 kcal/kg was of little value at the lowest levels of milk protein fed (5.5–9.0% of energy). N balance plateaued at about 140 kcal/kg with intermediate protein concentrations (9–12.5%). Although the rate of improvement in balance declined at the higher energy levels, N balance continued to improve with increased energy through the entire range tested with formulas providing more than 12½% of energy as protein. The data from this study are best described by a three-dimensional model linking protein utilization with protein and energy intake (Fig. 10.4). Quoting Calloway and Spector (1954) on N balance in young men "to the general principles set forth—that on a fixed adequate protein intake, energy level is the deciding factor in nitrogen balance and that with a fixed adequate caloric intake, protein level is the determi-

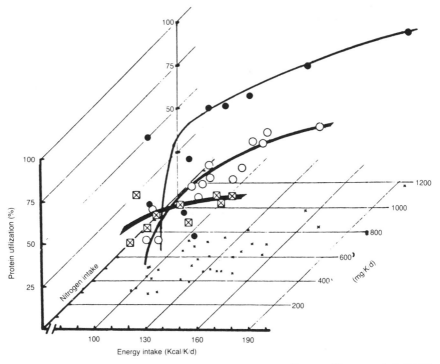

Source: Morasso et al. (1979)

FIG.10.4.   THREE-DIMENSIONAL MODEL SHOWING PROTEIN UTILIZATION AS A FUNCTION OF ENERGY INTAKE AND PROTEIN INTAKE
The curves are drawn according to three protein:energy ratios: □ = 5.5–9%, 0 = 9 to 12.5%, ● > 12.5%. Energy intake is represented on the x axis, nitrogen intake on the y axis and the z axis shows protein utilization. The P/E function would be a family of straight lines with an origin at zero, sweeping the x:y plane from 0 to 90 degrees. In the graph, experimental nitrogen:energy interaction points are shown by small crosses; their projection onto the z axis determines a series of planes on each of which protein utilization is a continuous function for varying energy intake at constant percentage of energy from protein.

nant—may be added a corollary. That is, at each fixed inadequate protein intake there is an individual limiting energy level beyond which increasing calories without protein or protein without calories is without benefit."

SUMMARY OF PROTEIN-ENERGY RELATIONSHIPS

In both children and adults, male and female, increasing the energy content of the diet from marginally sub-maintenance levels to luxus levels results in a more positive or less negative nitrogen balance. This effect requires the presence of sufficient protein of adequate quality to allow N balance. The literature is not entirely consistent as regards differential effects of added energy according to protein quality but the evidence tends in the main to support the view that the impact of energy is greater with proteins of better quality. In that case, the difference between proteins will appear to be greater at higher than at lower energy intakes.

ENERGY INTAKES OF AMERICANS

Data from the Health and Nutrition Examination Survey (HANES) conducted by the United States Department of Health, Education and Welfare in 1971–74 provide the best current information on energy intakes of individuals in various age, sex, race and income categories (DHEW, 1979). The energy intake data were calculated from tables of food composition applied to recalled food consumption for the previous day. The data are subject to the well-known methodologic problems but they should, in aggregate, provide a reasonable picture of consumer food habits. Average energy intakes of all groups generally met or exceeded average requirements levels through childhood but by the age of puberty (12–14 years in females, 15–17 years in males) average intakes by white males were 80–88% of standard, black males 66–71% and females of both races 72–76%.

Average intake figures centered around 80% of standard for adults aged 25–34 years (except for males in the "above poverty" class, for whom the figures were 93–94%), but average intakes of the elderly were very low: 71–79% of standard for whites and 64–71% for blacks.

Table 10.5 records energy intakes at the 10th, 50th and 90th centiles by age and sex. Energy intakes of female adults of all ages are alarmingly low as are intakes of elderly men. In fact, intakes are so low as to cast doubt on their validity as representing habitual intakes of these consumer groups. Three lines of evidence suggest that the figures are spuriously low. Firstly, evidence from a number of species suggests that the minimum energy intake compatible with body maintenance (except lying at rest) is about 1.5 times the basal metabolic expenditure (BMR) (FAO/WHO, 1973). The BMR of healthy older men is, for example, of the order of 1400–1600 kcal/d and the recorded 50th centile intake is only 1695 kcal. Secondly, habitual consumption of low intakes would result in low body weights, and yet obesity is more prevalent than undernutrition. Finally, studies in which subjects are given controlled diets under controlled conditions show requirements to be far higher than reported energy intakes. Referring again

TABLE 10.5.   ENERGY INTAKE (KCAL/DAY) BY SEX AND AGE, ALL RACES AND INCOME LEVELS, HANES DATA, 1971–1974[1]

Age (yrs)	10th	Centile 50th	90th
Males			
2–3	879	1478	2307
6–7	1314	1986	2936
15–17	1656	2791	4674
25–34	1546	2588	4130
65 and +	997	1695	2647
Females			
2–3	826	1342	2030
6–7	1056	1774	2623
12–14	1052	1859	2824
25–34	847	1547	2520
65 and +	728	1239	1943

[1] Source: DHEW (1979).

to the elderly, energy intakes required to maintain body weight with ½ to 1 hour of moderate physical activity per day in three separate Berkeley studies (Margen and Durkin, 1980, Schutz 1979, Zanni et al. 1979) of healthy men were 2554 ± 222 kcal, 2343 ± 247 and 3030 ± 324 kcal/d. For six older women (65–74 years) requirement was 2022 ± 65 kcal/d (Calloway 1980).

People do consume very little food on some days, and feast on others, so the occurrence of exceptionally high and low values is to be expected. However, if the real and usual food intake of older adults is as low as the HANES data indicate it must mean that average activity is less than ½ hour of moderate walking a day, that their body mass is decreasing over time, that their rate of metabolism has fallen as an adaptive mechanism, or some combination of these.

IMPLICATIONS OF ENERGY INTAKE FOR PROTEIN NEEDS

One way to tackle the problem presented by these low energy intake figures is to examine the ratios of protein to energy in the reported diets in relation to requirements. The NRC's 1979 Recommended Dietary Allowance (RDA) (NRC/FNB 1979) for protein is considered to be two standard deviations above the mean requirement for protein having an assumed NPU of 70. Thus, average protein requirement is 0.77 RDA. The RDA for energy is the average requirement and a range of usual requirements is provided in the latest revision of the RDA's. There is no evidence that individual requirements for energy and protein are linked, so a high requirer of protein may have average or low energy requirements. Table 10.6 provides P/E ratios based on three assumptions about requirements. If the individual's protein requirement is two standard deviations above the mean (the RDA) and energy needs are average (the RDA), P/E ratio required is 0.07 at ages 1–3 years, 0.08 for adolescents and adults and 0.11 for the elderly. For pregnant women the figure is 0.128. Average requirers of both protein and energy (0.77 RDA for protein, RDA for energy) need

TABLE 10.6. PROTEIN/ENERGY RATIOS REQUIRED IN THE DIET ACCORDING TO VARIOUS ASSUMPTIONS REGARDING REQUIREMENTS FOR EACH[1]

| | Protein, g/d | | P/E Ratios | | |
Category/Age (yrs)	RDA	.77RDA	RDA/ RDA	.77RDA/ RDA	RDA/ Low range
Children					
1-3	23	18	.071	.055	.102
Males					
15-18	56	43	.080	.061	.106
23-50	56	43	.082	.064	.097
76+	56	43	.109	.084	.136
Females					
11-14	46	35	.084	.064	.122
23-50	44	34	.088	.068	.110
76+	44	34	.110	.085	.147
Pregnant*	74	57	.128	.099	.156
Lactating*	64	49	.102	.078	.121

[1] RDA = Recommended Dietary Allowance (NRC/FNB 1979).

lower P/E ratios, ranging from 0.06 to 0.10. In the extreme case of high protein (RDA) and low energy requirement (lowest NRC tabulated figure), P/E ratios increase to a range of 0.10 to 0.15.

The HANES data on protein and energy intakes can be used to construct a similar series of P/E ratios based on assumptions about the co-occurrence of protein and energy in the diet. Table 10.7 presents ratios for matched centiles of intake (10th/10th, etc.) and two series of extreme assumptions, namely that the 10th centile consumer of protein is a 50th or 90th centile consumer of energy. The most likely diets appear to derive about 13 to 17% of energy from protein. Only the most extreme assumption (10th protein/ 90th energy) yields P/E ratios as low as .05-.06. In the case assumed, the higher intake of energy would offset the physiologic effect of lower protein intake.

TABLE 10.7. PROTEIN/ENERGY (KCAL/KCAL) RATIOS IN DIETS OF AMERICANS, HANES DATA, 1971-1974[1]

| | Intake centile P/E[2] | | | | |
Age, yrs	10/10	50/50	90/90	10/50	10/90
Males					
2-3	0.13	0.14	0.15	0.08	0.05
6-7	0.14	0.15	0.16	0.09	0.06
15-17	0.13	0.14	0.15	0.08	0.05
25-34	0.15	0.16	0.17	0.09	0.06
65 and +	0.15	0.16	0.17	0.09	0.06
Females					
2-3	0.13	0.15	0.16	0.08	0.05
6-7	0.14	0.14	0.16	0.08	0.06
12-14	0.14	0.15	0.16	0.08	0.05
25-34	0.14	0.16	0.17	0.07	0.05
65 and +	0.15	0.16	0.17	0.09	0.06

[1] Source: DHEW (1979).
[2] Protein intake (in g) × 4 ÷ energy intake (kcal).

In practice, energy and protein intakes increase together, as illustrated by data from studies of pregnant women (Fig. 10.5) (Calloway 1974). Protein and energy intake rose in parallel across the ranges of 5–25 g N and 1000–3500 kcal/d. N balances from this same series were used to estimate the impact of protein and energy intakes. The best fit of the data for 199 balances indicates that N balance (g/d) = $-2.31 + \{0.54 \times 10^{-3}$ Energy Intake (kcal/d)$\} + 0.204$ {N Intake (g/d)}. Thus 100 kcal and 280 mg N have the same effect on N balance, each addition of either increasing N balance by about 1 mg/kg of body weight. Maximum benefit would accrue from progressive energy intakes in which protein supplied about 7% of the calories.

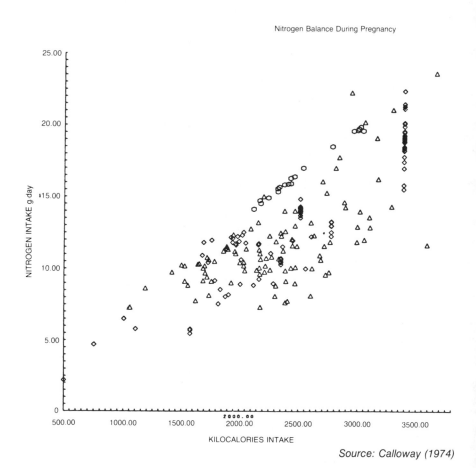

Source: Calloway (1974)

FIG. 10.5.   CORRESPONDENCE OF ENERGY AND PROTEIN CONTENT IN SELF-SELECTED DIETS OF PREGNANT WOMEN

IMPLICATIONS FOR FOOD STANDARDS AND LABELING

The implications of these protein-energy relationships for individual food standards and labeling are almost nil, because these relationships depend on the totality of intake, not on single components of the diet. The least-risk stance is, however, that a new or substitute food should provide as much protein per calorie of as good quality as the food it replaces. Protein quality for man might better be estimated by amino acid scores corrected for measured digestibility, or by well-controlled animal studies, than by continued routine assays in children who are available for study because they are poor and malnourished.

REFERENCES

ANDERSON, H.L., HEINDEL, M.B. and LINKSWRITER, H.   1969.   Effect on nitrogen balance of adult man of varying source of nitrogen and level of calorie intake. J. Nutr. 99:82.

BUTTERFIELD-HODGDON, G. and CALLOWAY, D.H.   1977.   Protein utilization in men under two conditions of energy balance and work. Federation Proc. 36:1166.

CALLOWAY, D.H.   1974.   Nitrogen balance during pregnancy. In "Nutrition and Fetal Development", M. Winick (Ed.). John Wiley & Sons, N.Y., p. 79.

CALLOWAY, D.H.   1975.   Nitrogen balance of men with marginal intakes of protein and energy. J. Nutr. 105:914.

CALLOWAY, D.H.   1980.   Unpublished data. University of California, Berkeley.

CALLOWAY, D.H. and CHENOWETH, W.L.   1973.   Utilization of nutrients in milk- and wheat-based diets by men with adequate and reduced abilities to absorb lactose. I. Energy and nitrogen. Am. J. Clin. Nutr. 26:939.

CALLOWAY, D.H., KING, J.C. and APPEL, J.A.   1980.   Nutritional balance studies in pregnant women. Proc. Symposium on Maternal Nutrition During Pregnancy and Lactation, Lausanne, 1979, p. 74.

CALLOWAY, D.H. and MARGEN, S.   1971.   Variation in endogenous nitrogen excretion and dietary nitrogen utilization as determinants of human protein requirement. J. Nutr. 101:205.

CALLOWAY, D.H. and SPECTOR, H.   1954.   Nitrogen balance as related to caloric and protein intake in active young men. Am. J. Clin. Nutr. 2:405.

FAO/WHO (FOOD AND AGRICULTURE ORGANIZATION/WORLD HEALTH ORGANIZATION).   1973.   Energy and protein requirements. World Health Organization Tech. Rep. No. 522, Geneva.

GARZA, C., SCRIMSHAW, N.S. and YOUNG, V.R.   1976.   Human protein requirements: the effect of variations in energy intake within the maintenance range. Am. J. Clin. Nutr. 29:280.

GRAHAM, G.G., CORDANO, A. and BAERTL, J.M.   1964.   Studies in infantile malnutrition. III. Effect of protein and calorie intake on nitrogen retention. J. Nutr. 84:71.

GRAHAM, G.G., MORALES, E., ACEVADO, G., BAERTL, J.M. and CORDANO, A. 1969. Dietary protein quality in infants and children. II. Metabolic studies with cottonseed flour. Am. J. Clin. Nutr. 22: 577.

HOLEMANS, K., LAMBRECHTS, A. and MARTIN, H. 1955. Nitrogen metabolism and fat absorption in malnutrition and in Kwashiorkor. J. Nutr. 55:477.

INOUE, G., FUJITA, Y. and NIIYAMA, Y. 1973. Studies on protein requirements of young men fed egg protein and rice protein with excess and maintenance energy. J. Nutr. 103:1673.

IYENGAR, A.K., NARASINGA RAO, B.S. and REDDY, V. 1979. Effect of varying protein and energy intakes on nitrogen balance in Indian preschool children. Br. J. Nutr. 42:417.

LEVERTON, R.M., GRAM, M.R. and CHALOUPKA, M. 1951. Effect of the time factor and calorie level on nitrogen utilization of young women. J. Nutr. 44:537.

MARABLE, N.L., HICKSON, J.F., JR., KORSLUND, M.K., HERBERT, W.G., DESJARDINS, R.F. and THYE, F.W. 1979. Urinary nitrogen excretion as influenced by a muscle-building exercise program and protein intake variation. Nutr. Reports Int. 19:795.

MARGEN, S. and DURKIN, N. 1980. Unpublished data. University of California, Berkeley.

MORASSO, M.C., D'ANDREA, S., OVANDO, M.T., RIO, M.E., CLOSA, S.J. and MEREDITH, C. 1979. Protein utilization in undernourished infants: effects of energy intake and protein: energy ratio. Nutr. Reports Int. 20:353.

MUNRO, H.N. and WIKRAMANAYAKA, T.W. 1954. Absence of a time factor in the relationship between level of energy intake and protein metabolism. J. Nutr. 52:99.

NAGESWARA RAO, C., NADAMUNI NAIDU, A. and NARASINGA RAO, B.S. 1975. Influence of varying energy intake on nitrogen balance in men on two levels of protein intake. Am. J. Clin. Nutr. 28:1116.

NRC/FNB (NATIONAL RESEARCH COUNCIL/FOOD AND NUTRITION BOARD). 1979. "Recommended Dietary Allowances," 9th edition. Natl. Acad. Sci., Washington, D.C.

RICHARDSON, D.P., WAYLER, A.H., SCRIMSHAW, N.S. and YOUNG, V.R. 1979. Quantitative effect of an isoenergetic exchange of fat for carbohydrate on dietary protein utilization in healthy young men. Am. J. Clin. Nutr. 32:2217.

ROSE, W.C. 1949. Amino acid requirements of man. Federation Proc. 8:546.

SCHUTZ, Y. 1979. Energy and protein metabolism in elderly men. Unpublished doctoral dissertation, University of California, Berkeley.

SCRIMSHAW, N.S., TAYLOR, Y. and YOUNG, V.R. 1973. Lysine supplementation of wheat gluten at adequate and restricted energy intakes in young men. Am. J. Clin. Nutr. 26:965.

DHEW (U.S. DEPT. HEALTH EDUCATION AND WELFARE). 1979. Dietary intake source data, United States 1971–1974. DHEW Publ. (PHS) 79-1221.

YOUNG, V.R. 1972. Protein energy relationships. Unpublished paper. Gordon Res. Conf. on Food and Nutr., Colby College, N.H.

ZANNI, E., CALLOWAY, D.H. and ZEZULKA, A.Y. 1979. Protein requirements of elderly men. J. Nutr. 109:513.

DISCUSSION

DR. YOUNG: That was a very nice overview of a very complex problem. In view of the fact that energy balance-energy intake considerations do seriously complicate the conduct and the interpretation of human metabolic experiments, particularly in the context of interpretation of nitrogen balance data, what suggestions do you have, Dr. Yates, as to how to minimize the confounding influences of energy intake levels relative to actual energy requirements? In other words, how do we get over this very complex problem and minimize the energy factor?

DR. YATES: I think one way obviously would be to come up with a standardized activity level expected in a metabolic study—such as an increment above the basal metabolic rate (a certain percentage that's agreed upon) or a certain amount of activity that would be reproducible in a number of laboratories. That's obviously the problem we have in trying to interpret information from different laboratories. I'm not sure that kind of uniformity could be derived. It obviously would take some time.

DR. MARGEN: I don't think we can answer your question, Dr. Young, and you know that perfectly well. I think that's why you asked the question. And the problems are two. The first is that until we can really measure energy output with sufficient accuracy and know at least what the variability is, we're going to continue to be faced with this horrible problem which is plaguing us. We know what the problems are of attempting to manipulate diets to maintain body weight and what the significance of that is. We generally never know how much body weight to allow the individual to vary before we attempt to change the caloric input and then, when we do that, we don't know whether the individual can alter his or her activity right under our eyes under the most controlled conditions. And when we attempt to control or measure output, even under the controlled conditions of the metabolic unit, we know that our errors are just much too great for the degree of sensitivity that we need. In the meantime, we just have to attempt as best we can, as Dr. Yates suggests, and as I'm certain Dr. Young and Dr. Scrimshaw are doing, to attempt to standardize both our diets in terms of the protein and energy and to attempt to standardize activity as best we can in various population groups where these studies are going to be done. We're probably going to have to rely primarily upon the maintenance of a steady body weight for some time. But even that presents real problems because if you are dealing with a relatively high protein intake, you're certainly going to get continued nitrogen retention, as we have demonstrated. You're going to get an increase in body weight, in spite of absolutely no change at all in terms of energy, when you switch from a low to a high protein intake. So we're really caught in a terrible bind at this point. I don't really have any definite answer or any solution.

DR. PELLETT: How realistic are these type of experiments which are done on well nourished individuals where there are adequate reserves of fat and other reserves of energy combined with relatively small energy deficits during the experimental period? In contrast, for a malnourished child of a less developed country, there may be virtually no reserves following chronic malnutrition and this may be combined with an extreme deficit which may be compounded by infections.

DR. YATES: Well, I think that's why there's such a large accretion of nitrogen in those cases and that's why it's hard to directly compare data from a malnourished child to a normally, healthy growing child. It depends on the kind of data you want: Do you want to have information on the child that is perhaps in the population that has the greatest need, comparable to a malnourished child in a developing country, or do you want to have an overall estimate of what a child of this age actually requires in terms of protein—that's a question that I'm sure FDA and a number of people are interested in.

DR. MARGEN: I think that's an extremely important answer. And the other important thing is the phenomenon which has not been discussed and that is this whole matter of adaptation. Over time, with excess or underfeeding (energy), the slope of the curves change, indicating changes in efficiency. There are changes both in the direction of alteration in weight and changes in the direction of nitrogen retention or loss, etc. So obviously there is some type of adaptation which we don't understand.

DR. SCRIMSHAW: The last few slides began to approach the question of percentage of protein calories for that portion of the population with very low levels of activity. There are, of course, two sources of variation. 1) Intakes may be low in individuals with *ad libitum* access to energy because of life-style, and 2) Factors in the social and economic situations may force lower levels of activity as an adaptation to low energy intakes.

What is your conclusion as to the percentage of protein calories there would need to be in an ordinary U.S. mixed diet for those individuals in the poverty group who are at the 2.5 percent confidence interval for protein requirement?

DR. YATES: I think the data on that second to the last slide probably indicated that in the range of around 16%, one would find an individual with low activity levels but high protein requirements. It's hard to tell whether, as you say, the decrease in their activity level is because of their social situation, or conversely, because of decreased intake, their activity is therefore decreased as a result. So which comes first, it's hard to tell.

DR. TORUN: I think that Dr. Yates put her finger on a very sensitive area when she mentioned that it's very important to standardize physical activity conditions to interpret some of these studies. I cannot add any more to what Dr. Yates and Dr. Margen answered to Dr. Young. I do want to point out that this problem of not standardizing physical activity may also in part explain some of the differences that we have found in recent studies of milk requirements for growing children. For the past five or six years, we have included as part of our standard protocol in all these studies, a program of games to stimulate our children to be physically active. Our results, compared to requirement studies done some twelve or fourteen years ago, are quite different and this difference in activity may be partly responsible for the difference.

I had some difficulties in interpreting the data from Dr. George Graham's studies because he was dealing with acutely malnourished children, some with marasmus and some with kwashiorkor. But the point is that his studies show that in some way there is a relationship between energy and protein. That relationship is more complex than what has already been shown here because things work both ways. We showed some years ago at MIT that protein intake also affects energy balance in adults (Am. J. Clin. Nutr. *30*: 1983; 1977). This was also very nicely shown by Dr. Graham a few months ago in children who were adequately nourished and with whom he was doing different protein quality assessments. He revised his data and showed that children, who were receiving different levels of proteins under isoenergetic conditions, gained more weight when their protein intake improved. We should think not only of standardizing physical activity and energy balance but also of standardizing in some way the ratio of protein to energy.

MS. MURPHY: In response to your comments questioning the energy values in HANES, the 1977 USDA Household Food Consumption Survey included a great number of families. Three-day diet records were obtained on some 37,000 individual people. The early indications are that they're getting an average intake for women of 1600 calories a day. I know Dr. Hegsted's group is trying to come up with recommended meal patterns for women at 1600 calories and men at 2400 calories. So those low levels that you were talking about from HANES may really be there.

DR. YATES: I think we get back, though, to the methodology in reporting 24-hour recalls and in estimating portion size: (1) the individuals that are interviewed are probably not as trained as some of us are in estimation; (2) many times there's a tendency to under-report rather than over-report food intake. Especially if you're a woman in today's society (which I can attest to), a number of things that are socially more acceptable tend to be what people *think* they're doing. When we've done balanced studies at Berkeley (although perhaps with individuals that might not be selected for the Hanes study), we've found higher requirements than the Hanes data indicate are necessary to maintain body weight. The elderly that we studied which had these higher calorie requirements and intake levels, were somewhat unusual for their age in the sense that they were willing to come and live in a metabolic unit and thus may be more active. We may, in fact, be looking at people that are more active to begin with, thus affecting our data; but at the same time, I think it's difficult to base requirements solely on 24-hour recalls. We *can* compare 24-hour recalls in 1977 to what was obtained in 1972 and 1973 to look at changes in consumption patterns, but the methodology is still, I think, questionable.

DR. MARGEN: I just want to second what Dr. Yates just said. I think that we really have to be very careful about the interpretation of those data because certainly those of us who work with people on metabolic wards find that the caloric intake of these people who are relatively inactive is quite different from the type of data which comes out of free-living population.

(*Editors' Note*: Dr. Allison Yates presented Dr. Calloway's paper.)

11

Effects of Variation in Protein Digestibility

Daniel T. Hopkins

The digestibility of nutrients in foods has been measured with humans for over one hundred years. According to Murlin and Mattill (1938), Mayer in 1871 and Rubner in 1879, both of whom were working in the laboratory of Carl Voit, made the first trustworthy reports on the apparent digestibility of dry substances in cereals. Both investigators reported studies in which human subjects consumed different types of wheat and rye breads with nothing additional other than butter and beer for periods of four days.

Another early report on investigations into the digestibility of foods by humans is summarized by Lusk (1928) as follows: "The year of 1882 was that of the "Bread Reform League" in England and was the time when Liebeg in Munich held the opinion that milling removed necessary salts and decreased the digestibility of bread. In 1883, Rubner showed Liebeg this was not so, that poor utilization was associated with the degree of milling above a rate which yielded 70% of flour and that the nitrogen in bran was very poorly utilized by human beings, although it was an excellent food for cattle and therefore should be used in agriculture." Thus, we see that the notion that the digestibility of protein of foods is an important consideration is not a new one.

At the turn of the century, Atwater (1902) developed digestibility factors to be applied to the proximate constituents of diets for the calculation of physiological fuel values, which were estimates of the available energy of the diet. The digestibility factors that Atwater suggested be used for the general classes of foods are given in Table 11.1. These digestibility factors have been reviewed and updated according to more recent information by Merrill and Watt (1955). The publication by Merrill and Watt contains an impressive tabulation of the results of 331 determinations of apparent digestibility of foods of plant origin. The authors compiled the results of human digestion experiments reported in the literature since 1875 and stated that although data published in languages other than English may

TABLE 11.1. COEFFICIENT OF APPARENT DIGESTIBILITY OF PROTEIN OF DIFFERENT FOODS FOR USE IN CALCULATING PHYSIOLOGICAL FUEL VALUES[1]

Food Group	%
Animal foods	97
Cereals	85
Legumes, dried	78
Vegetables	83
Fruits	85
Vegetable foods of mixed diet diet	84
Total food of mixed diet	92

[1] Atwater, 1902.

not be completely covered, they believe the greater portion has been reviewed. The Merrill and Watt publication is probably the best source of digestibility data published prior to 1950.

The digestibility coefficients summarized by Merrill and Watt were used for the calculation of energy values in Agriculture Handbook #8 (Watt and Merrill, 1963). These coefficients of digestibility have served as a reference point or a benchmark with which the digestibility of foods for humans are compared.

What is not generally recognized is that the digestibility coefficients reported by Merrill and Watt and in Handbook #8 are apparent digestibility values rather than "true" digestibility values, although this fact is stated clearly in Atwater's original publications as well as in the explanation of the tables by Merrill and Watt. Because apparent digestibility underestimates the actual or "true" digestibility of proteins, it is likely that the coefficients in Handbook #8 contain considerable and, perhaps, excessive safety margins if they are used to calculate the digestibility of proteins for humans. Apparent digestibility is calculated as follows:

Apparent Protein Digestibility =

$$\frac{\text{Total N Consumed-Total Fecal N}}{\text{Total N Consumed}} \times 100$$

In the calculation of apparent digestibility, no account is made for the portion of the fecal nitrogen that is metabolic in origin and is not derived from the undigested residue of food.

True digestibility is calculated as follows:

True Protein Digestibility =

$$\frac{\text{Total N Consumed-(Total Fecal N-Metabolic Fecal Nitrogen)}}{\text{Total N Consumed}} \times 100$$

The true protein digestibility of food is always higher than the apparent digestibility because true protein digestibility takes into account the metabolic fecal nitrogen which is not of dietary origin. The total fecal excretion of metabolic fecal nitrogen is relatively constant and more or less independent of food intake. At low levels of protein intake apparent digestibility deviates more from true digestibility than at high levels of intake because the metabolic fecal nitrogen makes up a larger proportion of the fecal nitrogen when the protein intake is low than when it is high. An illustration of this relationship is shown in Table 11.2 in which the apparent digestibility of a protein with a true digestibility of 100% is calculated at different levels of

TABLE 11.2. APPARENT DIGESTIBILITY OF A 100% DIGESTED PROTEIN AT VARIOUS LEVELS OF INTAKE

Intake/kg/Day		Apparent Digestibility, %	
Protein, g	Nitrogen, mg	MFN = 9[1]	MFN = 12
0.1	16	44	25
0.2	32	72	62
0.3	48	81	75
0.4	64	86	81
0.5	80	89	85
0.6	96	91	88
0.7	112	92	89
0.8	128	93	91
0.9	144	94	92
1.0	160	94	92
1.5	240	96	95
2.0	320	97	96

[1] Metabolic fecal nitrogen = 9 mg/kg/day or 12 mg/kg/day.

protein intake. In the calculation, the excretion of metabolic fecal nitrogen is assumed to be 9.0 or 12.0 mg N/kg body weight/day, figures that have been suggested by FAO/WHO (1973). Metabolic studies in which protein digestibility figures are obtained are frequently conducted at levels of protein intake between 0.2 and 0.8 g/kg/day. Apparent digestibility calculated in such studies would underestimate true digestibility by $10-15$ percentage units. Even in studies in which protein intakes are $1-1.5$ g/kg/day, apparent digestibility will underestimate true digestibility by 5 percentage units. It is unlikely that cereal and grain products containing relatively low levels of protein are fed to human subjects at levels above 1 g protein/kg/day, illustrating that apparent digestibility figures for such products of necessity must underestimate the true digestibility by at least 5%. An illustration using real data on whole egg collected by Kishi *et al.* (1978) is shown in Table 11.3. The data demonstrates the point that whole egg digestibility is considerably underestimated by apparent digestibility calculations and that the degree of underestimation is inversely related to the level of nitrogen intake.

The digestibility of products can be satisfactorily compared using apparent digestibilities provided that the level of nitrogen intake is similar. An

TABLE 11.3.   APPARENT VS TRUE DIGESTIBILITY OF EGG PROTEIN[1]

Intake	Nitrogen Fecal	Digestibility	
mg/kg/day		Apparent %	True, %[2]
31.9	13.5	57.6 (45−65)[3]	97.3 (85−105)[3]
63.4	15.1	76.2 (73−80)	96.2 (93−100)
79.7	13.8	82.7 (80−86)	98.7 (96−102)

[1] Calculated from Kishi et al. (1978).
[2] Metabolic fecal nitrogen assumed to be 12.7 mg/kg/day.
[3] Range of values.

illustration of this mathematical relationship is shown in Table 11.4 in which the apparent digestibility of proteins that have true digestibilities of 50, 75 and 100% are calculated using various levels of nitrogen intake. One can see from this table that the degree of underestimation of true digestibility is relatively constant provided that the level of nitrogen intake is constant. For example, at a level of 0.4 g/kg/day intake, the true digestibility is underestimated by 19 percentage units no matter what the true digestibility. At intakes of 1 g/kg body weight/day digestibility is underestimated by 8 percentage units irrespective of the level of true digestibility. This fact is important when considering literature values of apparent digestibility.

TABLE 11.4.   APPARENT DIGESTIBILITY VS TRUE DIGESTIBILITY AT SEVERAL LEVELS OF NITROGEN INTAKE[1]

Protein Intake	N Intake	True Digestibility, %		
		50	75	100
		−Apparent Digestibility, %		
0.1	16	−25	0	25
0.2	32	12	38	62
0.4	64	31	56	81
0.8	128	41	66	91
1.0	160	42	67	92
1.5	240	45	70	95
2.0	320	46	71	96

[1] Protein and N intakes expressed in gms/Kg body weight/day; for calculation of true digestibility, a value for metabolic fecal nitrogen of 12 mg/N/Kg body weight/day was used.

In order to evaluate the effect of variations in protein digestibility on the nutritional evaluation of foods, a number of reports in the literature from 1940 to present in which digestibility was reported were examined. Although not every study in which digestibility was measured was examined, sufficient reports were read and summarized to give a reasonable view of the ranges of protein digestibility that might be expected in the food supply. In examining the reports, true digestibility was calculated from the data where possible. In order to make the calculations, certain assumptions had to be made. When the authors did not specify the level of metabolic fecal nitrogen that should be used for subjects under their experimental conditions, the figures of 12, 9 and 31 mg/kg/day were used for adult males, adult females and children 2–12 years old, respectively (FAO/WHO, 1973). In most reports, the body weight was specified so metabolic fecal nitrogen could be calculated. However, in cases where body weight was not specified, it was assumed that males were 70 kg, females weighed 58 kg and the total metabolic fecal nitrogen was 840 and 522 mg/day for males and females, respectively. In summarizing the reports, a certain amount of pooling of data was necessary. Where possible, the mean values reported by the authors for digestibility were used. Where this was not possible, the apparent and true digestibility were calculated from mean nitrogen intakes and excretions of experimental groups with similar nitrogen intakes. When digestibility was measured with groups of subjects fed a protein at several different levels of intake, an overall average was calculated by averaging the digestibility coefficients from each of the groups fed each level after appropriate weighting to allow for differences in the number of subjects in each group.

The true digestibility of mixed diets by North Americans has been measured by several investigators and results typical of these studies are tabulated in Table 11.5. The average true digestibility of protein of the normal mixed diet is 92%, with values of individual studies ranging from 88 to 96%.

TABLE 11.5.   DIGESTIBILITY OF PROTEIN IN NORTH AMERICAN MIXED DIETS

| | Ref. | Digestibility, % | |
		Apparent	True
Diet mixed	(3)	81	88
Diet mixed	(2)	76	90
Diet mixed	(5)	88	95
Diet mixed	(1)	89	96
Diet mixed	(4)	84	89
Diet, lactovegetarian	(4)	83	88
Diet, vegetarian	(6)	82	87
Diet, vegetarian	(4)	77	100
Average		82	92

1) Bolourchi et al, 1968.
2) Bricker et al, 1945.
3) Crampton et al, 1960.
4) Register et al, 1967.
5) Smith et al, 1933.
6) Turk et al, 1973.

It is interesting to note that the protein of vegetarian diets in two studies was 94% digested; results comparable with that of the mixed diet. These protein digestibility coefficients are higher than might be expected from studies from other countries with subjects fed largely vegetarian diets (Table 11.6). Probably vegetarians from more affluent countries tend to select foods that are refined and quite digestible. Merrill and Watt (1955) reported that the average coefficient for apparent protein digestibility of ordinary mixed diets in 93 digestion experiments was 93.3 indicating again

TABLE 11.6. DIGESTIBILITY BY ADULTS OF PROTEIN OF MIXED DIETS FROM
SEVERAL DIFFERENT CULTURES

Country	Ref.	Diet	Digestibility, % Apparent	True
India	(4)	Rice, red gram dahl, milk powder, vegetables	61	73
India	(5)	Rice, red gram dahl, skimmed milk powder, vegetables	69	77
India	(6)	Ragi, red gram dahl, potatoes, whole milk powder, vegetables	48	54
Nigeria	(3)	Cassava, rice, dried fish, vegetables	61	91
Rural Guatemalan Diet[1]	(1)	Black beans, corn tortilla, rice, wheat rolls, cheese, whole eggs, vegetables	69	77
Ceylon	(2)	Polished and unpolished rice, meat, fish, dairy products, bread, fruits, vegetables	82	87

[1] Determined with North American subjects.
1) Calloway and Kretsch (1978).
2) Cullumbine (1950).
3) Nicol and Phillips (1976).
4) Pasricha et al (1965).
5) Rao et al (1975).
6) Subrahmanyan et al (1955).

that mixed diets are highly utilized. One can conclude that in a typical North American diet, protein digestibility is not an important consideration.

Studies on whole diets from other cultures show a somewhat different picture. Studies in India have shown that the protein of diets based on rice, dahl, milk powder and vegetables have true digestibilities of 73 and 77% (see Table 11.6). The protein of a typical diet based on the cereal, ragi, were even more poorly digested (54%). Somewhat higher true digestibility values of 87 and 91% were found in studies of typical diets in Ceylon and Nigeria, respectively. The low digestibility of protein in typical diets of India have also been demonstrated in studies with children (Table 11.7). In these studies, the protein of rice diets were found to be from 77 to 85% digested and the protein of diets based on kaffir corn and ragi had even lower true digestibility values of 66 and 68%, respectively. These types of observations show that although the protein of a typical North American diet is highly digestible, low digestibility may be a concern in diets in other cultures. Problems involved in the utilization of plant proteins in the diet by low-income groups in developing countries has been thoroughly reviewed by Swaminathan (1967).

The low digestibility of diets in other cultures is probably due largely to an inherent characteristic of the foods in the diet rather than to genetic factors in the population or to pathological problems with the digestive

TABLE 11.7.   DIGESTIBILITY OF PROTEIN IN DIETS BY INDIAN CHILDREN

	Ref.	Digestibility, % Apparent	True
Rice diet, India	(3)	75	85
Rice diet, India	(4)	68	84
Rice diet, India	(6)	64	83
Rice diet, India	(5)	59	77
Rice diet, India	(2)	71	83
Rice diet, India	(3)	75	88
Kaffir corn diet, India	(3)	55	66
Ragi diet, India	(1)	50	68
Ragi diet, India	(2)	53	65
Millet diet, India	(3)	53	63

1) Daniel et al (1965).
2) Joseph et al (1958).
3) Kurien et al (1960).
4) Narayanaswamy et al (1972).
5) Panemangalore et al (1964).
6) Parthasarathy et al (1964)B.

tract. Nasr et al. (1976) examined the feces of randomly selected Iranian villagers fed a hospital diet. Although enteropathy was found in nearly one-third of the subjects, it was not associated with malabsorption. Most subjects had a normal level of nitrogen (less than 2 g/day) in the feces indicating normal protein digestibility. Calloway and Kretsch (1978) fed American subjects in California diets containing either egg albumen or ingredients from a diet typical of rural Guatemala. The Guatemalan type diet contained ingredients such as black beans, corn tortillas, rice, wheat rolls, cheese, whole eggs and vegetables. Recalculation of their data suggest that the true protein digestibility of the egg albumen diet was 99%; whereas the comparable value for the protein of the rural Guatemalan diet was 77%. The digestibility coefficient of the protein of the Guatemalan diet was similar to the typical diets based on rice from India discussed above. The results suggest that the low digestibility of diets from certain cultures is a function of the diet itself and not due to differences in the digestive physiology of members of population.

Kies and Fox (1977) determined the fecal nitrogen of subjects fed a mixed diet in which the principal source of protein was peanuts to which was added several different levels of hemicellulose material derived from psyllium seeds. The fecal nitrogen excretion was increased significantly from 1.3 g/day to 1.7 g/day when the dietary level of hemicellulose was increased from 4 to 24 g/day (Table 11.8). This increase in fecal nitrogen would cause the digestibility of the diet to decrease by about 6 percentage units. The increase in fecal nitrogen from feeding hemicellulose suggests that fiber per se in the diet causes an increase in nitrogen excretion that, on a physiological basis, probably should be assigned to the metabolic fecal nitrogen rather than to the indigestible residue of the diet.

Similar observations have been made on the effect of dietary fiber on fecal nitrogen excretion by Southgate and Durnin (1970) and more recently by Kelsay et al. (1978). These investigators fed diets either low in residue or

TABLE 11.8.   THE EFFECT OF DIETARY FIBER ON THE APPARENT DIGESTIBILITY OF PROTEIN FROM A MIXED DIET

Fiber in Diet	Nitrogen, g/day Intake	Fecal	Apparent Digestibility
Low fiber, 4.2 g/day Hemicellulose[1]	6.8	1.28	—
Medium fiber, 14.2 g/day Hemicellulose	6.8	1.54	—
High fiber, 24.2 g/day Hemicellulose	6.8	1.66	—
Low fiber, 4 g/day[2] Neutral detergent fiber	14.9	1.35	90
High fiber, 20 g/day Neutral detergent fiber	15.2	2.58	81
Low fiber, 6 g/day[3] Pentosan & cellulose	13.1	1.03	92
Medium fiber, 16 g/day Pentosan & cellulose	13.9	1.27	91
High fiber, 32 g/day Pentosan & cellulose	14.7	2.17	85

[1] Kies & Fox (1977).
[2] Kelsay et al (1978).
[3] Southgate & Durnin (1970).

containing fiber from fruits and vegetables, and in the case of Southgate and Durnin, whole wheat bread. Subjects eating the diets with higher levels of dietary fiber had increased excretion of nitrogen. The apparent digestibility was reduced by as much as 9 percentage units in the study by Kelsay *et al.* (1978) and by lesser amounts in the study by Southgate and Durnin (Table 11.8). These studies show that nitrogen in diets high in dietary fiber can be expected to be less well digested than nitrogen in low fiber diets.

There is insufficient data to determine whether the increased fecal nitrogen in high fiber diets is due to indigestible residues, increased bacterial cells from bacterial fermentation or metabolic fecal nitrogen from the gut. As a practical manner, this is a moot point since the increase in fecal nitrogen, in most cases, can be assigned to the food and the food can be penalized when the true digestibility coefficient is assigned. If, with the development of *in-vitro* assays, this proves impractical, the increase in nitrogen excretion will have to be assigned to the protein requirement. As far as the normal North American diet, the effect of fiber on apparent protein digestibility does not appear to be of sufficient magnitude to be of concern.

The variation in the digestibilites of protein is an important consideration in considering the contribution of digestibility to protein quality. Although many investigators have measured digestibility of foods repeatedly in the same subjects, few have reported their data in such a way that the variance within subjects and between subjects can be estimated. Furthermore, variation in true digestibility between subjects is difficult to evaluate. Some of the variation may be due to differences in metabolic fecal nitrogen excretion and thus be more a reflection of variability of protein requirements than differences in digestibility.

It is difficult to determine how much protein digestibility varies between people or even with one person from time to time. Crampton and co-workers (1960) measured the digestibility of muffins and biscuits made from a mixture of ingredients that met all of the nutritional requirements. The subjects were senior college students and the digestibility study was conducted as part of a senior nutrition laboratory course. Digestibility was measured by the indirect method using a chromic oxide procedure. Calculation of the standard deviation of the digestibility data collected over a three-year period on 96 students gave a value of 7.3%. Other workers have shown somewhat lower levels of variation in measuring digestibility of protein and typical standard deviations are given in Table 11.9. Again, due to the nature of the procedures used to estimate digestibility, one must suspect that a large part of the variability may be contributed by experimental techniques.

TABLE 11.9. EXAMPLES OF STANDARD DEVIATIONS OBSERVED IN TYPICAL DIGESTIBILITY STUDIES

Protein Source	Subjects	Ref.	Standard Deviation of Digestibility Coeff.
Mixed diet	College students	(1)	7.3
Mixed diet	College students	(6)	1.3
Egg albumen	College students	(3)	2.7
Various rice diets	Indian children	(4)	2.1
Various rice diets	Indian children	(5)	2.3
Rice and ragi diet	Indian children	(2)	5.2

(1) Crampton et al (1960).
(2) Joseph et al (1958).
(3) Kishi et al (1978).
(4) Narayanaswarmy et al (1972).
(5) Parthasarathy et al (1964)B.
(6) Register et al (1967).

The digestibility of protein in individual ingredients by adults has been reported by many investigators. The apparent and true digestibility of food proteins, calculated from the data on nitrogen intake and excretion, where necessary, are tabulated in Table 11.10. These data are presented primarily as reference and the data in Table 11.10 is further summarized in Table 11.11 in which average digestibility values and ranges are tabulated.

The digestibilites of animal proteins (Table 11.11) are well over 90%. The common assumption that whole egg is completely digested is borne out by the true digestibility figure of 97%. Whole corn generally seems to have a true digestibility of approximately 87%. In one study, autoclaved whole corn had a true digestibility value as low as 75%; however, the conditions and preparations of this corn may have had some effect on the value. Opaque-2 corn seems to have a somewhat higher value than 87%, perhaps due to the soft nature of the endosperm. An experimental Opaque-2 corn with a high amylose content (thus, a flint-type endosperm) had lower digestibility. Ready-to-eat cereals generally had true digestibilities in the range of 70%. The results summarized in this table indicate that the true digestibilites of proteins in products derived from corn, with the exception of ready-to-eat cereals, can be expected to be 85% or better.

TABLE 11.10.   DIGESTIBILITY OF PROTEIN IN VARIOUS FOODS BY ADULTS

| Protein Source | Ref. | Digestibility, % | |
		Apparent	True
Meat			
Beef	(15)	—	92
Beef	(38)	—	98
Beef steak	(33)	—	97
Beef tenderloin, roasted	(2)	78	91
Beef, canned, infant	(49)	82	99
Beef, low fat, ground	(45)	73	91
Beef, powdered, defatted	(14)	77	97
Pork loin & tenderloin	(46)	90	98
Tuna fish, canned	(4)	76	90
Turkey breast, roasted	(2)	77	91
Dairy Products			
Casein	(14)	78	94
Casein	(3)	86	97
Casein	(29)	71	96
Cottage cheese	(4)	85	99
Casein lactalbumin	(28)	83	100
Lactalbumin	(3)	82	93
Lactalbumin	(29)	69	95
Milk, whole	(7)	77	94
Milk, whole	(31)	69	90
Milk, whole	(41)	77	98
Milk, whole, powdered	(41)	80	99
Milk, whole, powdered	(42)	75	90
Egg and Egg Products			
Egg, whole	(21)	73	99
Egg, whole	(34)	73	106
Egg, whole	(48)	80	96
Egg, whole	(47)	76	97
Egg, whole	(33)	—	96
Egg, whole	(30)	86	100
Egg, whole	(41)	78	98
Egg, whole	(6)	77	93
Egg, whole	(46)	89	97
Egg, scrambled	(12)	86	95
Egg, scrambled	(16)	79	96
Egg, guinea fowl	(37)	67	98
Egg, spray dried	(4)	78	92
Egg, dried	(13)	84	97
Egg, dried	(45)	80	98
Egg, flakes	(42)	79	92
Egg, powdered, defatted	(14)	61	100
Egg, albumen	(14)	72	102
Egg, albumen	(3)	82	93
Cereals (Corn and Corn Products)			
Corn, whole	(18)	74	92
Corn, whole	(43)	77	86
Corn meal	(39)	72	84
Corn, in various stages of maturity	(8)	74	86

TABLE 11.10.   *(Cont'd.)*

Protein Source	Ref.	Digestibility, % Apparent	True
Corn, field, autoclaved 15#, 60 minutes	(22)	64	75
Corn, whole opaque 2	(9)	80	92
Corn, whole opaque 2	(47)	74	92
Corn, degerminated, opaque 2	(18)	75	92
Corn, degerminated, opaque 2	(47)	74	95
Corn, whole, opaque 2 high amylose	(18)	52	70
Corn, Corn Chex	(19)	50	62
Corn, Corn Flake	(19)	50	62
Corn, Corn Flake	(22)	67	77
Corn, Kix	(19)	62	73
Corn, Kix	(22)	68	78
Cereals (Wheat & Wheat Products)			
Wheat, whole, Canadian	(26)	85	88
Wheat, hard spring	(22)	76	86
Wheat, whole	(49)	80	93
Wheat, whole, hot cereal	(30)	73	87
Wheat, whole, hot cereal	(30)	66	80
Wheat, whole, hot cereal	(31)	68	88
Wheat, whole, English	(26)	74	80
Wheat flour, Canadian 90% extra	(25)	87	92
Wheat flour, English 90% extra	(25)	80	86
Wheat flour, Canadian 80% extra	(25)	91	95
Wheat flour, English 80% extra	(25)	83	90
Wheat flour, white	(5)	88	96
Wheat flour, white	(7)	82	97
Bread, white	(40)	85	101
Bread, white	(35)	91	97
Bread	(15)	—	96
Bread	(32)	78	95
Bread	(24)	92	97
Bread, whole wheat	(24)	87	92
Bread, coarse brown	(35)	84	91
Wheat gluten	(17)	55	100
Wheat gluten	(14)	85	104
Wheat gluten	(3)	87	96
Wheat gluten	(4)	82	96
Wheat germ	(15)	—	81
Wheat endosperm (Farina)	(31)	82	102
Wheat endosperm (Farina)	(41)	78	96
Wheat endosperm (Farina)	(30)	85	99
Wheat, puffed wheat	(22)	73	84
Wheat, puffed wheat	(30)	69	83
Wheat, shredded	(19)	45	53
Wheat, shredded	(30)	66	80
Wheat, shredded	(32)	68	85
Wheat, Wheaties	(19)	58	67
Wheat, Wheaties	(30)	73	88
Wheat, Wheaties	(22)	74	85
Wheat, 40% Bran	(19)	60	69

TABLE 11.10.   *(Cont'd.)*

Protein Source	Ref.	Digestibility, % Apparent	True
Cereals (Rice and Rice Products)			
Rice	(36)	62	91
Rice, polished	(10)	77	89
Rice, polished	(23)	81	93
Rice, polished	(16)	68	82
Rice, high protein	(10)	78	85
Rice, Rice Krispies	(19)	65	77
Rice, Special K	(19)	58	63
Cereals (Oats and Oat Products)			
Oats, quick oatmeal	(30)	78	92
Oats, quick oatmeal	(31)	67	87
Oat, oatmeal	(27)	70	76
Oat, oatmeal	(22)	78	88
Oats, Cheerios	(22)	78	89
Oats, Cheerios	(19)	56	63
Oat flakes, Frosted Flakes	(19)	62	67
Oat flakes, Life	(19)	65	70
Legumes & Oilseed Products			
Peas, Alaskan Field	(12)	79	88
Peanut butter	(46)	86	95
Peanut flour	(3)	82	91
Peanut flour	(4)	76	91
Peanut flour	(14)	78	98
Soy flour	(15)	—	75
Soy flour	(7)	70	92
Soy flour, defatted extruded	(20)	79	90
Soy flour, defatted extruded	(45)	66	84
Soy flour, defatted extruded	(3)	76	88
Soy protein, isolated	(3)	82	93
Soy protein, isolated	(38)	—	97
Soy protein, isolated	(4)	81	95
Soy protein, spun	(28)	83	101
Soy protein, spun	(44)	88	107
Sunflower seed, flour	(6)	—	90
Cottonseed	(33)	—	78
Cottonseed, flour, glandless	(1)	90	98
Cottonseed, flour, deglanded	(1)	88	95
Miscellaneous Foods			
Millet	(20)	65	79
Triticale	(20)	76	90
Metrecal	(11)	90	101
Algae, extruded	(11)	58	67
Algae, lyophilized	(11)	68	78
Yeast, kitchen	(33)	—	87
Wheat, meat analogue	(45)	78	95

(1) Alford and Onley (1978).
(2) Arnold et al (1968).
(3) Bodwell (1979).
(4) Bodwell et al (1980).
(5) Bolourchi et al (1968).
(6) Bricker and Smith (1951).
(7) Bricker et al (1945).
(8) Chen et al (1966).
(9) Clark et al (1967).
(10) Clark et al (1971).
(11) Dam et al (1965).
(12) Esselbaugh et al (1952).
(13) Garza et al (1977).
(14) Hawley et al (1948).
(15) Hegsted et al (1946).
(16) Inoue et al (1973).

TABLE 11.10.   *(Cont'd.)*

(17) Inoue et al (1974).
(18) Kies and Fox (1972).
(19) Kies and Fox (1973).
(20) Kies et al (1975).
(21) Kishi et al (1978).
(22) Kuether and Myers (1948).
(23) Lee et al (1971).
(24) Macrae et al (1942).
(25) McCance and Widdowson (1947).
(26) McCance and Walsham (1948).
(27) McCance and Glaser (1948).
(28) Morse et al (1972).
(29) Mueller and Cox (1947).
(30) Murlin et al (1938).
(31) Murlin and Mattill (1938).
(32) Murlin et al (1941).
(33) Murlin et al (1944).

(34) Navarrete et al (1977).
(35) Newman et al (1912).
(36) Nicol and Phillips (1976)A.
(37) Nicol and Phillips (1976)B.
(38) Scrimshaw and Young (1979).
(39) Sherman and Winters (1918).
(40) Sherman (1920).
(41) Sumner et al (1938).
(42) Sumner and Murlin (1938).
(43) Truswell and Brock (1962).
(44) Turk et al (1973).
(45) Vemury et al (1976).
(46) Watts et al (1959).
(47) Young et al (1971).
(48) Young et al (1973).
(49) Young et al (1975).

TABLE 11.11.   SUMMARY OF THE TRUE DIGESTIBILITY BY ADULTS OF PROTEIN IN SOME COMMON FOODS

Protein Source	Number of Reports	Digestibility, % Mean	Digestibility, % Range
Meat, poultry, fish	10	94	90–99
Milk, casein, lactalbumen	12	95	90–100
Egg, egg albumen	19	97	92–106
Corn, whole	4	87	84–92
Corn, opaque 2	4	93	92–95
Corn, opaque 2, high amylose	1	70	—
Corn, ready-to-eat cereal	5	70	62–78
Whole wheat	6	87	80–93
Wheat flour, Canadian, 90% extraction	1	92	—
Wheat flour, Canadian, 80% extraction	1	95	—
Wheat flour, white	2	96	96–97
Bread, coarse, brown or whole wheat	2	92	91–92
Bread, white	5	97	95–101
Wheat gluten	4	99	96–104
Wheat germ	1	81	—
Farina (wheat endosperm)	3	99	95–102
Wheat, ready-to-eat	9	77	53–88
Rice, polished	4	89	82–91
Rice, ready-to-eat cereals	3	75	77–85
Oatmeal	4	86	76–92
Oats, ready-to-eat cereals	4	72	63–89
Millet	1	79	—
Triticale	1	90	—
Cottonseed	3	90	70–98
Peas, Alaskan field	1	88	—
Peanuts	4	94	91–98
Soy flour	5	86	75–92
Soy protein, isolate	3	95	93–97
Soy protein, isolate, spun	2	104	101–107
Sunflower seed, flour	1	90	—

Over the years a considerable amount of research has been reported on the digestibility of wheat protein. Whole wheat seems to have a true digestibility of approximately 87%. White flour is almost completely digested and has a true digestibility coefficient of 96%, a value comparable to wheat gluten (99%) and farina (99%). Flours of intermediate extraction between that of whole wheat and white flour likewise have an intermediate digestibility as can be seen from Table 11.11. As would be expected, bread manufactured from white flour is highly digested (97%), whereas brown bread or whole wheat bread have somewhat lower true digestibility (82%). The digestibility of protein of ready-to-eat cereals based on wheat is somewhat depressed. The digestibility of protein from rice, oatmeal and millet is also summarized in Table 11.11. The average true digestibility of protein from polished rice and oatmeal is 89 and 86%, respectively, values not too dissimilar to those for whole wheat and corn. Digestibility of the ready-to-eat cereals is again lower. In one study, the true digestibility of protein of millet was 79%. The digestibility of protein in millet or sorghum is low; however, these are not grains that are widely consumed in the United States. True digestibility of protein from several oil seeds and legumes are also summarized on Table 11.11. Unrefined legumes such as field peas and soy flour have true digestibilities of about 86 to 88%. When the products are refined and the proteins are isolated, the digestibility is improved and, in fact, comparable to that of animal proteins. Cottonseed flour, peanuts and sunflower seed products all have true digestibilities exceeding 90%.

The digestibility of proteins in food has also been measured with children by DeMaeyer and Vanderborght (1961), Bressani and co-workers (1963), a group in the Mysore Laboratories in India and others. These data are tabulated in Table 11.12. The digestibility of proteins by children generally seem to be similar to that by adults although the digestibility of certain foods, most noticeably dried skim milk, may be somewhat reduced. This may be due to an actual difference between adults and children, perhaps due to the nature of the skim milk itself, or perhaps due to lactose intolerance.

The true digestibility coefficients for children seem to be considerably less than that of adults (Table 11.13). The lower values may be due to differences in technique. The estimation of metabolic fecal nitrogen is more difficult with children than for adults. However, it is possible that children do not absorb protein as well as adults. In any event, the difference which is approximately 5% is not large enough to be of concern. Digestibility values obtained with adults can be satisfactorily applied to diets for children.

The key digestibility data from children and adults for the more important classes of protein are summarized in Table 11.14. The information given shows that for the most part the true digestibility of proteins that are found in the American diet are quite high whether measured with adults or children. The true digestibility of animal products is very high (95%) while the digestibility of legumes and cereals is somewhat lower (perhaps as low as 85%). As legumes and cereals are refined so as to become more palatable,

TABLE 11.12.   DIGESTIBILITY OF PROTEIN IN FOODS BY CHILDREN

Protein Source	Ref.	Digestibility, %	
		Apparent	True
Egg, whole, fresh	(7)	—	97
Egg	(15)	—	92
Egg, whole, powdered	(9)	70	89
Milk, whole	(3)	85	92
Milk	(15)	—	93
Milk, powdered, skim	(10)	69	87
Milk, powdered, skim	(11)	68	86
Milk, powdered, skim	(6)	68	88
Milk, powdered, skim	(5)	70	88
Milk, powdered, skim	(4)	70	86
Milk, powdered, skim	(7)	—	91
Milk, human, lyophilized	(7)	—	90
Fish flour	(7)	—	82
Fish flour	(12)	76	89
Corn	(15)	—	82
Corn	(14)	54	62
Wheat, white flour	(13)	74	85
Wheat, white flour	(2)	82	90
Wheat, white flour	(15)	—	93
Wheat, endosperm (Farina)	(1)	86	94
Rice, polished	(15)	—	85
Rice, polished	(14)	77	88
Rice, polished	(8)	72	89
Kaffir corn	(5)	56	74
Soy flour	(7)	—	88
Soy flour	(7)	—	88
Soy flour	(10)	65	84
Soy protein, isolated	(3)	85	93
Soy protein, isolated	(7)	—	95
Soy protein, isolated	(15)	—	92
Peanut flour	(7)	—	92
Cottonseed flour	(7)	—	87
Sesame flour	(7)	—	88

(1) Barness et al, 1961
(2) Bressani et al, 1963
(3) Bressani et al, 1967
(4) Daniel et al, 1970
(5) Daniel et al, 1966
(6) Daniel et al, 1965
(7) Demaeyer and Vanderborght, 1961
(8) Joseph et al, 1963
(9) Parthasarathy et al, 1963
(10) Parthasarathy et al, 1964A
(11) Parthasarathy et al, 1964B
(12) Rao et al, 1964
(13) Reddy, 1971
(14) Tasker et al, 1962
(15) Viteri et al, 1971

TABLE 11.13.   A COMPARISON OF COEFFICIENTS OF TRUE DIGESTIBILITY OF PRO-
TEINS BY ADULTS AND CHILDREN

Protein Source	Children	Adults
Egg	93	97
Milk	89	95
Corn	82	87
Wheat, white flour	89	96
Wheat, endosperm	94	99
Rice, polished	87	89
Soy flour	87	86
Soy protein, isolated	93	95
Peanuts	92	94

TABLE 11.14.   SUMMARY OF TRUE DIGESTIBILITY OF PROTEIN OF MAJOR OR PO-
TENTIAL MAJOR SOURCES OF PROTEIN IN THE NORTH AMERICAN DIET

Protein source	True digestibility, %
Animal protein	95
Corn	87
Wheat	87
Wheat, flour, white	96
Bread, white	97
Rice, polished	89
Oatmeal	86
Peas, beans	88
Soy flour	86
Soy protein, isolated	95
Peanut	94
Cottonseed	90
Sunflower	90

their digestibility again approaches that of animal proteins. Although unrefined grains and legumes probably have protein that is 10% less well
digested than that of animal protein, it seems unlikely that digestibility
is a major consideration in the American diet. It is conceivable that special
cases might arise where digestibility of American foods could become a consideration in assessing protein nutritional value. Examples of such cases
might be the introduction of large quantities of millet, sorghum, and other
special ingredients into the diet. However, these can be dealt with on a
case by case basis. For this purpose, and for screening new foods for problems associated with digestibility, in-vitro or animal assays should be
sufficient.

ACKNOWLEDGEMENTS

I wish to acknowledge the invaluable assistance of Dr. C.E. Bodwell,
who first suggested the topic of this paper and assisted in the gathering
of the many literature references. I also wish to acknowledge the assistance of Thomas M. Hopkins who assisted in the recalculation of the true
digestibility coefficients.

REFERENCES

ALFORD, B.B. and ONLEY, K.   1978.   The minimum cottonseed protein
    required for nitrogen balance in women.   J. Nutr. 108:506.

ARNOLD, T.S., RITCHEY, S.J. and MOORE, M.E.   1968.   Nitrogen balance
    and plasma amino acids on low-protein diets.   J. Am. Diet. Assoc. 52:135.

ATWATER, W.O.   1902.   Principles of nutrition and nutritive value of food.
    Farmers Bulletin 142:5.

BARNESS, L.A., KAYE, R. and VALYASEVI, A.   1961.   Lysine and potas-
    sium supplementation of wheat protein.   Am. J. Clin. Nutr. 9:331.

BODWELL, C.E.   1979.   Unpublished data. USDA, Beltsville, MD.

BODWELL, C.E., SATTERLEE, L.D. and HACKLER, L.R. 1980. Protein digestibility of the same protein preparations by human and rat assays and by *in-vitro* enzymic digestion methods. Am. J. Clin. Nutr. *33*:677.

BOLOURCHI, S., FRIEDEMANN, C.M. and MICKLESEN, O. 1968. Wheat flour as a source of protein for adult human subjects. Am. J. Clin. Nutr. *21*:827.

BRESSANI, R., WILSON, D., BEHAR, M., CHUNG, M. and SCRIMSHAW, N.S. 1963. Supplementation of cereal proteins with amino acids. Lysine supplementation of wheat flour fed to young children at different levels of protein intake in the presence and absence of other amino acids. J. Nutr. *79*:333.

BRESSANI, R., VITERI, F., ELIAS, L.G., DE ZAGHI, S., ALVARADO, J. and ODELL, A.D. 1967. Protein quality of a soybean protein textured food in experimental animals and children. J. Nutr. *93*:349.

BRICKER, M., MITCHELL, H.H. and KINSMAN, G.M. 1945. The protein requirements of adult human subjects in terms of the protein contained in individual foods and food combinations. J. Nutr. *30*:269.

BRICKER, M.L. and SMITH, J.M. 1951. A study of the endogenous nitrogen output of college women, with particular reference to use of the creatinine output in the calculation of the biological values of the protein of egg and of sunflower seed flour. J. Nutr. *44*:553.

CALLOWAY, D.H. and KRETSCH, M.J. 1978. Protein and energy utilization in men given a rural Guatemalan diet and egg formulas with and without added oat bran. Am. J. Clin. Nutr. *31*:1118.

CHEN, C.S., FOX, H.M., PEO, E.R. Jr., KIES, C., BLUNN, C.T. and COLVILLE, W.L. 1966. Utilization of environmentally produced high nitrogen corn by weanling rats and adult humans. J. Nutr. *90*:295.

CLARK, H.E., ALLEN, P.E., MEYERS, S.M., TUCKETT, S.E. and YAMAMURA, Y. 1967. Nitrogen balances of adults consuming Opaque-2 maize protein. Am. J. Clin. Nutr. *20*:825.

CLARK, H.E., HOWE, J.M. and LEE, C.J. 1971. Nitrogen retention of adult human subjects fed a high protein rice. Am. J. Clin. Nutr. *24*:324.

CRAMPTON, E.W., FARMER, F.A., McKIRDY, H.B., LLOYD, L.E., DONEFER, E. and SCHAD, D.J. 1960. A statistical study of apparent digestibility coefficients of the energy-yielding components of a nutritionally adequate mixed diet consumed by 103 young human adults. J. Nutr. 72:177.

CULLUMBINE, H. 1950. Nitrogen balance studies on rice diets. Brit. J. Nutr. *4*:129.

DAM, R., LEE, S., FRY, P.C. and FOX, H. 1965. Utilization of algae as a protein source for humans. J. Nutr. *86*:376.

DANIEL, V.A., LEELA, R., DORAISWAMY, T.R., RAJALAKSHMI, D., RAO, S.V., SWAMINATHAN, M. and PARPIA, H.A.B. 1965. The effect of supplementing a poor Indian ragi diet with L-lysine and DL-threonine on the digestibility coefficient, biological value and net utilization of the proteins and on nitrogen retention in children. J. Nutr. Dietet. *2*:138.

DANIEL, V.A., LEELA, R., DORAISWAMY, T.R., RAJALAKSHMI, D., RAO, S.V., SWAMINATHAN, M. and PARPIA, H.A.B. 1966. The effect of supplementing a poor kaffir corn (sorghum vulgare) diet with L-lysine and DL-threonine on the digestibility coefficient, biological value and net utilization of proteins and retention of nitrogen in children. J. Nutr. Dietet. 3:10.

DANIEL, V.A., DORAISWAMY, T.R., SWAMINATHAN, M. and RAJAKSHMI, D. 1970. The effect of dilution of milk proteins with non-essential amino acids (L-alanine and L-glutamic acid) on nitrogen retention and biological value of the proteins in children. J. Nutr. 24:741.

DeMAEYER, E.M. and VANDERBORGHT, H.L. 1961. Determination of the nutritive value of different foods in the feeding of African children. In "Progress in meeting protein needs of infants and preschool children," Pub. 843, p. 143, NAS-NRC, Washington, D.C.

ESSELBAUGH, N.C., MURRAY, H.C., HARDIE, L.W. and HARD, M.M. 1952. The replacement value of the Alaska field pea (pisum sativum) for human subjects. J. Nutr. 46:109.

FAO/WHO. 1973. Energy and protein requirements. Tech. Rept. Series No. 522, World Health Organization, Geneva.

GARZA, C., SCRIMSHAW, N.S. and YOUNG, V.R. 1977. Human protein requirements: a long-term metabolic nitrogen balance study in young men to evaluate the 1973 FAO/WHO safe level of egg protein intake. J. Nutr. 107:335.

HAWLEY, E.E., MURLIN, J.R., NASSET, E.S. and SZYMANSKI, T.A. 1948. Biological values of six partially purified proteins. J. Nutr. 36:153.

HEGSTED, D.M., TSONGAS, A.G., ABBOTT, D.B. and STARE, F.J. 1946. Protein requirements of adults. J. Lab. and Clin. Med. Vol. 31:261.

INOUE, G., FUJITA, Y., and NIIYAMA, Y. 1973. Studies on protein requirements of young men fed egg protein and rice protein with excess and maintenance energy intakes. J. Nutr. 103:1673.

INOUE, G., FUJITA, Y., KISHI, K., YAMAMOTO, S. and NIIYAMA, Y. 1974. Nutritive values of egg protein and wheat gluten in young men. Nutr. Reports Int. 10:201.

JOSEPH, K., KURIEN, P.P., SWAMINATHAN, M. and SUBRAHMANYAN, V. 1958. The metabolism of nitrogen, calcium, and phosphorous in undernourished children. 5. The effect of partial or complete replacement of rice in poor vegetarian diets by ragi (Eleusine coracana) on the metabolism of nitrogen, calcium and phosphorous. Brit. J. Nutr. 13:213.

JOSEPH, K., TASKER, P.K., NARAYANARAO, M., SWAMINATHAN, M., SREENIVASAN, A. and SUBRAHMANYAN, V. 1963. The effect of supplements of ground nut flour or ground nut protein isolate fortified with calcium salts and vitamins or of skim-milk powder on the digestibility coefficient, biological value and net utilization of the proteins of poor Indian diets given to undernourished children. Brit. J. Nutr. 17:13.

KELSAY, J.L., BEHALL, K.M. and PRATHER, E.S.   1978.   Effect of fiber from fruits and vegetables on metabolic responses of human subjects. I. Bowel transit time, number of defecations, fecal weight, urinary excretions of energy and nitrogen and apparent digestibilities of energy, nitrogen and fat.   Am. J. Clin. Nutr. *31*:1149.

KIES, C. and FOX, H.M.   1972.   Protein nutritional value of Opaque-2 corn grain for human adults.   J. Nutr. *102*:757.

KIES, C. and FOX, H.M.   1973.   Comparisons of dry breakfast cereals as protein resources: human biological assay at equal intakes of cereal.   Cer. Chem. Vol. *50*:233.

KIES, C., FOX, H.M. and NELSON, L.   1975.   Triticale, soy-TVP, and millet based diets as protein suppliers for human adults.   J. Fd. Sci. *40*:90.

KIES, C. and FOX, H.M.   1977.   Dietary hemicellulose interactions influencing serum lipid patterns and protein nutritional status of adult men.   J. Fd. Sci. *42*:440.

KISHI, K., MIYATANI, S. and INOUE, G.   1978.   Requirement and utilization of egg protein by Japanese young men with marginal intakes of energy. J. Nutr. *108*:658.

KUETHER, C.A. and MYERS, V.C.   1948.   The nutritive value of cereal proteins in human subjects.   J. Nutr. *35*:651.

KURIEN, P.P., NARAYANARAO, M., SWAMINATHAN, M. and SUBRAHMANYAN, V.   1960.   The metabolism of nitrogen, calcium, and phosphorus in undernourished children.   6.   The effect of partial or complete replacement of rice in poor vegetarian diets by kaffir corn (sorghum vulgare). Brit. J. Nutr. *14*:339.

KURIEN, P.P., SWAMINATHAN, M. and SUBRAHMANYAN, V.   1961. The metabolism of nitrogen, calcium and phosphorus in undernourished children.   Brit. J. Nutr. *15*:345.

LEE, C.J., HOWE, J.M., CARLSON, K. and CLARK, H.E.   1971.   Nitrogen retention of young men fed rice with or without supplementary chicken. Am. J. Clin. Nutr. *24*:318.

LUSK, G.   1928.   "The Elements of the Science of Nutrition," 4th ed., W.B. Saunders Company, Philadelphia, PA.

MACRAE, T.F., HUTCHINSON, J.C.D., IRWIN, J.O., BACON, J.S.D. and McDOUGALL, E.I.   1942.   Comparative digestibility of wholemeal and white breads and the effect of the degree of fineness of grinding on the former.   J. Hyg. *42*:423.

McCANCE, R.A. and WIDDOWSON, E.M.   1947.   The digestibility of English and Canadian wheats with special reference to the digestibility of wheat protein by man.   J. Hygiene *45*:59.

McCANCE, R.A. and WALSHAM, C.M.   1948.   The digestibility and absorption of the calories, proteins, purines, fat and calcium in wholemeal wheaten bread.   Brit. J. Nutr. *2*:26.

McCANCE, R.A. and GLASER, E.M.   1948.   The energy value of oatmeal and the digestibility and absorption of its protein, fats and calcium.   Brit. J. Nutr. 2:221.

MERRILL, A.L. and WATT, B.K.   1955.   "Energy Value of Foods," Agriculture Handbook No. 74, Human Nutrition Research Branch, ARS-USDA, Washington, DC.

MORSE, E.H., MERROW, S.B., KEYSER, D.E. and CLARK, R.P.   1972. Comparative utilization of casein-lactalbumin and spun soy protein diets by human subjects.   Am. J. Clin. Nutr. 25:912.

MUELLER, A.J. and COX, W.M., Jr.   1947.   Comparative nutritive value of casein and lactalbumin for man.   J. Nutr. 34:285.

MURLIN, J.R. and MATTILL, H.A.   1938.   Digestibility and nutritional value of cereal proteins in the human subject.   J. Nutr. 16:15.

MURLIN, J.R., NASSET, E.S. and MARSH, M.E.   1938.   The egg-replacement value of the proteins of cereal breakfast foods with a consideration of heat injury.   J. Nutr. 16:249.

MURLIN, J.R., MARSHALL, M.E. and KOCHAKIAN, C.D.   1941.   Digestibility and biological value of whole wheat breads as compared with white bread.   J. Nutr. 22:573.

MURLIN, J.R., EDWARDS, L.E. and HAWLEY, E.E.   1944.   Biological values and true digestibilities of some food proteins determined on human subjects.   J. Biol. Chem. 156:785.

NARAYANASWAMY, D., DORAISWAMY, T.R., DANIEL, V.E., SWAMINATHAN, M. and PARPIA, H.A.B.   1972.   Effect of supplementing poor rice diet with low cost protein food, chick pea or skim milk powder on nitrogen retention and net protein utilization in children.   Nutr. Rpts. Int. 5:171.

NASR, K., HAGHIGHI, P., ABADI, P., LAHIMGARZADEH, A., HEDAYATI, H., HALSTEAD, J.A. and REINHOLD, J.D.   1976.   Idiopathic enteropathy: an evaluation in rural Iran with an appraisal of nutrient loss.   Am. J. Clin. Nutr. 29:169.

NAVARRETE, D.A., LOUREIRO DE DAQUI, V.A., ELIAS, L.G., LaCHANCE, P.A. and BRESSANI, R.   1977.   The nutritive value of egg protein as determined by the nitrogen balance index.   Nutr. Reports Int. 16:695.

NEWMAN, L.F., ROBINSON, G.W., HALNAN, E.T. and NEVILLE, H.A.D. 1912.   Some experiments on the relative digestibility of white and wholemeal breads.   J. Hygiene 12:9.

NICOL, B.M. and PHILLIPS, P.G.   1976A.   The utilization of dietary protein by Nigerian men.   Brit. J. Nutr. 36:337.

NICOL, B.M. and PHILLIPS, P.G.   1976B.   Endogenous nitrogen excretion and utilization of dietary protein.   Brit. J. Nutr. 35:181.

PANEMANGALORE, M., PARTHASARATHY, H.N., JOSEPH, K., SANKARAN, A.N., RAO, M.N. and SWAMINATHAN, M.   1964.   The metabolism of nitrogen and the digestibility coefficient and biological value of the proteins and net protein utilization in poor rice diet supplemented with methionine-fortified soya flour or skim milk powder.   Canadian J. Biochem. 42:641.

PARTHASARATHY, H.N., DORAISWAMY, T.R., RAO, S.V., RAO, M.N., SWAMINATHAN, M., SREENIVASAN, A. and SUBRAHMANYAN, A. 1963. The digestibility coefficient, biological value and net protein utilization of egg proteins and protein blends having amino acid composition similar to FAO reference pattern and an "ideal" reference protein pattern as determined in children. Fd. Sci. *12*:168.

PARTHASARATHY, H.N., DORAISWAMY, T.R., PANEMANGALORE, M., RAO, M.N., CHANDRASEKHAR, B.S., SWAMINATHAN, M., SREENIVASAN, A. and SUBRAHMANYAN, V. 1964A. The effect of fortification of processed soya flour with DL-methionine hydroxy analogue or DL-methionine on the digestibility, biological value, and net protein utilization of the proteins as studied in children. Canadian J. Biochem. *42*:377.

PARTHASARATHY, H.N., JOSEPH, K., DANIEL, V.A., DORAISWAMY, T.R., SANKARAN, A.N., RAO, M.N., SWAMINATHAN, M., SREENIVASAN, A. and SUBRAHMANYAN, V. 1964B. The effect of supplementing a rice diet with lysine, methionine, and threonine on the digestibility coefficient, biological value, and net protein utilization of the proteins and on the retention of nitrogen in children. Canadian J. Biochem. *42*:385.

PASRICHA, S., RAO, N., MOHANRAM, K. and GOPALAN, C. 1965. Nitrogen balance studies on women in India. J. Am. Diet. Assoc. *47*:269.

RAO, C.N., NAIDU, A.N. and RAO, B.S.N. 1975. Influence of varying energy intake on nitrogen balance in men on two levels of protein intake. Am. J. Clin. Nutr. *28*:1116.

RAO, S.V., DORAISWAMY, T.R., MOORJANI, M.N. and SWAMINATHAN, M. 1964. The digestibility coefficient, biological value and net utilization of the proteins of fish flour from oil sardine (clupea longiceps) in children. J. Nutr. Dietet. *1*:178.

REDDY, V. 1971. Lysine supplementation of wheat and nitrogen retention in children. Am. J. Clin. Nutr. *24*:1246.

REGISTER, U.D., INANO, M., THURSTON, C.E., VYHMEISTER, I.B., DYSINGER, P.W., BLANKENSHIP, J.W. and HORNING, M.C. 1967. Nitrogen-balance studies in human subjects on various diets. Am. J. Clin. Nutr. *20*(7):753.

SCRIMSHAW, N.S., BRESSANI, R., BEHAR, M. and VITERI, F. 1958. Supplementation of cereal proteins with amino acids. J. Nutr. *66*:485.

SCRIMSHAW, N.S. and YOUNG, V.R. 1979. Soy protein in adult human nutrition. In Soy Protein and Human Nutrition, H.L. Wilcke, D.T. Hopkins, and D. H. Waggle (Editors). Academic Press, NY, p. 121.

SHERMAN, H.C. and WINTERS, J.C. 1918. Efficiency of maize protein in adult human nutrition. J. Biol. Chem. *35*:301.

SHERMAN, H.C. 1920. Protein requirement of maintenance in man and the nutritive efficiency of bread protein. J. Biol. Chem. *41*:97.

SMITH, A.H., ANDERSON, W.E., BROOKE, R.O. and GORDON, W.G. 1933. The comparative digestibility of the doughnut. J. Am. Diet. Assoc. *9*:6.

SOUTHGATE, D.A.T. and DURNIN, J.V.G.A.   1970.   Calorie conversion factors. An experimental reassessment of the factors used in the calculation of the energy value of human diets.   Brit. J. Nutr. 24:517.

SUBRAHMANYAN, V., NARAYANARAO, M., RAMARAO, G. and SWAMINATHAN, M.   1955.   The metabolism of nitrogen, calcium and phosphorus in human adults on a poor vegetarian diet containing ragi (eleusine coracana).   Brit. J. Nutr. 9:350.

SUMNER, E.E., PIERCE, H.B. and MURLIN, J.R.   1938.   The egg replacement value of several proteins in human nutrition.   J. Nutr. 16:37.

SUMNER, E.E. and MURLIN, J.R.   1938.   The biological value of milk and egg protein in human subjects.   J. Nutr. 16:141.

SWAMINATHAN, M.   1967.   Availability of plant proteins. In Newer Methods of Nutritional Biochemistry with Applications and Interpretations, A. Albanese (Editor). Vol. III.   Academic Press, NY, p. 197.

TASKER, P.K., DORAISWAMY, T.R., NARAYANARAO, M., SWAMINATHAN, M., SREENIVASAN, A. and SUBRAHMANYAN, V.   1962.   The metabolism of nitrogen, calcium and phosphorus in undernourished children.   Brit. J. Nutr. 16:361.

TRUSWELL, A.S. and BROCK, J.F.   1962.   The nutritive value of maize protein for man.   Am. J. Clin. Nutr. 10:142.

TURK, R.E., CORNWELL, P.E., BROOKS, M.D. and BUTTERWORTH, C.E., Jr.   1973.   Adequacy of spun-soy protein containing egg albumin for human nutrition.   J. Am. Diet. Assoc. 63:519.

VEMURY, M.K.D., KIES, C. and FOX, H.M.   1976.   Comparative protein value of several vegetable protein products fed at equal nitrogen levels to human adults.   J. Fd. Sci. 41:1086.

VITERI, F., BRESSANI, R. and ARROYAVE, G.   1971.   Fecal and urinary nitrogen losses. As cited in PAG, Energy and protein requirements—recommendations by a joint FAO/WHO informal gathering of experts.   PAG Bulletin Vol. V, 35, United Nations, NY, 1975.

WATT, B.K. and MERRILL, A.L.   1963.   "Composition of Foods," Agriculture Handbook No. 8, Consumer and Food Economics Research Division, ARS-USDA, Washington, DC.

WATTS, J.H., BOOKER, L.K., McAFEE, J.W., WILLIAMS, E.G., WRIGHT, W.G. and JONES, F.   1959.   Biological availability of essential amino acids to human subjects.   J. Nutr. 67:483.

YOUNG, V.R., OZALP, I., CHOLAKOS, B.V. and SCRIMSHAW, N.S.   1971.   Protein value of Colombian Opaque-2 corn for young adult men.   J. Nutr. 101:1475.

YOUNG, V.R., TAYLOR, Y.S., RAND, W.M. and SCRIMSHAW, N.S.   1973.   Protein requirements of man: efficiency of egg protein utilization at maintenance and submaintenance levels in young men.   J. Nutr. 103:1164.

YOUNG, V.R., FAJARDO, L., MURRAY, E., RAND, W.M. and SCRIMSHAW, N.S.   1975.   Protein requirements of man: comparative nitrogen balance response within the submaintenance to maintenance range of intakes of wheat and beef proteins.   J. Nutr. 105:534.

DISCUSSION

DR. SARWAR: At least in animal studies, the amino acid digestibility of different sources of protein is not the same as nitrogen digestibility. For example, in the case of legumes, the true digestibility of protein in field peas would be around 90% and the methionine digestibility could be as low as 75%. Have you seen anything in the literature with humans where they have found this difference? In cereal proteins, will you find that lysine is less digestible than other amino acids?

DR. HOPKINS: I have not seen anything in the literature in which the digestibility of individual amino acids has been studied. Without doubt, based on animal studies, in certain cases the availability of certain amino acids could be reduced. An example would be the browning reaction with lysine.

DR. TORUN: In the paper that we'll be discussing tomorrow, in the context of the use of amino acid scoring patterns to predict nutritive value, we have considered the possibility that not all the amino acid nitrogen in the diets is absorbed uniformly. We'll give an example to illustrate this point where there is a poor correlation between protein quality assessed by chemical scoring (corrected for digestibility) and actual *in vivo* studies. If anybody here has some information to answer the question about the differences in absorption of amino acids, it would be very useful.

I would also like to make another very brief comment and this is really a corollary of some of the data that Dr. Hopkins presented. The same diets that were tested by Dr. Calloway and coworkers with students at Berkeley were tested by Schneider et al. at INCAP with Guatemalan peasants. These studies pointed out that not just the diet but also other "adaptation" factors, probably having to do with intestinal flora, affect digestibility. The digestibility of the Guatemalan rural diet was better in the Guatemalans than that observed in the Berkeley students. The egg-based diet was absorbed better by the Guatemalan men than the rural Guatemalan diet but to a lesser degree than the way it was absorbed by the U.S. students.

DR. BRESSANI: I question the significance of true digestibility. I think that apparent digestibility is more meaningful, as expressing NR as a function of NI rather than as a function of NA.

Secondly, you haven't said anything about the effects of processing on digestibility and this is very important. It can go either way. For example, lime treatment of corn will improve the digestibility of maize. On the other hand, roasting processes will decrease digestibility. Cooking increases digestibility of beans, but excess cooking will decrease it.

Then there are many instances where natural compounds will react with essential amino acids. For example, gossypol will bind lysine in cottonseed. Some of the polyphenolic compounds in many foods will also bind essential amino acids. Digestibility of natural diets, particularly those of developing

countries, is a very important factor in determining quality and requirements of protein, particularly when beans are present.

DR. HOPKINS: The polyphenolic compounds do react with protein and
there are genetic factors in grains that will affect digestibility. Genetic
factors influence the storage protein, how the storage proteins are packed,
and the ratio of storage proteins to structural proteins. The sorghums and
the millets are probably the best examples. Nonetheless, my feeling is that
as far as the American diet is concerned, we're dealing with fairly well
defined food ingredients, and are not going to see sorghum and millets
coming into the diet. For regulatory purposes, as far as the protein quality is
concerned, digestibility probably is not a big factor. When we are talking
about new foods or problems with some of the largely vegetarian diets in
some of the developing countries, it is a different story and protein digestibility is of concern.

DR. MARGEN: We really don't have that much information on this
matter of the relative digestibility of protein and absorbability of amino
acids. We can be lead quite astray in making "corrections" for either apparent or true digestibility, in interpreting protein quality handling in the
intestine. Since almost all of the fecal nitrogen is bacteria, it is almost
impossible to know if it was not digested or absorbed. We're not observing
protein as such, only material which has been "processed."

DR. SATTERLEE: Dr. Hopkins, in using the true digestibility, you are
using a constant factor for metabolic fecal nitrogen. Because of this, I would
tend to second Dr. Bressani's concern, which is, are you really sure obligatory nitrogen is nine or twelve mg N/Kg body wt/day? It may be that
apparent digestibility is giving us a better indicator of the variation in
digestibility, without being nullified by a constant obligatory nitrogen
value.

DR. HOPKINS: I feel that the protein requirement should account for
the metabolic fecal nitrogen. We should be interested in what goes in the
mouth and how much is getting through the gut. The true digestibility
figures, by and large over the long run, should be more constant between
laboratories than apparent digestibility figures and the use of true digestibility data should counteract the effects of the level of protein in the diet.

DR. EGGUM: Regarding the metabolic nitrogen excretion, I think that
this is more related to dry matter consumption. Anyway, this is what we
find in both pigs and rats. The question was also raised here if the amino
acids were absorbed to a different degree compared to total nitrogen. We
find, for instance, in the cereals, that the lysine in general has a significantly lower digestibility than total nitrogen. I think this can be explained
by the fact that the lysine is in the aleurone layer which has a thick,
cellulosic cell wall so it's more difficult to digest. The prolamine fraction is

almost completely digestible but is very low in lysine. So I think it is a fact that the amino acids are absorbed to a different degree compared to total nitrogen.

DR. BODWELL: A comment about apparent versus true digestibility. Indeed, we, by convention, usually use factors for obligatory nitrogen losses to go from apparent to true. And indeed, surely, this probably changes with the level of protein intake that we feed. This is, the underlying actual losses of fecal and urinary obligatory nitrogen probably differ according to the intake of protein. However, this just casts doubt on the validity of apparent digestibility. We don't know what the obligatory nitrogen losses underlying those values for apparent digestibility really are. I'm not sure that we know what apparent digestibility means.

Part III In Vitro Methods for Assessing Protein Nutritional Value

12

Collaborative Studies of Amino Acid Analyses: A Review and Preliminary Observations from a Nine-Laboratory Study

M.L. Happich, C.E. Bodwell and J.G. Phillips

The use of amino acid composition data has become increasingly important in the assessment of protein nutritional quality of food products, particularly of new food products. The contents of individual essential amino acids in foods are used in calculating Food and Agriculture Organization (FAO) amino acid scores and chemical scores and predicting protein efficiency ratios by use of mathematical equations (Alsmeyer *et al.*, 1974; Happich *et al.*, 1975; Lee *et al.*, 1978; Satterlee *et al.*, 1979). Essential amino acid composition data may be used as a basis to enrich foods with individual amino acids and to combine protein sources for complementation and mutual supplementation of the proteins, with an enhancement of nutritional value (Bressani, 1977). These and other uses of amino acid data will increase in the future. In particular, amino acid composition data are potentially important for use in assay methods for nutritional labeling and demonstrating compliance with FDA and USDA product standards (See Chapters 4, 5, and 6). The potential importance of amino acid determination in food analysis requires standardization of procedures for the preparation of protein food samples for amino acid analysis, and knowledge of the inter- and intralaboratory variation of procedures used to hydrolyze the protein in food sources and of the subsequent analysis of individual amino acids. These requirements are particularly important if amino acid analysis data are used to calculate the protein nutritional value of food sources for nutritional labeling, regulatory purposes, fortification of foods, or mutual supplementation of foods.

Reference to a brand or firm name does not constitute endorsement by the U.S. Department of Agriculture over others of a similar nature not mentioned.

In the United States there are no official methods for sample preparation prior to hydrolysis or for hydrolysis of the proteins prior to amino acid analysis. Each laboratory is free to use available methods and procedures of choice and to modify them as desired. The only official methods for determining amino acid composition in food in the United States are a titration method for free amino acids in lemon juice (AOAC, 1980; Sec. 22.106), one for proline in honey (AOAC, 1980; Sec. 31.116), and one for available lysine which includes hydrolysis of the protein (AOAC, 1980; Sec. 43.224).

STUDIES ON ION-EXCHANGE AMINO ACID ANALYSES

Collaborative studies on variation of the methods for determining amino acids have been reported by Porter *et al.*, 1968; Derse, 1969; Knipfel *et al.*, 1971; Cavins *et al.*, 1972; Westgarth and Williams, 1974; Williams *et al.*, 1979; and most recently Sarwar *et al.* (see Chapter 13).

Porter and colleagues in Great Britain conducted assays on a standard mixture of amino acids in solution and on two test proteins, gelatin and freeze-dried cod muscle. Their objective was to assess the precision of the ion-exchange column chromatographic technique, including the manual procedure, as operated in seven different laboratories. They concluded that the automatic amino acid analyzers afford better precision than the manual procedure, but that experienced workers could obtain broadly similar results with the manual procedure. The absolute mean deviations were lower for the standard mixture of amino acids than for the test proteins, gelatin and cod muscle. Seven of the nine collaborators recovered 81.5 to 88% of the total Kjeldahl nitrogen of the cod muscle as amino acid nitrogen. For gelatin, methionine and tyrosine had high mean absolute deviations (>±10%).

Derse (1969) reported a study in which a sample of 50% protein soybean meal was issued to 12 collaborating laboratories proficient in the use of amino acid analyzer techniques. The results of the test, which were considered good, illustrated differences in percent of each amino acid found in the 50% protein soybean meal. Methionine and cystine showed the greatest differences between laboratories. Methionine ranged from a trace to 0.72% and cystine ranged from 0.44% to 1.36%. Standard deviations and coefficients of variation were not calculated.

Knipfel *et al.* (1971) studied the analysis of 2 hydrolyzates of each of 3 proteins (casein, soy flour, and fish flour) prepared by Knipfel and sent to five cooperating laboratories. Duplicate analyses of individual hydrolysates within laboratories produced coefficients of variation of 3.7 to 4.4% (Knipfel *et al.*, 1971). There was a significant hydrolyzate-amino acid interaction for soy flour and casein. The concentration of some amino acids was higher in one hydrolyzate and the concentration of others was higher in the second hydrolyzate. The methionine content of casein showed a difference of about 30% between the two hydrolyzates. Practically identical results could be produced from a standard sample tested in the different laboratories by varying analytical procedures. Adjustment of the mean values for amino acid concentrations for each protein source within individual laboratories to constant total recovery of 90g/16g N reduced the interlaboratory variation considerably.

Kwolek and Cavins (1971) examined variability in published amino acid data on plant seed or seed products from 18 selected references. Cysteine, tryptophan, tyrosine, and methionine had higher relative standard deviations than the other 14 amino acids. A mean relative standard deviation of about 8% (ranging from 5.9 to 14.1, depending on the amino acid) was observed between samples.

Cavins et al. (1972) conducted an interlaboratory study with five laboratories on the amino acid analysis of soybean meal. The analysts evaluated four methods of hydrolysis: in sealed ampoules, in sealed flasks, and with two acid-to-sample ratios by the reflux method. Later, they evaluated the effect of mesh size (40-, 80-, over 70-, and under 270-mesh) of the soybean meal on the amino acid analysis results. Interlaboratory variations were significant for twelve amino acids and ammonia in the hydrolysis study and for all (18) amino acids and ammonia in the study on mesh size. Tryptophan results obtained by four procedures on 80-mesh samples were in good agreement. Analysis of the 80-mesh samples for cystine by two procedures, one an oxidation and the other a reduction followed by a derivatization, produced values that did not agree. Normalization of results to 95% nitrogen recovery had only a small effect on the statistical analysis.

Westgarth and Williams (1974) compared the methods of Miller (1967) and of Spies and Chambers (1949) for the determination of tryptophan on groundnut, soya bean, and cottonseed meals at eight laboratories. Tryptophan values obtained by both methods were similar. The collaborators preferred the Miller method because it appeared to provide a satisfactory estimation of tryptophan in feedstuffs.

Williams et al. (1979) reported determinations of "cyst(e)ine" in feedstuffs in eight laboratories by the Moore (1963) performic acid oxidation (PAO) method and the Spencer and Wold (1969) dimethyl sulfoxide (DMSO) method. Casein, fishmeal, extracted leaf protein concentrate, groundnut, soya bean, and wheat were the protein sources analyzed. The authors state that the "variation between duplicate hydrolyses within laboratories was similar by both methods (mean coefficients of variation 4.3 and 5.0%) and smaller than the variation between laboratories (mean coefficients of variation 14.8 and 14.3%)." Apparently both methods were considered to be satisfactory for estimating "cyst(e)ine" in the samples tested except for casein. Six of the eight laboratories found significantly more "cyst(e)ine" in casein by the DMSO method. Although the DMSO method was faster, the PAO method was preferred by the collaborators because methionine could be estimated at the same time.

Past collaborative studies were important at the time they were conducted and are historically important now. However, they were usually either limited in scope, did not include hydrolysis procedures for all amino acids, did not rigorously standardize procedures used or did not analyze a variety of protein sources.

The collaborative study on amino acid analysis recently completed in Canada by Sarwar *et al.* is described in Chapter 13.

METHODOLOGY CONSIDERATIONS

The preparation of the sample, hydrolysis methods, hydrolysis time, resolution of some amino acids and/or amino acid derivatives, and a reference protein for amino acid analysis required consideration when a protocol was developed for the current amino acid analysis collaborative studies. Methodology selected for the current collaborative studies is given below.

Sample Preparation

Chemical analyses and particularly amino acid analyses of protein food sources must be made on homogeneous material. Meats, meat products, and variety meats which have a high content of water and fat should be lyophilized, defatted (the defatting solvent thoroughly removed), ground finely, mixed thoroughly, and sampled for proximate analysis and amino acid analyses at the same time. Petroleum ether is the solvent of choice to defat samples prior to amino acid analysis in the ERRC (Eastern Regional Research Laboratory) Meat Laboratory. We compared petroleum ether with acetone and diethyl ether for defatting of beef. We found no appreciable differences in amino acid analyses after the samples were defatted with any one of these three solvents (unpublished data). Removal of carbohydrate from protein food sources prior to amino acid analysis, although recommended by Block and Weiss (1956a), is not widely practiced today.

Hydrolysis of Proteins

The use of different procedures or conditions for the hydrolysis of proteins by different investigators is a well-known source of variability in the analyses (Block and Weiss, 1956b; Blackburn, 1978). Hydrolysis with 6 N HCl is a widely used method of choice except for the determination of tryptophan. Although some analysts determine cystine and methionine in the usual routine acid hydrolyzates, it is not the preferred method. Great care must be taken to prevent oxidation during hydrolysis with hydrochloric acid, removal of hydrochloric acid after hydrolysis and evaporation of the hydrolyzate to dryness. Hackler (1974) suggested that greatest care should be taken in the removal of air from the hydrolysis sample and tube. Even so, cysteine is subject to oxidation and cystine is racemized to meso-cystine and DL-cystine during acid hydrolysis and chromatography (Friedman and Noma, 1975). Cysteine and cystine react with tryptophan, carbohydrates and other substances commonly present during hydrolysis (Friedman and Noma, 1975). Methionine and cystine contents of a protein sample can be determined fairly accurately by oxidation with performic acid, followed by hydrolysis with 6 N HCl, and analyzed by an ion-exchange procedure (Walker *et al.*, 1975; Shram *et al.*, 1954; Neumann *et al.*, 1962; Moore, 1963). This method is preferred for the determination of methionine and cystine.

Tryptophan is extensively destroyed during hydrochloric acid hydrolysis of proteins, producing ammonia as the only recognizable end product (Friedman and Finley, 1971; Friedman and Finley, 1975). Basic hydrolysis of proteins is used by most analysts for the determination of tryptophan. By the method of Hugli and Moore (1972) (hydrolysis with 4.2 N sodium hydroxide), partially hydrolyzed potato starch is added as an antioxidant to protect the tryptophan (Dreze, 1960; Hugli and Moore, 1972). Oelshlegel *et al.* (1970) suggested the use of a polypropylene or polyallomer tube, resistant to hot alkali, inside a glass tube for the hydrolysis.

Hydrolysis Time

Hydrolysis time, another potential variable, has been selected by various analysts rather arbitrarily. For convenience, the times most often used have been 18, 21, 22, or 24 hr (with 6 N HCl). Cherry (1978; 1979) found that variations in the release of amino acids from cottonseed protein products are a function of hydrolysis time. When samples were hydrolyzed with 6 N HCl for 8, 16, 24, 32, 48, and 72 hr, the release of amino acids increased between 8 and 16 hr and decreased between 16 and 24 hr, leveled off between 24 and 32 hr, and then increased to optimum levels at 48 hr. Little change or a decline in amino acid content was observed between 48 and 72 hr. The amount of change varied greatly among amino acids.

Hackler (1971) studied the release of methionine from pea beans, beef round, and gelatin by hydrolysis with 6 N HCl for 5½, 11, 22, 44, or 66 hr. Twenty-two hr of hydrolysis produced the highest value for methionine from pea beans; there were only small differences in the methionine released from beef round in 11, 22, 44, or 66 hr; the methionine values obtained for gelatin were highest (and similar) after either 44 or 66 hr of hydrolysis.

Hackler *et al.* (unpublished data) also studied the release of amino acids from the proteins of rice (Table 12.1) by hydrolysis with 6 N HCl for 6, 12, 24,

TABLE 12.1. SPECIFIC AMINO ACID CONTENT (GRAMS AMINO ACID/16G NITROGEN) OF TWO PROTEIN SOURCES AS AFFECTED BY VARYING HYDROLYSIS TIME

| Amino acid | Lyophilized defatted beef[1] | | | | Rice[2] | | | | |
| | Hydrolysis time, hr | | | | Hydrolysis time, hr | | | | |
	18	21	24	48	6	12	24	48	72
Histidine	3.8	3.8	3.8	3.9	2.1	2.2	2.1	2.3	2.2
Isoleucine	4.8	4.9	4.7	4.9	3.1	3.4	4.0	4.5	4.6
Leucine	8.4	8.2	8.2	8.3	7.7	8.2	8.1	9.0	8.7
Lysine	8.9	8.9	9.1	8.8	3.0	3.3	3.3	3.5	3.4
Methionine	3.5	3.4	3.0	2.6	2.3	2.4	2.4	2.4	2.4
Phenylalanine	4.4	4.5	4.4	4.4	4.4	4.7	5.1	5.3	5.1
Threonine	4.6	4.3	4.5	4.3	3.0	3.0	3.4	3.1	2.9
Tyrosine[3]	3.8	3.7	3.8	3.9	2.6	2.9	3.1	3.0	3.0
Valine	5.0	5.0	5.0	5.2	4.8	5.4	6.1	6.6	6.4

[1]M.L. Happich, unpublished data; acid hydrolyzate.
[2]L.R. Hackler et al., unpublished data; acid hydrolyzate.
[3]Tyrosine is not an essential amino acid but may replace part of the phenylalanine requirement for humans (Rose and Wixon, 1955).

48 and 72 hr. Valine and isoleucine showed the greatest overall differences in recovery with variation in hydrolysis time. The highest values were recovered by 48 or 72 hr, respectively, and were 8 and 15% higher, respectively, than values obtained after a 24 hr hydrolysis. Methionine recovery varied least reaching the highest value by 12 hours hydrolysis and remaining constant thereafter. Threonine and tyrosine recoveries were best from a 24 hr hydrolysis. Forty-eight hr hydrolysis showed the highest values for histidine, leucine, lysine and phenylalanine, i.e., 9.5, 11.1, 6.0 and 3.9% higher, respectively, than the values recovered by 24 hr hydrolysis.

Although not part of the collaborative study, a time study is underway at ERRC, USDA, to evaluate the release of amino acids from the proteins of the food sources in the collaborative study after 18, 21, 24, and 48 hr of hydrolysis. Results for lyophilized, defatted beef (Table 12.1), indicate that, in general, the largest differences occurred in the 48 hr samples. Methionine and threonine show a decrease of 25.7 and 6.5%, respectively, after 48 hr hydrolysis as compared with values after 18 hr. Valine shows an increase of 4.0%. Changes in the other amino acids varied from 1.1 to 2.6%, values within experimental error.

Reference Protein for Amino Acid Analysis

Most analysts use an amino acid reference standard. Commonly, a reference protein is used to determine the accuracy of the recovery of amino acids from the protein and of the analytical methods used. The selection of a reference protein for amino acid analysis of food sources is difficult because the protein content of a food source is usually a mixture of individual proteins. These individual proteins are seldom available in purified form and are often of unknown proportions. Because other constituents such as carbohydrates, mineral salts, other naturally occurring materials, and materials added during manufacture may produce reactions with, or changes in, the hydrolyzed amino acids, a good general standard is lacking for protein food sources.

CURRENT COLLABORATIVE STUDY

Nine laboratories participated in the current collaborative study: Campbell Soup Company; General Mills, Inc.; Kraft, Inc.; Mead Johnson; Northern Regional Research Center, (USDA); Procter and Gamble Company; Quaker Oats Company; Ralston Purina Company; and the University of Nebraska. In a tenth laboratory, Nabisco, Inc., amino acid compositions were determined on the same pretest and test samples by gas chromatography for comparison with results from the automated ion-exchange chromatographic analyzers.

Objectives and Design

The objectives of the study were:

1. Standardize the preparation of protein food sources for amino acid analysis, including the hydrolysis of proteins.
2. Compare two different approaches for normalizing amino acid analysis data.
3. Determine the inter- and intralaboratory variation for the analysis of individual amino acids in selected protein food sources.

The experimental protocol is set forth in Table 12.2.

Sample Preparation and Storage

The samples for the current collaborative study were purchased or prepared as indicated: The pretest samples, tuna fish and peanut flour, were originally obtained for use in studies by Bodwell *et al.* (1978); textured soy protein was obtained in the spring of 1979 and ground; casein was a sample of the ANRC (Animal Nutrition Research Council) casein purchased in January, 1979. These four samples were stored at $-20°C$ until shipped to ERRC (USDA) in August, 1979. The non fat dried milk powder (low heat, spray dried, containing no added nutrients); the wheat flour (commercial bakers' bread flour of hard, red spring wheat, enriched with niacin, iron, thiamine, and riboflavin); and the sample of lean beef (semitendinosus muscle) were all obtained in September–November, 1979. The beef was lyophilized, defatted with petroleum ether, and ground in a Wiley Mill to pass a 2-mm screen. All samples were thoroughly mixed several times; fifteen 30g samples were placed in air tight bottles, sealed for use in the amino acid analysis, and stored just above freezing at about 4°C or less. One sample was taken for proximate analysis on each protein source. Nitrogen was determined by the macro Kjeldahl method in quintuplicate (AOAC, 1980; Sec. 24.027; Sec. 2.057).

Hydrolysis of Proteins

The hydrolysis procedures for the current study were specified in much detail to minimize interlaboratory variation. Either the stopcock style glass flask assembly or a drawn glass tube was allowed as the container for the hydrolysis. The contents (protein source + 6 N HCl) were frozen by placing the lower portion of the flask or tube in a dry ice-acetone bath. The flask or tube was evacuated of air by a high vacuum pump, the vacuum was then closed off and the contents were thawed. These steps were repeated twice. When the flask assembly was used, the stopcock also was evacuated of air. Finally, the flask was closed by the stopcock or the tube was sealed by use of a Bunsen burner. Based on the work of Hackler (1971) and others, 22 hr hydrolysis time at an oven temperature of 110°C was selected. For tryptophan analyses, the samples were hydrolyzed with 4.2 N NaOH by the method of Hugli and Moore (1972) in a stopcock style test-tube assembly or a drawn test tube. A polypropylene tube was inserted inside the tube to hold the protein sample and other hydrolysate components. Otherwise the samples were treated the same as for the hydrochloric acid hydrolyses for evacuation of air, hydrolysis time and temperature (Table 12.2). Collaborators were requested to run duplicate hydrolyses on each protein source.

TABLE 12.2.   EXPERIMENTAL PROTOCOL FOR AMINO ACID ANALYSIS COLLABORATIVE STUDY

Protein Sources

Pretest Samples:	Tuna
	Peanut flour
Test Samples:	ANRC casein
	Freeze-dried defatted beef (semitendinosus muscle)
	Non fat dried milk
	Wheat flour (commercial bakers' bread flour)
	Textured soy protein

Hydrolysis (in duplicate; under vacuum):

6 N HCl[1]	17 Common amino acids,
30–40 mg protein (N × 6.25)	Hydroxylysine,
10 ml 6 N HCl	Hydroxyproline,
Performic acid pretreatment (Moore, 1963) followed by 6 N HCl[1]	Methionine as methionine sulfone
	Cysteine as cysteic acid
4.2 N NaOH (Hugli and Moore, 1972)	Tryptophan
8–10 mg protein (N × 6.25)	
1.0 ml 5 N NaOH	
1.0 ml isopropyl alcohol	
Partially hydrolyzed potato starch (25 mg; for casein, beef, milk, and tuna only)	

Time: 22 hr

Temperature: 110°C

Analyzer: Ion-exchange chromatography—9 collaborators (each by use of own system)
Gas chromatography—special collaborator

Buffers: Na$^+$ citrate
or
Li$^+$ citrate

Amino acid hydrolyzates analysis: analyzed in duplicate

Calibration standards:
For 17 common amino acids and ammonia (collaborator's own standard and a common standard)

For cysteic acid and methionine sulfone (collaborator's own standard and a common standard)

Reporting of data: as grams amino acid/16g N (each collaborator's standard use as a reference)

[1]Hydrolyzed by same conditions.

Amino Acid Analysis

Each collaborator used the analyzer and buffer system normally used in his/her laboratory, a recognized variable we could not change. We requested collaborators to run duplicate amino acid analyses on each hydrolyzate so that an estimate could be made of the within instrument variability of each analyzer. The data were calculated as grams amino acid per 16g of nitrogen and returned to us. The choice of 16g nitrogen is not meant to imply that this is equivalent to 100g of protein in the case of each protein source.

Resolution of Methionine Sulfone

We found that the current collaborators had a preference for using lithium acetate rather than sodium acetate buffers to resolve methionine sulfone in the performic acid oxidized 6 N HCl hydrolyzates. The difficulty with the sodium system lies in its incomplete resolution of methionine sulfone, aspartic acid, and threonine. The chromatogram obtained by a 24-hr separation procedure, Figure 12.1, was taken from Moore *et al.* (1958). The elution times for aspartic acid, methionine sulfone, and threonine were well separated with methionine sulfoxide eluted prior to aspartic acid. However, in a 90-min analysis procedure by Greenberg at ERRC

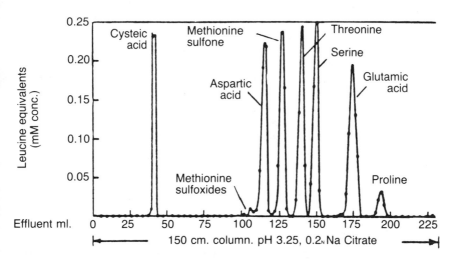

FIG. 12.1. CHROMATOGRAPHIC FRACTIONATION OF MIXTURE OF STANDARD AMINO ACIDS ON COLUMN OF AMBERLITE IR-120. REPRINTED WITH PERMISSION FROM: ANALYTICAL CHEMISTRY 30(7): 1185, 1958, MOORE, S., SPACKMAN, D.H., AND STEIN, W.H. CHROMATOGRAPHY OF AMINO ACIDS ON SULFONATED POLYSTYRENE RESINS. COPYRIGHT BY THE AMERICAN CHEMICAL SOCIETY

(Figure 12.2), aspartic acid and threonine eluted only 2.5 min apart, making it difficult to place methionine sulfone well resolved between the two. Talley, Plant Products Laboratory, ERRC, regularly and successfully analyzes for methionine sulfone and methionine sulfoxide using a sodium citrate buffer at pH 3.25 in a 3¾-hr procedure. The sulfoxide is eluted about 2 min before the aspartic acid. He keeps his conditions precisely the same from sample to sample to obtain good separations consistently.

At the ERRC Meat Laboratory, we regularly use a sodium citrate buffer at pH 2.9 to elute hydroxyproline at 38–39 min, which is about 6 min before aspartic acid emerges. The column is 0.9 cm in diameter and contains 31.5 cm of a cation exchange resin. With this buffer, methionine sulfone coelutes with aspartic acid. Although it is possible to elute methionine sulfone from the column before aspartic acid, it must precede methionine sulfoxide and not interfere with hydroxyproline. Each laboratory had to work out this separation problem. A lithium acetate buffer system was recommended if a laboratory had an extra column available. The run could be aborted as soon as the amino acids required were eluted.

Normalization of Data

ANRC casein, one of the proteins in this test, was selected as the standard protein for this study. Upon completion of the study, amino acid analysis data will be normalized against a "best" analysis of casein and against total Kjeldahl nitrogen recovery. Determined ammonia will be included in the calculation of total nitrogen recovery. Inter- and intralaboratory variation will be determined before and after casein or nitrogen normalization.

Calibration Standards

As suggested by Yates, Campbell Soup Company, a calibration standard for the 17 common amino acids and ammonia and a calibration standard for cysteic acid and methionine sulfone were sent to all collaborators. The analysis of a common standard will allow for interpretation of the data upon completion of the study. Each standard was to be analyzed against the appropriate calibration standard solution each collaborator used for analysis of the test samples. These data will allow us to determine whether the standard was a source of variation in the analysis and will allow for possible adjustment of the data obtained by a common standard.

Statistical Analyses

Standard deviations and coefficients of variation were determined for the lysine and valine data obtained from the two pretest samples (tuna fish and peanut flour) and from the five test samples (ANRC casein, textured soy protein, wheat flour, defatted lyophilized beef, and non fat dried milk powder), with and without identified outliers. Outliers were identified by

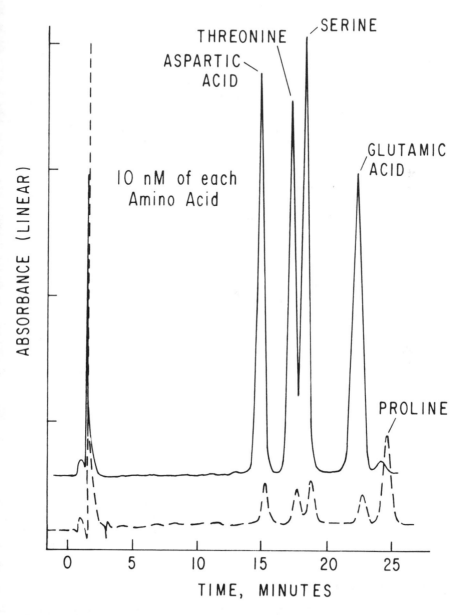

FIG. 12.2. PARTIAL CHROMATOGRAM OF AMINO ACID CALIBRATION STANDARD FROM 90-MIN AMINO ACID ANALYSIS. (R., GREENBERG, EASTERN REGIONAL RESEARCH CENTER, USDA).

the method of Anscombe and Tukey (1963). An analysis of variance was done on the data for all amino acids from the two pretest samples and for lysine from the five test samples.

Results from the Pretest Samples

The collaborators are enthusiastic and are making an effort to return the data promptly. During this study we have encountered the usual problems of amino acid analysis. Several of the collaborators who do not commonly use the performic acid-6 N HCl procedure for methionine and the 4.2 N NaOH hydrolysis procedure for tryptophan spent time becoming familiar with the method, the primary reason for sending out two pretest samples. The return of the data from the pretest samples took longer than anticipated. To date, we have received pretest data from eight laboratories and the special collaborator. Only three laboratories have submitted results for methionine sulfone and three for tryptophan. Six laboratories reported the analysis of cystine as cysteic acid (Table 12.3). At least two of those who had analyzed for methionine sulfone used lithium acetate buffers and the third used sodium acetate buffers.

TABLE 12.3.   METHIONINE, CYSTINE, AND TRYPTOPHAN DATA FOR PRETEST SAMPLES[1]

	Grams amino acid/16g nitrogen						
	Laboratories						
Amino acid	2	3	4	5	6	8	S.D.
	Tuna						
Methionine	2.40	—	2.84	—	—	—	—
½ Cystine	0.91	1.07	0.92	0.69	0.49	1.05	0.22
Tryptophan	—	—	1.35	1.07	0.89	—	0.23
	Peanut flour						
Methionine	1.15	—	1.32	—	—	1.09	0.12
½ Cystine	1.14	1.64	1.28	1.01	1.74	1.58	0.30
Tryptophan	—	—	0.58	0.72	0.48	—	0.12

[1]The data have not been normalized (see text).

In the determination of lysine in the two pretest samples, tuna fish and peanut flour (Table 12.4), one laboratory (Laboratory 5) had very high values for tuna fish and another (Laboratory 3) had low values for peanut flour. Statistical data (Table 12.5) show that the interlaboratory standard deviations (S.D.) for reproducibility of the lysine determinations were 0.95 from tuna and 0.37 from peanut flour. The coefficient of variation (C.V.) for each source is similar: 11.3% for tuna and 11.1% for peanut flour. Interlaboratory precision (reproducibility) for the determination of lysine in the pretest is lower than desirable. However, when the high lysine value for tuna (Laboratory 5) is treated as an outlier, the C.V. is lowered to 6.3% (Table 12.5). Likewise, removing the low value for lysine in peanut flour (Laboratory 3) lowers the C.V. to 7.6%. Both the very high and the very low values for lysine are outliers by the method of Anscombe and Tukey (1963).

TABLE 12.4.   LYSINE DATA FOR PRETEST SAMPLES[1]

	Hydrolyzate[2]			
	Grams amino acid/16g nitrogen			
	Tuna		Peanut flour	
Laboratory	1	2	1	2
1	8.52	8.55	3.18	3.21
2	7.43	7.79	3.74	3.61
3	8.25	8.87	2.56	2.66
4	8.55	8.35	3.20	3.15
5	10.48[3]	10.43[3]	3.87[3]	3.78[3]
6	7.39	7.15	3.49	3.32
7	7.85	7.78	3.41	3.59
8	8.2	8.1	3.15	3.1

[1]The data have not been normalized (see text).
[2]Each value is the average of two analyses with the exception of those from Laboratory 5.
[3]Single analyses.

The intralaboratory precision (repeatability) of lysine determinations for tuna fish and peanut flour is high as indicated by the values shown in Table 12.5.

The interlaboratory precision of the valine determination for tuna fish and peanut flour is affected by the very low values obtained by one laboratory (Table 12.6). The C.V.'s for these protein sources were 20.8% and 19.0%, respectively (Table 12.7). Treating the very low results from Laboratory 6 as outliers (outliers by the method of Anscombe and Tukey, 1963) lowers the C.V.'s to 13.5 and 14.3%, respectively. The intralaboratory precision (S.D.) for the valine determinations is high, however, the relative precision (C.V.) is lower than that for lysine (Tables 12.5 and 12.7).

The analysis of variance for the pretest samples (two protein sources, duplicate 6 N HCl hydrolyses, and duplicate analyses of each hydrolyzate, no outliers) indicates that overall interlaboratory (reproducibility) S.D.'s for lysine and for valine (Table 12.8) were different from those of either tuna fish or peanut flour individually (Tables 12.5 and 12.7). In general the interlaboratory variation for all amino acids was high. Intralaboratory

TABLE 12.5.   INTER- AND INTRALABORATORY VARIATION FOR LYSINE DETERMINATION OF PRETEST SAMPLES[1]

	Tuna		Peanut flour	
	8 Laboratories	Outlier removed	8 Laboratories	Outlier removed
Range, g AA[2]/16g N	7.15−10.48	7.15−8.87	2.56−3.87	3.1−3.87
Mean, g AA/16g N	8.36	8.05	3.31	3.44
S.D., g AA/16g N, (inter-)	0.95	0.51	0.37	0.26
C.V., % (inter-)	11.3	6.3	11.1	7.6
S.D., g AA/16g N, (intra-)	0.21	0.22	0.08	0.08
C.V., % (intra-)	2.5	2.8	2.4	2.3

[1]The data have not been normalized (see text).
[2]AA = amino acid.

TABLE 12.6.   VALINE DATA FOR PRETEST SAMPLES[1]

	Hydrolyzate[2]			
	Grams amino acid/16g nitrogen			
	Tuna		Peanut flour	
Laboratory	1	2	1	2
1	5.18[3]	5.14	3.49[3]	3.54
2	4.52	4.53	3.41	3.31
3	3.45	3.92	2.53	2.47
4	5.15	5.15	3.35	3.10
5	5.56[3]	6.06[3]	3.58	4.00
6	2.84	2.60	2.21	2.19
7	4.79	4.87	4.12	4.02
8	5.35	5.30	3.75	3.55

[1]The data have not been normalized (see text).
[2]Each value the average of two analyses.
[3]Single analyses.

TABLE 12.7.   INTER- AND INTRALABORATORY VARIATION FOR VALINE DETERMINATIONS OF PRETEST SAMPLES[1]

	Tuna		Peanut flour	
	8 Laboratories	Outlier removed	8 Laboratories	Outlier removed
Range, g AA[2]/16g N	2.60−6.06	3.45−6.06	2.19−4.12	2.47−4.12
Mean, g AA/16g N	4.65	4.93	3.29	3.44
S.D., g AA/16g N, (inter-)	0.97	0.66	0.63	0.49
C.V., % (inter-)	20.8	13.5	19.0	14.3
S.D., g AA/16g N (intra-)	0.18	0.17	0.15	0.16
C.V., % (intra-)	3.9	3.4	4.6	4.6

[1]The data have not been normalized (see text).
[2]AA = amino acid.

variation (repeatability) was also determined for each amino acid during the overall analysis of variance (Table 12.8). The C.V.'s for arginine, histidine, proline, and isoleucine ranged from 4.3 to 11.9%. Cystine had an extremely high C.V. of 31.6%. The C.V.'s of other amino acids were 3.8% or below. The analysis of variance showed that the laboratory-hydrolysis interaction was significant ($p < 0.05$) for several amino acids (i.e., aspartic acid, leucine, tyrosine, lysine, glutamic acid, alanine, and valine). However, only one, glutamic acid, showed a serious difference (~7%) between duplicate hydrolyses at a particular laboratory, while the other amino acids showed significant laboratory-by-hydrolysis interactions because of a difference in levels between laboratories. Laboratory-by-protein source interactions were significant ($p < 0.01$) for all amino acids except proline and were not significant for ammonia. These are due largely to the different amounts of amino acids found in the two samples and the variance of differences between laboratories.

TABLE 12.8.   STANDARD DEVIATIONS (S.D.) AND COEFFICIENTS OF VARIATION (C.V.) FOR EACH AMINO ACID FROM DATA FOR TWO PRETEST SAMPLES COMBINED[1]

Amino acid	6 N HCl hydrolysis			
	Interlaboratory		Intralaboratory	
	S.D. g AA[2]/16g N	C.V.[3] %	S.D. g AA[2]/16g N	C.V.[3] %
Aspartic acid	1.19	12.4	0.24	2.5
Threonine	0.50	15.1	0.10	3.1
Serine	0.33	8.2	0.13	3.3
Glycine	0.50	10.4	0.18	3.6
Proline	0.44[4]	12.8	0.37[4]	10.7
Glutamic	1.26	8.6	0.26	1.7
Alanine	0.47	10.6	0.12	2.7
Valine	0.69	17.6	0.11	2.7
Isoleucine	0.79	22.8	0.41	11.9
Leucine	0.62	9.8	0.12	2.0
Tyrosine	0.38	10.8	0.14	3.8
Phenylalanine	0.35	8.2	0.15	3.4
Lysine	0.88	15.4	0.14	2.5
Histidine	0.53	6.6	0.29	8.4
Arginine	0.85	10.5	0.35	4.3
Ammonia	0.62[5]	52.3	0.47[5]	39.1
Cystine	0.46[5]	77.5	0.19[5]	31.6
Methionine	0.31[6]	15.3	0.07[6]	3.7

[1]From analysis of variance for 8 laboratories unless noted; no values (outliers) were excluded.
[2]AA = amino acid.
[3]C.V. (coefficient of variation) is standard deviation ÷ the mean value for each amino acid.
[4]Analysis of variance for six laboratories.
[5]Analysis of variance for five laboratories.
[6]Analysis of variance for four laboratories.

Preliminary Results from Five Test Samples

To date we have received amino acid analysis data from four collaborators on the five test samples (ANRC casein, textured soy protein, wheat flour, freeze-dried defatted beef, and non fat dried milk powder) hydrolyzed with 6 N HCl. Table 12.9 presents the statistical data for lysine. The highest interlaboratory variability (reproducibility) was observed for casein and defatted beef, principally due to the low lysine values obtained from one laboratory (Laboratory 4) for both protein sources (Fig. 12.3). The variability for the other three protein sources is under 5%. The statistical analyses were done on the data without any adjustments or normalization (nitrogen recovered or standard protein).

Intralaboratory variability (repeatability) shows moderate to high precision for each protein source (Table 12.9). The coefficient of variation of these five samples varied from 1.8 to 4.9%. All are acceptable values for high precision except the 4.9% value for the textured soy protein.

An analysis of variance was performed to determine the effect of laboratory, protein source, and hydrolysis, and their interactions on the determination of lysine. Significant interactions were found between laboratory and protein source and between protein source and hydrolysis. Laboratory-by-protein source interactions were apparent (Figure 3). The significant

TABLE 12.9. INTER- AND INTRALABORATORY VARIATION FOR LYSINE DETERMINATIONS OF FIVE TEST PROTEIN SOURCES[1,2]

Summary	Casein	Textured soy protein	Wheat flour	Defatted beef	Non fat dried milk powder
Range, g AA[3]/16g N	6.35–8.40	5.34–6.18	1.62–1.91	6.95–8.53	6.81–7.84
Mean, g AA/16g N	7.75	5.88	1.82	7.78	7.38
S.D. g AA/16g N (inter-)	0.65	0.25	0.09	0.55	0.36
C.V., % (inter-)	8.4	4.2	4.9	7.0	4.9
S.D. g AA/16g N (intra-)	0.23	0.29	0.07	0.27	0.13
C.V., % (intra-)	3.0	4.9	3.8	3.5	1.8

[1]The data have not been normalized (see text).
[2]Data from four laboratories.
[3]AA = amino acid.

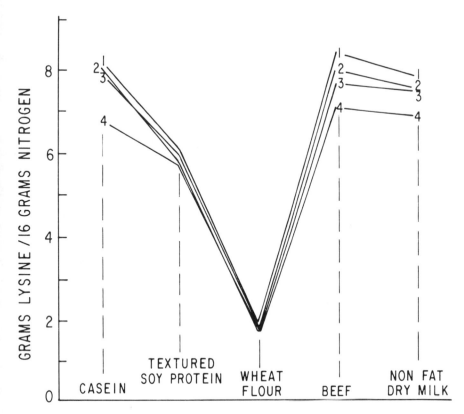

FIG. 12.3. LABORATORY-BY-PROTEIN SOURCE INTERACTIONS FOR LYSINE; AV-
ERAGE VALUES OVER DUPLICATE HYDROLYSES AND DUPLICATE DETERMINA-
TIONS OF EACH HYDROLYSIS (4 VALUES/POINT) FOR FIVE PROTEIN SOURCES.
NUMBERED LINES REPRESENT THE FOUR LABORATORIES

protein source-by-hydrolysis interaction was due to the relatively large
mean difference for lysine between hydrolyses for casein (Fig. 12.4).

The interlaboratory standard deviation for lysine over the five protein
sources is 0.66, with a C.V. of 10.6%. The intralaboratory S.D. over the five
protein sources was 0.13, and the C.V. was 2.1.

Inter- and intralaboratory variation data for valine (Table 12.10) is high
for all five protein sources. There was one laboratory which had consistently
low results on four of the sources. When those results were treated as
outliers (outliers by the method of Anscombe and Tukey, 1963), the inter-
laboratory variation was lowered by about 50%. In all cases, the intralaboratory variation was low, indicating high precision. The precision among
these four laboratories is higher for the test samples than it was among the
eight laboratories for the pretest samples.

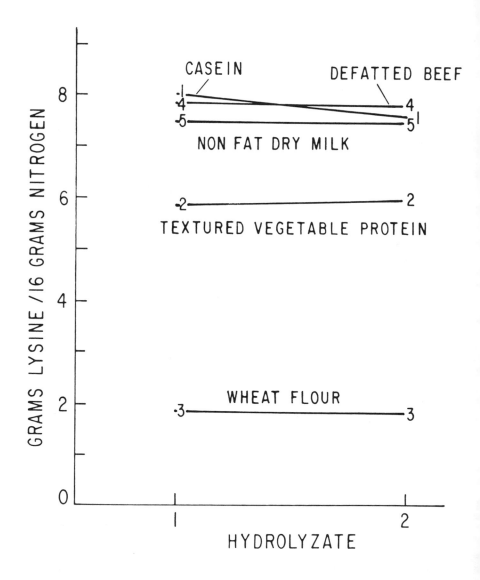

FIG. 12.4.   PROTEIN SOURCE-BY-HYDROLYSIS INTERACTIONS FOR LYSINE; AV-
ERAGED OVER DUPLICATE HYDROLYSES AND FOUR LABORATORIES (8 VALUES/
POINT) FOR FIVE PROTEIN SOURCES

TABLE 12.10. INTER- AND INTRALABORATORY VARIATION FOR VALINE DETERMINATIONS OF FIVE TEST PROTEIN SOURCES[1,2]

	Casein		Textured soy protein		Wheat flour	Defatted beef		Non fat dried milk powder	
	4 Labs	Outlier removed	4 Labs	Outlier removed	4 Labs	4 Labs	Outlier removed	4 Labs	Outlier removed
Range, g AA[3]/16g N	4.26– 6.80	5.20– 6.80	3.00– 4.33	3.68– 4.33	2.90– 3.60	3.37– 4.70	3.80– 4.70	4.28– 5.80	5.06– 5.80
Mean, g AA/16g N	5.75	6.13	3.89	4.06	3.28	4.18	4.42	5.16	5.50
S.D., g AA/16g N (inter-)	0.82	0.47	0.40	0.21	0.28	0.50	0.30	0.61	0.25
C.V., % (inter-)	14.42	7.66	10.21	5.19	8.54	11.92	6.71	11.84	4.52
S.D., g AA/16g N (intra-)	0.13	0.15	0.08	0.09	0.04	0.08	0.09	.09	0.11
C.V., % (intra-)	2.26	2.45	2.06	2.22	1.22	1.91	2.04	1.74	2.00

[1]Data have not been normalized (see text).
[2]Data from four laboratories, except where outlier from one laboratory was removed. Outlier data are from three laboratories.
[3]AA = amino acid.

SUMMARY

A new collaborative study on amino acid analysis was initiated with three objectives: to standardize the preparation of protein food sources for amino acid analysis, including the hydrolysis of proteins; to compare two different approaches for normalizing amino acid analysis data; and to determine the inter- and intralaboratory variation for the analysis of individual amino acids in selected protein food sources. Nine laboratories participated in the study. Seven protein sources, two as pretest samples (peanut flour and lyophilized tuna fish) and five as test samples (ANRC casein, textured soy protein, bakers' commercial wheat flour for bread, lyophilized defatted beef, and non fat dried milk powder), were prepared and analyzed. Three procedures were used for hydrolysis of the protein: 6 N hydrochloric acid for analysis of 17 common amino acids, hydroxylysine and hydroxyproline; performic acid pretreatment followed by 6 N hydrochloric acid for analysis of cystine and methionine as cysteic acid and methionine sulfone, respectively; 4.2 N sodium hydroxide for the analysis of tryptophan by the method of Hugli and Moore. Preliminary data on the two pretest samples from eight laboratories indicate the interlaboratory variation for lysine to be higher than desirable, with coefficients of variation (C.V.) of 11.3% for tuna and 11.1% for peanut flour; however, these are 6.3% and 7.6%, respectively, when one outlier is removed from the data for each of the pretest samples. The intralaboratory C.V.'s for lysine were 2.5% for tuna and 2.4% for peanut flour (8 laboratories) and 2.8% and 2.3%, respectively, with the outlier values removed. Preliminary statistical analyses of data for five test samples from four laboratories were done. These data were not normalized to nitrogen recovery or a standard amino acid profile for the reference protein casein. This will be done after data from all collaborators are received. Interlaboratory standard deviations (S.D.), expressed as grams amino acid per 16g nitrogen, for lysine were: casein, 0.65; textured soy protein, 0.25; bakers' commercial bread wheat flour, 0.09; lyophilized, defatted beef, 0.55; non fat dried milk powder, 0.36; C.V.'s were 8.4%, 4.2%, 4.9%, 7.0%, and 4.9%, respectively. The interlaboratory precision on some protein sources, therefore, was lower than desired. The intralaboratory precision was generally high with C.V.'s for lysine of: 3.0%, 4.9%, 3.8%, 3.5%, and 1.8%, respectively, for the five test samples above. The interlaboratory standard deviation for lysine determined by an analysis of variance over the five protein sources and the four laboratories was 0.66 with a C.V. of 10.6%, and the intralaboratory S.D. was 0.13 with a C.V. of 2.1%. Collaborative studies on amino acid analysis in the literature are reviewed briefly.

REFERENCES

ALSMEYER, R.H., CUNNINGHAM, A.E., and HAPPICH, M.L.   1974. Equations predict PER from amino acid analysis.   Food Tech. 28(7):34.

AOAC.   1980.   "Official Methods of Analysis," 13th ed.   Association of Official Analytical Chemists, Washington, D.C.

ANSCOMBE, F.J. and TUKEY, J.W.   1963.   The examination and analysis of residuals.   Technometrics 5:141.

BLACKBURN, S.   1978.   Amino acid analysis: An important technique.   In "Amino Acid Determination," 9th ed. by S. Blackburn, p. 1.   Marcel Dekker, Inc., New York.

BLOCK, R.J. and WEISS, K.W.   1956.   (a) Preparation of the Sample for Analysis, p. 14, (b) Introduction, p. 6.   In "Amino Acid Handbook," Charles C. Thomas, Springfield, Illinois.

BODWELL, C.E., WOMACK, M., SCHUSTER, E.M., and BROOKS, B.   1978.   I. Biochemical indices in humans of protein nutritive value. II. Postprandial plasma urea nitrogen.   Nutritional Reports Inter. 18(5):579.

BRESSANI, R.   1977.   Protein supplementation and complementation.   In "Evaluation of Proteins for Humans" ed. By C.E. Bodwell, p. 204.   The AVI Publishing Company, Inc., Westport, Connecticut.

CAVINS, J.F., KWOLEK, W.F., INGLETT, G.E., and COWAN, J.C.   1972.   Amino acid analysis of soybean meal: Interlaboratory study.   JAOAC 55(4): 686.

CHERRY, J.P.   1978 and 1979.   Progress report on topic, amino acids. Unpublished reports to AOAC.

DERSE, P.H.   1969.   Amino acid analysis employing microbiological assays and automated analyzers.   Feedstuffs Feb. 8, 42.

DREZE, A.   1960.   Dosage of tryptoptophan in natural medium. II-Stability of tryptophan during alkaline hydrolysis carried out in presence of carbohydrates.   Bull. Soc. Chim. Biol. 42(4):407.

FRIEDMAN, M. and FINLEY, J.W.   1971.   Methods of tryptophan analysis.   J. Ag. Food Chem. 19:626.

FRIEDMAN, M. and FINLEY, J.W.   1975.   Evaluation of methods for tryptophan analysis in proteins.   In "Protein Nutritional Quality of Foods and Feeds." Part 1 ed. by Mendel Friedman, p. 423.   Marcel Dekker, Inc., New York.

FRIEDMAN, M. and NOMA, A.T.   1975.   Methods and problems in chromatographic analysis of sulfur amino acids.   In "Protein Nutritional Quality of Foods and Feeds." Part 1 ed. by Mendel Friedman, p. 521.   Marcel Dekker, Inc., New York.

HACKLER, L.R.   1971.   Methods of analysis for the sulfur amino acids.   Feedstuffs, Feb. 27, 18.

HACKLER, L.R. et al.   Unpublished data.

HACKLER, L.R.   1974.   Amino acid analysis of foods with emphasis on cystine and methionine.   Presented at the 34th Annual IFT Meeting, June, New Orleans.

HAPPICH, M.L.   1980.   Unpublished data.

HAPPICH, M.L., SWIFT, C.E., and NAGHSKI, J.   1975.   Equations for predicting PER from amino acid analysis—a review and current scope of application.   In "Protein Nutritional Quality of Foods and Feeds," Part I, ed. by Mendel Friedman, p. 125.   Marcel Dekker, Inc., New York.

HUGLI, T.E. and MOORE, S. 1972. Determination of the tryptophan content of proteins by ion exchange chromatography of alkaline hydrolysates. J. Biol. Chem. 247(9):2828.

KNIPFEL, J.E., AITKEN, J.R., HILL, D.C., McDONALD, B.E., and OWENS, B.D. 1971. Amino acid composition of food proteins: inter- and intralaboratory variation. JAOAC 54(4):777.

KWOLEK, W.F. and CAVINS, J.F. 1971. Relative standard deviations in determinations of amino acids. JAOAC 54(6):1283.

LEE, Y.B., ELLIOTT, J.G., RICKANSRUD, D.A., and HAGBERG, E.C. 1978. Predicting protein efficiency ratio by the chemical determination of connective tissue content in meat. J. Food Sci. 43:1359.

MOORE, S., SPACKMAN, D.H., and STEIN, W.H. 1958. Chromatography of amino acids on sulfonated polystyrene resins. Anal. Chem. 30(7):1185.

MOORE, S. 1963. On the determination of cystine as cysteic acid. J. Biol. Chem. 238(1):235.

MILLER, E.L. 1967. Determination of the tryptophan content of feedingstuffs with particular reference to cereals. J. Sci. Fd. Agric. 18:381.

NEUMANN, N.P., MOORE, S., and STEIN, W.H. 1962. Modification of the methionine residues in ribonuclease. Biochemistry 1(1):68.

OELSHLEGEL, F.J., Jr., SCHROEDER, J.R., and STAHMANN, M.A. 1970. A simple procedure for basic hydrolysis of proteins and rapid determination of tryptophan using a starch column. Anal. Biochem. 34(2):331.

PORTER, J.W.G., WESTGARTH, D.R., and WILLIAMS, A.P. 1968. A collaborative test of ion exchange chromatographic methods for determining amino acids. Br. J. Nutr. 22:437.

ROSE, W.C., and WIXOM, R.L. 1955. The amino acid requirements of man. XIV. The sparing effect of tyrosine on the phenylalanine requirement. J. Biol. Chem. 217:95.

SATTERLEE, L.D., MARSHALL, H.F., and TENNYSON, J.M. 1979. Measuring protein quality. JAOCS 56:103.

SHRAM, E., MOORE, S., and BIGWOOD, E.J. 1954. Chromatographic determination of cystine as cysteic acid. Biochem. J. 57(1):33.

SPENCER, R.L. and WOLD, F. 1969. A new convenient method for estimation of total cystine-cysteine in proteins. Anal. Biochem. 32:185.

SPIES, J.R. and CHAMBERS, D.C. 1949. Chemical determination of tryptophan in proteins. Anal. Chem. 21:1249.

WALKER, H.G., Jr., KOHLER, G.O., KUZMICKY, D.D., and WITTS, S.C. 1975. Problems in analysis for sulfur amino acids in feeds and foods. In "Protein Nutritional Quality of Foods and Feeds," Part 1, ed. by Mendel Friedman. p. 549. Marcel Dekker, Inc., New York.

WESTGARTH, D.R. and WILLIAMS, A.P. 1974. A collaborative study on the determination of tryptophan in feedingstuffs. J. Sci. Fd. Agric. 25:571.

WILLIAMS, A.P., HEWITT, D., and COCKBURN, J.E. 1979. A collaborative study on the determination of cyst(e)ine in feedingstuffs. J. Sci. Fd. Agric. 30:469.

DISCUSSION

DR. PELLETT: A brief question on detail. In this procedure were internal standards permitted, such as norleucine and amino β guanidino propionic acid, and if so, were they used regularly to test for ninhydrin deterioration?

MS. HAPPICH: No, we did not suggest or recommend the use of internal standards.

DR. PELLETT: Was there a special reason for that? Did you expect that the ninhydrin would stay constant or was it agreed that comparisons would be run using new reagents? In ordinary analysis, one has to take care of ninhydrin deterioration and internal standards seem to be the only way of doing this.

MS. HAPPICH: We expected that the ninhydrin would be freshly prepared and would be used soon there after. We did not expect the collaborators to use ninhydrin which was old or was deteriorated.

DR. PELLETT: I certainly think it needs to be looked at because ninhydrin doesn't only deteriorate, it often increases in its potency when freshly made, than it plateaus for a bit, and finally decreases rapidly in potency. Thus, time factors for age of ninhydrin in amino acid analysis can be important considerations.

MS. HAPPICH: They are important considerations. A calibration standard should be run, of course, each time analyses are run.

DR. STAHMANN: I want to commend Mrs. Happich and Dr. Bodwell on a very careful study which demonstrates that amino acid analysis by chromatography can be very precise and accurate. I also want to comment a bit about the tryptophan analysis. The tryptophan analysis that Moore and his group developed was based on earlier studies in which we showed you could do basic hydrolysis in cheap plastic centrifuge tubes enclosed in a glass envelope to exclude air. We also showed that tryptophan could be separated by two methods. One was to change the buffer used for elution so the tryptophan follows all other acids but this required mixing up a new buffer. A simpler way that may be somewhat easier was use of a starch column as we did; then you can use the same buffers that are used in regular amino acid analysis. Tryptophan then follows well behind all other amino acids. This does not require making up a new buffer.

As to amino acid analysis, and the question of using internal standards, this may depend upon how many standards you run. If you run a standard before and after every analysis, then you may not need an internal standard. It also may depend on how carefully the ninhydrin solution was made and particularly upon how much peroxide is present in the methyl cellosolve in which the ninhydrin is dissolved. Good methyl cellosolve from a drum which was filled directly from the line in which it was produced gives fairly stable solutions. Old methyl cellosolve with peroxides in it give very unstable ninhydrin solutions. Thus, the need for internal standards may depend upon how the ninhydrin solution is made up.

MS. HAPPICH: That's true. Thank you, Dr. Stahmann. We in our laboratory use DMSO (dimethylsulfoxide), not methyl cellosolve, to make up the ninhydrin solution. I think that among our collaborators there may be some variation. They're probably not all using methyl cellosolve but some of them may be.

DR. SATTERLEE: One of the things that occurs during the tryptophan analysis is that lysinoalanine is formed during sodium hydroxide hydrolysis. The various collaborators should be very careful to separate it out from underneath the tryptophan peak, or else it can cause some very high tryptophan values. The starch methodology that Dr. Stahmann refers to does not have the problem of lysinoalanine interference.

MS. HAPPICH: Yes, the ion exchange method does have that problem. We did send lysinoalanine to those collaborators who requested it. They were then able to make sure that it was separated on their particular column. We run this on a short column as a very short procedure with the ion-exchange chromatographic analyzer and we do know that lysinoalanine is separated from tryptophan. Tryptophan comes off first and lysinoalanine follows, well separated. We gave the collaborators the benefit of our experience by telling them just how we separated tryptophan and lysinoalanine.

13

Health Protection Branch Organized Collaborative Study on Amino Acid Analysis

G. Sarwar, D.A. Christensen, A.J. Finlayson, M. Friedman, L.R. Hackler, P.L. Pellett and R. Tkachuk

Inter- and intra-laboratory variation in amino acid composition of seven protein sources (casein, soy assay protein, pea flour, whole wheat flour, egg white solids, minced beef and rapeseed protein concentrate) was studied with seven laboratories collaborating. The samples were subjected to three separate hydrolyses (6N HCl, performic acid + 6N HCl, and 4.2N NaOH or treatment with p-dimethylaminobenzaldehyde) in each laboratory. All amino acids (except colorimetric determination of tryptophan in one laboratory) were then determined by ion-exchange chromatography using automatic amino acid analysers. For most of the protein sources studied, the inter-laboratory determination of isoleucine, leucine, lysine, phenylalanine, threonine and valine (coefficients of variation, CV, <10%) was less variable than that of tryptophan (CV of 14−20%), cystine and methionine (CV of 10−17%). Between laboratories CV values for tyrosine in four samples (casein, soy protein, egg white and minced beef) were less than 10% but were 13, 14 and 16% in rapeseed protein concentrate, wheat flour and pea flour, respectively. In general, variation between duplicate hydrolysates within laboratories was smaller than the variation between laboratories. For most amino acids, casein (CV <5%), soy protein (CV <6%) and minced beef (CV <7%) exhibited lower within laboratory variation than pea flour and rapeseed protein concentrate (CV of not more than 13%) while wheat flour and egg white (CV <10) occupied an intermediate position. Influence of the variation in amino acid analysis on the accuracy of the *in vitro* protein quality methods was estimated by calculating amino acid scores.

INTRODUCTION

In recent years there has been considerable interest in developing a rapid and suitable method for monitoring protein quality of foods for humans. Among the many chemical and biological methods described for this purpose in the literature, the computed protein efficiency ratio (C-PER) method of Satterlee et al. (1979), which is based on essential amino acid content plus in vitro protein digestibility, shows some promise. However, a major problem associated with this and other in vitro methods, which utilize amino acid profiles is that of obtaining reliable amino acid composition data. Variation in the analytical estimates of amino acid composition of various protein sources may result in very different, or even unreliable, protein quality estimates. Therefore, determination of variation in amino acid analysis becomes important in evaluating acceptability of the in vitro indices for estimating protein quality.

A review of the recent literature indicates that most amino acids in proteins can be satisfactorily determined by the use of an automatic amino acid analyser after properly controlled acid hydrolysis of the protein (Blackburn, 1978). Unfortunately, certain amino acids such as cyst(e)ine, tryptophan and methionine (to some extent) are destroyed during hydrolysis, necessitating separate hydrolyses for these amino acids. Cyst(e)ine and methionine are more accurately determined after oxidation by performic acid to their stable forms (cysteic acid and methionine sulfone, respectively) by the method of Moore (1963). Alkaline hydrolysis is the most feasible way of quantitative determination of tryptophan in proteins when using an automatic analyser (Blackburn, 1978). In this respect, Hugli and Moore (1972) developed a simple procedure (in which proteins are hydrolyzed with NaOH containing starch as an antioxidant) which permits an accurate determination of tryptophan.

Although suitable methods for determining all amino acids have been developed, information on inter-laboratory comparison of these methods is both limited and incomplete. Knipfel et al. (1971), in determining amino acid compositions of casein, soybean and fish flour, reported mean coefficients of variation (CV) between laboratories of up to 7%. However, much higher CV values have been reported for cystine (up to 38%) in casein, fish meal, leaf protein concentrate, peanut meal, soybean meal and wheat (Williams et al. 1979), and for tryptophan (up to 53%) in soybean, peanut and cottonseed protein concentrates (Westgarth and Williams, 1974). The lower variability noted by Knipfel et al. (1971) might have been due to the fact that all laboratories analysed the same hydrolysates which were prepared in one laboratory. In routine work, however, laboratories would prepare their own hydrolysates. Therefore, results of Knipfel et al. (1971) do not provide a realistic estimate of the error. Moreover, cystine and tryptophan were not studied by these workers.

The Health Protection Branch considered that more information on inter-laboratory variation in the determination of the amino acid composition of protein sources was required. Therefore, this collaborative study on amino acid analysis including appropriate hydrolyses for sulfur amino acids and tryptophan was undertaken. Seven protein sources (casein, soy assay protein, pea flour, whole wheat flour, egg white solids, minced beef and rapeseed protein concentrate) were subjected to three separate hydrolyses (6N HCl, performic acid + 6N HCl, and 4.2N NaOH or treatment with p-dimethylaminobenzaldehyde) in seven laboratories. All amino acids (except colorimetric determination of tryptophan in one laboratory) were then determined by ion-exchange chromatography using an automatic amino acid analyser.

EXPERIMENTAL

Materials

All samples were prepared at the organizing laboratory (laboratory I). The casein was ANRC Reference Protein (Humko Sheffield Chemical, Kraftco Corporation, Madison, Wisconsin, U.S.A.). Soy assay protein and egg white solids were purchased from Teklad Mills, ARS (Sprague-Dawley, Division of the Mogul Corporation, Madison, Wisconsin, U.S.A.). The samples of rapeseed protein concentrate and pea flour were kindly provided by Dr. J.D. Jones, FRI, Agriculture Canada, Ottawa, and Griffith Laboratories, Toronto, respectively. Whole wheat flour (Five Roses brand) and minced beef (lean grade) were purchased from a supermarket in Ottawa. The meat sample was freeze-dried and defatted.

All the samples were ground to a fine powder (40-mesh) and mixed in a Hobart Mixer for 30 minutes before subsampling and distributing to the participating laboratories. A vial (5.5 ml) of Beckman Calibration Standard for amino acid analysers (No. 31220) was also shipped to each of the collaborating laboratories.

Each laboratory analysed the samples for dry matter and nitrogen contents, and these values were used in calculating the amino acid results. Laboratory I also analysed the samples for ether extract, ash, and carbohydrates (by difference). Moisture, nitrogen, ether extract and ash were determined by the AOAC (1975) procedures.

For each sample, three hydrolyses (6N HCl, performic acid + 6N HCl and 4.2N NaOH) in duplicate were carried out by all the collaborating laboratories except laboratory 5 where the alkaline hydrolysis was replaced by p-dimethylaminobenzaldehyde treatment of the intact proteins.

Direct hydrolysis of samples in the 6N HCl was carried out to obtain hydrolysates suitable for analysis of all amino acids except cystine + cysteine and tryptophan (in all laboratories) and methionine (in some laboratories). The 6N HCl hydrolyses were carried out by each laboratory's normal procedure (Table 13.1).

Samples previously oxidized with performic acid were also hydrolyzed with 6N HCl to obtain hydrolysates suitable for the determination of cystine + cysteine as cysteic acid (all laboratories) and of methionine as

TABLE 13.1. TECHNIQUES IN INTERLABORATORY COMPARISON OF AMINO ACID ANALYSIS

Laboratory No.	Sample weight (mg protein)	6N HCl used (ml)	Method of hydrolysis	Temperature (°C)	Time (h)	Method of removal of 6N HCl after hydrolysis	Column System(s)	Amino acid analyser
1	50	175	Sealed flask under N_2	110	22	Evaporated under N_2 at 45°C in analytical evaporator	Two	Beckman 121MB
2	30–180	80	Sealed jar under N_2	105	20	Evaporated at 45°C under vacuum	One	Technicon TSM-IR
3	90	50	Sealed flask under N_2	110	24	Evaporated at 40°C under vacuum	One	Beckman 119BL
4	5	5	Evacuated and sealed tube	110	22	Evaporated at 40°C under vacuum	One	Durrum D500
5	30–35	10	Evacuated and sealed flask	110	22	Evaporated in vacuum desiccator over NaOH	Two	Beckman Spinco 120
6	75	300	Refluxed under N_2	110	24	Evaporated in rotary evaporator	Two	Beckman 120C
7	5	2	Evacuated and sealed tube	110	22	Evaporated in vacuum desiccator over NaOH	Two	Beckman 121

methionine sulfone (laboratories 1,2,4,5 and 6) by the procedure of Moore (1963).

Tryptophan was determined after the 4.2N NaOH hydrolysis using the procedure of Hugli and Moore (1972) in all collaborating laboratories except laboratory 5, in which unhydrolyzed samples were treated with p-dimethylaminobenzaldehyde by the method of Spies and Chambers (1949).

All amino acids, except tryptophan in laboratory 5, were determined by ion-exchange chromatography using an automatic amino acid analyser (Table 13.1). Tryptophan in laboratory 5 was determined colorimetrically by using a Bausch and Lomb Spectronic 20.

Statistical Analyses

Due to time constraints, only results for ten amino acids (isoleucine, leucine, lysine, methionine, phenylalanine, threonine, tryptophan, valine cystine and tyrosine) and nitrogen recovery in each protein source were subjected to preliminary statistical analyses and are presented in this discussion. The amino acid concentrations in the seven protein sources were transformed ($T = \log_e$) to achieve normality of residuals (an in-depth verification of the success of the transformation has not yet been completed). The transformed data for the ten amino acids as well as untransformed nitrogen recovery values were subjected to analysis of variance. As a result of this analysis, it was decided that data analysis should proceed for each protein source separately. Further statistical analysis to date has been limited to the computation of means, standard errors (SE), and coefficients of variation (CV) for both transformed and untransformed data.

RESULTS AND DISCUSSION

Means, SE and CV (between laboratories) for nitrogen recovery in each protein source are given in Table 13.2. Casein, soy protein and rapeseed protein concentrate seemed to give higher nitrogen recovery values than other protein sources. Inter-laboratory variation in nitrogen recovery appeared higher for egg white and pea flour than for other samples. Part of the variation was probably due to differences in the determination of nitrogen

TABLE 13.2. MEANS, STANDARD ERRORS (SE) AND COEFFICIENTS OF VARIATION (CV) BETWEEN LABORATORIES OF RECOVERY OF NITROGEN VALUES (%) IN SEVEN PROTEIN SOURCES.

Protein Source	Mean, %	SE	CV%
Casein	99	3	2.9
Soy assay protein	98	4	3.7
Pea flour	95	5	5.8
Whole wheat flour	94	4	4.3
Egg white solids	95	8	8.4
Minced beef	94	4	4.5
Rapeseed protein concentrate	97	4	3.7

in the samples which were analysed independently in each laboratory (Table 13.3). The use of different nitrogen values for a given sample may also have influenced estimates of inter-laboratory variation of amino acids obtained in this investigation (Tables 13.4–13.6). It is customary in tests of amino acid methodology to supply collaborators with nitrogen analyses for samples. In routine work, however, laboratories would carry out their own nitrogen analyses. Therefore, this study gives a more realistic estimate of the error.

Means, SE and CV (between laboratories) for ten amino acids in each protein source (calculated from untransformed data) are given in Tables 13.4 and 13.5. Inter-laboratory variation (expressed in terms of CV) for isoleucine, leucine, lysine, phenylalanine, threonine and valine in most protein sources was less than 10%. However, the CV values for some other amino acids, such as tryptophan and cystine in all protein sources; tyrosine in pea flour, wheat flour and rapeseed protein concentrate; and methionine in all protein sources except casein and egg white, were higher than 10%, with the values for tryptophan being as high as 20%.

The between laboratories CV values for cystine in casein (17.6%), soy protein (11.4%) and wheat flour (14.2%) obtained in this study were lower than those reported for these protein sources (casein = up to 38%, soybean = up to 18%, wheat = up to 23%) by Williams et al. (1979). Similarly inter-laboratory variation for tryptophan observed in this investigation (CV values of up to 20%) was lower than noted in Trial I (CV values of up to 53%) of Westgarth and Williams (1974). However, these workers were able to reduce the inter-laboratory variation by repeating the collaborative study, and CV values for tryptophan in the second trial were below 10%. In our study, one laboratory (No.5) determined tryptophan by the method of Spies and Chambers (1949) while the other six laboratories used Hugli and Moore's (1972) method (two of these laboratories did not have experience with this method). Therefore, repeating the study for tryptophan determination by the Hugli and Moore (1972) method might reduce variation between laboratories for this amino acid (Tables 13.4 and 13.5).

According to the transformed data, inter-laboratory variation (expressed as CV values) for the ten amino acids in all seven protein sources was not more than 5% (Table 13.6). The CV values for tryptophan (3.3–5.0%), cystine (2.1–5.0%) and methionine (0.8–3.8%) were higher than the values for other amino acids which were in most cases below 2%. Our CV values (Table 13.6) compare favorably with those observed by Knipfel et al. (1971). These workers employed a similar transformation in analysing their data, and reported average CV values of 2.1–7.4% for casein, soybean flour and fish flour. This favorable comparison was quite encouraging but unexpected because in the Knipfel et al. (1971) study, all laboratories analysed the same hydrolysates which were prepared in one laboratory as compared to independently prepared hydrolysates in this investigation. The lower variation obtained in this study may be attributed to the use of improved amino acid analysers.

TABLE 13.3. PROXIMATE COMPOSITION (% DRY BASIS) OF SEVEN PROTEIN SOURCES AS DETERMINED IN LABORATORY 1 AND RANGE OF NITROGEN VALUES OBTAINED BY ALL LABORATORIES

Protein Source	Nitrogen	Protein (N × 6.25)	Ether extract	Ash	Carbohydrates (by difference)
Casein	15.04(14.62 – 15.49)[1]	94.0	0.2	0.4	5.4
Soy assay protein	15.31(14.95 – 15.86)	95.7	0.2	1.8	2.3
Pea flour	4.82 (4.69 – 5.05)	30.1	1.1	3.8	65.0
Whole wheat flour	2.60 (2.55 – 2.71)	16.2	2.1	1.5	80.2
Egg white solids	13.83(13.83 – 14.39)	86.4	0.2	5.7	7.7
Minced beef, defatted	15.24(14.81 – 15.97)	95.2	0.4	3.9	0.5
Rapeseed protein concentrate	11.06(10.46 – 11.33)	69.1	0.9	8.0	22.0

[1]Range of nitrogen values obtained by 7 collaborating laboratories.

TABLE 13.4. MEANS, STANDARD ERRORS (SE) AND COEFFICIENTS OF VARIATION (CV; BETWEEN LABORATORIES) OF TEN AMINO ACIDS IN FOUR PROTEIN SOURCES (CALCULATED FROM UNTRANSFORMED DATA)

Amino acid (mg/gN)	Casein			Soy assay protein			Pea flour			Whole wheat flour		
	Mean	SE	CV%	Mean	SE	CV%	Mean	SE	CV%	Mean	SE	CV%
Isoleucine	335	16	4.8	295	10	3.3	267	21	7.9	209	15	7.2
Leucine	635	28	4.3	532	25	4.7	479	34	7.1	428	20	4.6
Lysine	528	27	5.0	396	23	5.8	480	33	6.8	166	12	7.3
Methionine	189	8	4.2	78	9	11.3	69	11	16.1	103	13	12.2
Phenylalanine	342	24	7.1	351	24	6.9	312	31	9.8	304	26	8.6
Threonine	290	20	7.0	240	21	8.8	260	14	5.4	183	18	9.7
Tryptophan	85	12	14.5	75	15	20.6	54	11	20.3	76	11	14.2
Valine	428	19	4.4	314	20	6.5	310	31	10.1	267	22	8.2
Cystine	28	5	17.6	74	8	11.4	94	13	14.1	139	20	14.2
Tyrosine	378	26	6.9	252	11	4.4	218	36	16.4	182	26	14.5
Recovery of nitrogen, %	99	3	2.9	98	4	3.7	95	5	5.8	94	4	4.3

TABLE 13.5. MEANS, STANDARD ERRORS (SE) AND COEFFICIENTS OF VARIATION (CV; BETWEEN LABORATORIES) OF TEN AMINO ACIDS IN THREE PROTEIN SOURCES (CALCULATED FROM UNTRANSFORMED DATA)

Amino acid (mg/gN)	Egg white solids			Minced beef			Rapeseed protein concentrate		
	Mean	SE	CV%	Mean	SE	CV%	Mean	SE	CV%
Isoleucine	330	37	11.2	261	9	3.4	260	24	9.0
Leucine	547	46	8.4	484	29	5.9	490	25	5.2
Lysine	436	39	8.8	496	47	9.5	356	21	5.9
Methionine	239	20	8.4	141	15	10.5	123	16	12.8
Phenylalanine	388	36	9.3	242	21	8.6	265	25	9.5
Threonine	293	29	9.9	263	11	4.0	262	22	8.3
Tryptophan	95	19	19.8	63	11	16.9	93	17	18.6
Valine	423	40	9.5	283	16	5.5	328	22	6.7
Cystine	176	19	11.1	64	8	12.8	154	27	17.5
Tyrosine	275	21	7.7	196	14	7.0	176	24	13.5
Recovery of nitrogen, %	95	8	8.4	94	4	4.5	97	4	3.7

TABLE 13.6.   BETWEEN LABORATORIES COEFFICIENTS OF VARIATION (%) FOR TEN AMINO ACIDS IN SEVEN PROTEIN SOURCES (CALCULATED FROM TRANSFORMED DATA)

Protein Source	Casein	Soy assay Protein	Pea flour	Whole wheat flour	Egg white solids	Minced beef	Rapeseed protein concentrate
Amino acid							
Isoleucine	0.8	0.6	1.4	1.3	2.0	0.6	1.6
Leucine	0.7	0.7	1.1	0.7	1.4	0.9	0.9
Lysine	0.8	1.0	1.1	1.4	1.6	1.5	1.0
Methionine	0.8	2.8	3.8	2.6	1.5	2.2	2.6
Phenylalanine	1.2	1.2	1.8	1.5	1.7	1.5	1.7
Threonine	1.3	1.7	0.9	1.9	1.8	0.7	1.6
Tryptophan	3.3	4.9	5.0	3.3	5.0	4.0	4.4
Valine	0.7	1.1	1.7	1.4	1.6	1.0	1.1
Cystine	5.0	2.8	3.1	3.0	2.1	3.2	3.6
Tyrosine	1.2	0.8	3.0	2.8	1.4	1.3	2.7

Means, standard deviations (SD) and CV (within laboratories) for ten amino acids in each protein source (calculated from untransformed data) are given in Tables 13.7 and 13.8. Within laboratories variation has been reported to be greater in laboratories using a single-column analyser than those using a two-column analyser (Porter et al., 1968). Therefore, in this initial statistical analysis for within laboratories variation, laboratories were grouped accordingly. Furthermore, laboratories 2 and 6 were, by examination, considerably more variable than the other laboratories. Hence, estimates of within laboratory variability (Tables 13.7–13.9) were based on the following grouping: Group No. 1, laboratories 3 and 4; group No. 2, laboratory 2; group No. 3, laboratories 1,5 and 7; group No. 4, laboratory 6.

In most cases, the within laboratories variation was much smaller than the variation between laboratories (Tables 13.4, 13.5, 13.7 and 8). For most amino acids, casein (CV values of less than 5%), soy protein (CV values of less than 6%) and minced beef (CV values of less than 7%), exhibited lower within laboratory variation than pea flour (CV values of 6–19%) and rapeseed protein concentrate (CV values of 8–18%) while wheat flour and egg white (CV values of less than 8%) occupied an intermediate position in this respect.

Based on the transformed data, the within laboratories CV values for the ten amino acids, were, in general, quite small (Table 13.9). The values for most amino acids in casein, pea flour, egg white, minced beef, soy protein and rapeseed protein concentrate were less than 1.0, 2.0, 2.0, 2.5, and 2.5%, respectively.

In order to estimate the influence of variation in amino acid analysis on the accuracy of *in vitro* protein quality indices, amino acid scores were calculated from the data obtained in this investigation (Table 13.10). The lysine-based scores were less variable (laboratory-to-laboratory difference of 10–14 percentage units) than those calculated from methionine + cystine (difference of 11–25 percentage units) or threonine (difference of 17–18 percentage units) content.

The variability of amino acid scores (Table 13.10) was compared with the variation in some of the bioassay methods for evaluating protein quality which was reported in a recent collaborative study (McLaughlan et al. 1980; Table 13.11). In general, amino acid scores appeared comparable to relative NPR and somewhat better than relative PER in terms of inter-laboratory variability (Tables 13.10 and 13.11).

Many correlations have been made over the years between chemical (amino acid) scores and biological methods in establishing acceptability of *in vitro* quality methods. This relationship can be summarized by saying that general agreement is sufficiently high to conform the basic validity of the approach but is not yet high enough for absolute predictive purposes (Pellett, 1978). Unlike the bioassay methods, amino acid scores do not take into account digestibility of protein and availability of amino acids. Moreover, uncertainty exists in selecting the most appropriate reference amino acid pattern used as a basis of calculating scores in predicting protein quality of humans.

TABLE 13.7. WITHIN LABORATORIES STANDARD DEVIATIONS (SD) AND COEFFICIENTS OF VARIATION (CV) FOR TEN AMINO ACIDS AND RECOVERY OF NITROGEN IN CASEIN, SOY ASSAY PROTEIN AND PEA FLOUR (CALCULATED FROM UNTRANSFORMED DATA)

Amino acid (mg/gN)	Casein			Soy assay protein			Pea flour		
	Mean[1]	SD[1]	CV%[1]	Mean[1]	SD[1]	CV%[1]	Mean[1]	SD[1]	CV%[1]
Isoleucine	320–362	2–13	0.9–3.5	289–312	1–26	0.2–8.4	250–301	3–31	1.1–10.3
Leucine	599–661	3–7	0.6–1.2	506–578	2–34	0.7–5.9	450–535	3–54	0.6–10.2
Lysine	505–568	1–7	0.2–1.4	377–438	1–14	0.3–4.8	455–500	8–42	1.7–9.3
Methionine	182–197	1–4	0.4–2.1	73–86	2–7	2.7–6.6	59–86	1–14	1.7–19.4
Phenylalanine	316–383	3–16	0.9–4.9	331–389	3–16	0.8–4.2	277–341	2–20	0.8–5.8
Threonine	262–308	5–8	1.6–2.7	211–253	1–8	0.2–3.1	249–275	1–9	0.2–3.8
Tryptophan	82–91	1–2	1.4–1.6	58–96	2–6	3.6–5.7	39–72	1–6	1.8–10.3
Valine	409–450	4–14	0.9–3.1	299–313	2–10	0.5–3.2	283–336	4–24	1.4–7.2
Cystine	25–38	1–2	1.8–8.0	71–81	1–5	0.7–5.2	85–113	1–13	1.0–13.4
Tyrosine	345–419	4–11	0.8–2.9	239–262	2–13	0.9–5.1	198–274	3–18	1.3–6.7
Recovery of nitrogen, %	96–100	1–2	0.4–1.9	95–102	1–4	0.6–3.6	90–99	1–2	0.8–2.7

[1]Range of values for laboratories 3,4; 2; 1,5,7; and 6.

TABLE 13.8. WITHIN LABORATORIES STANDARD DEVIATIONS (SD) AND COEFFICIENTS OF VARIATION (CV) FOR TEN AMINO ACIDS AND RECOVERY OF NITROGEN IN WHOLE WHEAT FLOUR, EGG WHITE SOLIDS, MINCED BEEF, AND RAPESEED PROTEIN CONCENTRATE (CALCULATED FROM UNTRANSFORMED DATA)

Amino acid (mg/gN)	Whole wheat flour			Egg white solids			Minced beef			Rapeseed protein concentrate		
	Mean[1]	SD[1]	CV%[1]	Mean[1]	SD[1]	CV%[1]	Mean[1]	SD[1]	CV%[1]	Mean[1]	SD[1]	CV%[1]
Isoleucine	198–231	3–11	1.6– 5.0	303–350	2–26	0.5– 7.5	259–267	2–5	0.7– 1.9	244–303	2–38	0.7–12.6
Leucine	419–460	5–20	1.3– 4.5	507–576	7–28	1.3–5.2	467–541	3–23	0.5– 4.7	455–513	1–14	0.1– 2.8
Lysine	154–189	2–12	1.4– 7.3	400–452	2–37	0.3–8.1	476–558	1–16	0.2– 3.3	338–393	3–35	0.8–10.4
Methionine	90–122	2–8	2.8– 7.8	228–261	2–7	0.5–3.2	124–155	1–4	0.4– 3.6	104–150	1–6	0.8– 4.2
Phenylalanine	277–352	3–8	1.0– 2.5	358–420	4–6	0.8–1.7	229–288	2–9	0.8– 3.5	234–300	3–8	1.2– 3.6
Threonine	161–204	2–7	1.1– 3.5	256–315	8–18	2.5–5.9	249–275	2–19	0.6– 6.9	239–278	2–23	0.6– 9.0
Tryptophan	70–90	1–5	0.9– 5.5	88–111	0–2	0–2.4	53–78	1–2	2.7– 3.9	86–105	2–8	2.0– 8.2
Valine	252–301	5–8	1.7– 2.8	382–449	4–19	1.0–4.2	277–300	2–19	0.8– 6.8	324–355	2–28	0.6– 7.8
Cystine	129–161	1–9	0.9– 6.5	162–198	1–6	0.7–3.4	59–72	1–4	0.9– 6.6	121–181	1–22	0.7–18.0
Tyrosine	167–219	1–25	0.9–11.6	272–282	7–19	2.4–6.8	187–226	1–8	0.3– 3.7	149–202	2–17	1.2–11.4
Recovery of nitrogen, %	91–96	1–2	0.4– 2.3	90–100	1–3	1.4–3.4	92–100	1	0.9– 1.2	93–99	1	0.5– 1.5

[1]Range of values for laboratories, 3,4; 2; 1,5,7; and 6.

TABLE 13.9. WITHIN LABORATORIES COEFFICIENTS OF VARIATION[1] (%) FOR TEN AMINO ACIDS IN SEVEN PROTEIN SOURCES (CALCULATED FROM TRANSFORMED DATA)

Protein source	Casein	Soy assay protein	Pea flour	Whole wheat flour	Egg white Solids	Minced beef	Rapeseed protein concentrate
Amino acid							
Isoleucine	0.11-0.31	0.03-1.46	0.19-1.82	0.29-0.84	0.09-1.28	0.12-0.34	0.29-0.94
Leucine	0.09-0.18	0.08-0.92	0.10-1.62	0.20-0.94	0.20-0.83	0.08-0.76	0.20-0.75
Lysine	0.02-0.22	0.05-0.80	0.28-1.52	0.28-1.43	0.06-1.33	0.04-1.71	0.28-1.43
Methionine	0.07-0.41	0.59-2.38	0.39-0.65	0.61-4.56	0.10-0.62	0.09-0.71	0.61-1.63
Phenylalanine	0.16-0.84	0.12-0.70	0.14-1.00	0.20-0.44	0.14-0.32	0.15-0.64	0.17-0.44
Threonine	0.27-0.48	0.04-0.55	0.04-0.68	0.23-0.65	0.42-1.05	0.11-1.23	0.23-0.65
Tryptophan	0.31-0.36	0.89-1.84	0.46-0.65	0.19-2.58	0.00-0.73	0.62-1.01	0.19-1.22
Valine	0.18-0.51	0.10-1.79	0.26-1.23	0.34-0.50	0.17-0.70	0.14-1.21	0.34-0.50
Cystine	1.10-2.44	0.18-1.45	0.22-1.32	0.17-2.95	0.13-0.63	0.28-1.59	0.17-1.33
Tyrosine	0.25-0.50	0.16-0.92	0.28-1.20	0.19-1.22	0.46-1.20	0.05-0.69	0.19-2.16

[1]Range of values for laboratories 3,4; 2; 1,5,7; and 6.

TABLE 13.10. AMINO ACID SCORES CALCULATED FROM DATA REPORTED IN TABLES 4, 5, 7 AND 8 BY USING FAO WHO (1973) REFERENCE SCORING PATTERN

Protein Source	Most limiting amino acid	Amino acid score, %	
		Mean[1]	Range[2]
Wheat flour	Lysine	49	45–55
Wheat flour + egg white (60:40 protein)	Lysine	80	74–86
Wheat flour + soy protein (60:40 protein)	Lysine	76	71–85
Wheat flour + rapeseed concentrate (50:50 protein)	Lysine	77	72–85
Casein	Met + Cys	99	94–107
Minced beef	Met + Cys	93	83–103
Pea flour	Met + Cys	74	65–90
Soy protein	Met + Cys	69	65–76
Wheat flour + L-lysine (200 mg/gN)	Threonine	73	64–82
Soy protein + L-methionine (100 mg/gN)	Threonine	96	84–101

[1]Calculated from amino acid data reported in Tables 13.4 and 13.5.
[2]Calculated from amino acid data reported in Tables 13.7 and 13.8.

TABLE 13.11. RELATIVE PROTEIN EFFICIENCY RATIO (RPER) AND RELATIVE NET PROTEIN RATIO (RNPR) VALUES (CASEIN + METHIONINE = 100) FOR FIVE SAMPLES (ABSTRACTED FROM MCLAUGHLAN ET AL., 1980)[1]

Sample	RPER, %		RNPR, %	
	Mean	Range	Mean	Range
Lactalbumin	81	66–91	87	77–96
Egg white	95	78–106	97	84–105
Wheat gluten	7	0–11	32	25–38
Soy assay protein	47	31–54	59	47–66
Soy protein + wheat gluten	54	46–65	64	60–73

[1]Results of an interlaboratory study involving six laboratories.

ACKNOWLEDGEMENTS

The authors acknowledge the advice and encouragement of Dr. J.L. Beare-Rogers, the technical assistance of H.G. Botting and F.J. Noel, and the statistical advice of Dr. G. Jarvis and S. Malcolm.

REFERENCES

AOAC 1975. Official Methods of Analysis, 12th ed., Association of Official Analytical Chemists, Washington, D.C.

BLACKBURN, S. 1978. Sample preparation and hydrolytic methods. *In* Amino Acid Determination, Methods and Techniques, 2nd ed., S. Blackburn (Ed.), Marcel Dekker, Inc. New York, pp. 7–38.

FAO/WHO (Food and Agriculture Organization/World Health Organization). 1973. Energy and protein requirements. Report of a joint FAO/WHO ad hoc expert committee on energy and protein requirements. Food and Agric. Org. Nutr. Report. Series No. 52, Rome.

HUGLI, T.E. and MOORE, S. 1972. Determination of the tryptophan content of alkaline hydrolysates. J. Biol. Chem. *247*:2828.

KNIPFEL, J.E., AITKEN, J.R., HILL, D.C., McDONALD, B.E. and OWEN, B.D. 1971. Amino acid composition of food proteins: Inter- and intra-laboratory variation. J.A.O.A.C. *54*:777.

McLAUGHLAN, J.M., ANDERSON, G.H., HACKLER, L.R., HILL, D.C., JANSEN, G.R., KEITH, M.O., SARWAR, G. and SOSULSKI, F.W. 1980. Assessment of rat growth methods for estimating protein quality: Interlaboratory study. J.A.O.A.C. *63:462*.

MOORE, S. 1963. On the determination of cystine as cysteic acid. J. Biol. Chem. *238*:235.

PELLETT, P.L. 1978. Protein quality evaluation revisited. Food Technol. *32*:60.

PORTER, J.W.G., WESTGARTH, D.R. and WILLIAMS, A.P. 1968. A collaborative test of ion-exchange chromatographic methods for determining amino acids. Br. J. Nutr. *22*:437.

SATTERLEE, L.D., MARSHALL, H.F. and TENNYSON, J.M. 1979. Measuring protein quality. J. Am. Oil Chemists' Soc. *56*:103.

SPIES, J.R. and CHAMBERS, D.C. 1949. Chemical determination of tryptophan in proteins. Anal. Chem. *21*:1249.

WESTGARTH, D.R. and WILLIAMS, A.P. 1974. A collaborative study on the determination of tryptophan in feeding stuffs. J. Sci. Food Agric. *25*:571.

WILLIAMS, A.P., HEWITT, D. and COCKBURN, J.E. 1979. A collaborative study on the determination of cyst(e)ine in feeding stuffs. J. Sci. Food Agric. *30*:469.

DISCUSSION

DR. SAMONDS: Could you tell us how you calculated the coefficients of variation for between lab variability?

DR. SARWAR: The amino acid data were subjected to analysis of variance and between laboratory standard errors (SE) were calculated from laboratory mean squares. Between laboratory CV were then calculated:

$$CV = \frac{SE \times 100}{\text{mean (between laboratory)}}$$

DR. ROTRUCK: I noticed in your study that you allowed investigators to use different times of hydrolysis whereas Mrs. Happich and Dr. Bodwell attempted to standardize their study by using the same time of hydrolysis. I was wondering if someone, either you or other people in the audience, could comment on the variability or the effect of time of hydrolysis on amino acid recovery. I'm a little uncomfortable with using a standardized time for all diet samples because of the potential effect of variables like carbohydrates on hydrolysis time. I know the experience in our laboratory has been that we can see different results depending on the type of sample. We feel that we have to use different times of hydrolysis.

DR. SARWAR: I'll just mention that we also have a time hydrolysis study which was part of this collaborative assay where we used different times (11 hours, 22 hours, 44 hours, and 68 hours). Our results are in but I did not have time to do a statistical analysis.

DR. STEINKE: In answer to your question, Dr. Rotruck, we have recently been doing some time studies with different food products and there are amino acids that increase with hydrolyses time, particularly isoleucine and valine. It seemed to be fairly consistent among the food products we've tested. There are others that decrease to some degree. I think a single hydrolyses time may not be adequate to optimize the values for all amino acids.

DR. STAHMANN: I was going to comment on the same thing. When very precise amino acid analysis are needed for structure determinations of pure isolated proteins, time studies are used because you do get some destruction of threonine and a few other amino acids. This destruction is small and I think that with food samples it's really not necessary to have that high degree of accuracy.

One other comment on a technique that might be useful to people doing amino acid analyses. For some fifteen years or more, we have mixed up our own buffers in 40 liter bottles connected to two or three amino acid analyzers. The buffers are made stable for months by adding a small amount of methiolate. This saves buying expensive buffers and you can make up as much as you want and hold it for a long time by adding a small amount of methiolate as a preservative.

DR. BODWELL: Dr. Sarwar, if I understand you correctly, you're saying that the labs used their own nitrogen analysis for the calculation of the compositions?

DR. SARWAR: Yes, that's true.

DR. BODWELL: Eventually, you're going to correct those to a constant nitrogen recovery so you get rid of that variable?

DR. SARWAR: Yes, that is our intention. When I get the final analysis from the statisticians and when we have the results of the time hydrolysis study, we will express the results as received and as calculated from corrected nitrogen recoveries.

DR. BODWELL: Eventually, when both of these collaborative studies are completed and all the statistics are done, I think we're going to have some very interesting comparisons, because they are two very different studies. You're sort of seeing "what is" and we're trying to see what "could be" with as much standardization as we can get (we're not getting quite as much as we'd like). As Ms. Happich and I have become aware, scientists (chemists especially) are very independent souls. On the pretest, for instance, one group decided they would compare methods while they were doing the pretest tryptophan analyses, so their tryptophan values came in as determined by three different methods. That was their three values. But the point is, we were trying to tighten things up as much as possible. I think you're seeing "what is" but I hope we don't take the values you show as what can be possible.

DR. SARWAR: Well, I agree with you. That was not the objective of our assay. The amino acid analysis has been around for a while now and our objective was that if somebody comes along and suggests that protein quality regulation (PER method) should be replaced by some kind of amino acid method, what is the variability? Now this was the first step. The second step is to correct the factors which were responsible for this variation.

DR. BODWELL: If we're really serious about using amino acids directly instead of rat assays for nutritional labeling, one of the first things we have to do is standardize the amino acid analysis. From your data, it looks like we also have to standardize nitrogen analysis which I thought had been done a long time ago!

DR. SARWAR: When you are doing a collaborative study, I think you know and I know, that you send out instructions and people do what they like. It's better that they are honest and they tell you what they've done and you record it.

14

Individual Amino Acid Levels and Bioavailability

K.J. Carpenter

When consideration of the protein quality of mixed diets is a serious matter, and it can only be estimated from the characteristics of the individual foods eaten, calculation has to based on the adequacy of each of several amino acids—the sulfur amino acids, lysine, and one or more of threonine, isoleucine and tryptophan. Complementarity between different proteins may be so considerable that calculations based on the NPU (Net Protein Utilization) of different individual foods can seriously underestimate the value of the whole.

The feed mix industry for pigs and poultry provides a "real world" model of such a situation. There, the commercial practice of companies who make their living by selling mixes giving competitive "growth per dollar," do their formulation in terms of 3 or 4 amino acids. They would consider the NPU value of a single ingredient almost irrelevant, except as an indicator of the availability of its first limiting amino acid.

With regard to bioavailability their custom is to apply a fixed "factor" to each class of material—the same factor for each amino acid, based on published assays for other samples of that class or on some in-house experience.

Presumably it is non-controversial that ideally the values we want to use for calculating nutritional value are "available" ones rather than "total" values from acid-hydrolysis of proteins. The question marks attached to the idea in relation to human nutrition are:

(i) whether the difference between the two for the normal range of human foods is ever of any considerable magnitude *and*

(ii) whether we have practicable ways of measuring the degree of availability.

In the present context we are concerned only with "in vitro" methods. They are, of course, the only ones that are at all practicable for routine application to large numbers of samples, but the *in vitro* methods must first be calibrated against true assays for bio-availability. Most such assays have used chicks or rats (e.g. Carpenter and Woodham, 1974). A much smaller number have been with dogs or pigs. I have no knowledge of any calibration of an *in vitro* test against bioassays with humans. We will return to this point. In practice it has been an implicit assumption that the young rat can be regarded as a fair model for humans in this respect.

Certainly there is no evidence that ordinary cooking, i.e. boiling, frying, roasting has any serious effect on the protein quality of foods (as tested with rats). In bread it is only the crust, or slices that are toasted, that show significant damage (Hutchinson *et al.* 1960). Cereals cooked by "popping" or some kinds of extrusion process have been shown to be damaged (reviewed by Liener, 1958). Milk can also be damaged in drying, especially by the roller process (Bujard and Finot, 1978). A further insidious deterioration can occur when milk powder is stored, after leaving the manufacturing plant, at a moisture content of 7% or more either at a high temperature or for a long period (Henry & Kon, 1950). This has been a real-life situation in the past (Table 14.1; Henry *et al.* 1948). The biological value of the powder may be halved and, although solubility is also reduced, a consumer may not appreciate the significance of the change.

TABLE 14.1.   CHANGES IN THE PROPERTIES OF SKIMMED MILK POWDER STORED IN NITROGEN AT 37° AT DIFFERENT MOISTURE CONTENTS[1]

Test Material	Amino groups (Van Slyke) mg NH$_2$ N /g total N	Protein quality for rats		
		Digestibility	Biological value	Protein efficiency ratio (PER)
untreated, as control	59	0.92	0.85	2.77
stored at 4.7% moisture content for 182 days	53	0.89	0.84	2.56
stored at 7.3% moisture content for 60 days	22	0.86	0.69[2]	1.82[2]

[1] Data from Henry *et al.* (1948); the results from rat experiments 5, 6 and 7 being pooled.
[2] With supplementary lysine the biological value was 0.83 and the PER was 2.48.

Milk powder used as a high-protein supplement in cookies for infants can also have its biological value virtually destroyed during their baking even though the product is tasty and pleasantly golden rather than "burnt" in flavor or appearance (Carpenter and March, 1961). In this case it was because of unfavorable results with children receiving cookies made with milk powder, compared with results obtained with milk powder added to cookies cooked separately and crumbled, that the extent of the damage to the protein in this product was investigated (R.F. Dean, unpublished results; cited by Clegg, 1960).

Where significant damage has been found to occur with relatively mild processing or cooking of foods it seems to be due consistently to the Maillard reactions that can occur in the "dry" state (7−20% moisture content) between the free ϵ-NH_2 groups of lysine units in protein and carbonyl groups present in another food component; these reactions result in the lysine becoming unavailable. In milk powder it is the lactose that is responsible for the reactivity; in cottonseed it is gossypol with its two aldehyde groups (reviewed by Carpenter and Booth, 1973). But these are exceptions. Few natural foods contain levels of free reducing sugars or other sources of carbonyl groups at levels that can block a significant proportion of the lysine present.

Similar reactions will occur with the complex phenols present when foods are heavily 'smoked' (Dvorak and Vognarova, 1965; Clifford et al. 1979) and also with formaldehyde that may be present in some 'smokes' or be added deliberately as a preservative (Hurrell and Carpenter, 1978).

Great numbers of "model" experiments have been done with mixtures of food proteins and glucose. Glucose and other monosaccharides are not found in foods at significant levels. Sucrose is widely present at high levels in fruits and in many cake mixes and processed foods. It will cause negligible damage to lysine at the mildest conditions needed for glucose to react, but with more severe conditions that allow inversion of the sugar, the glucose and fructose formed will react. Of course, this will also occur at lower temperatures if yeast is present to catalyze the hydrolysis of the sucrose (Hurrell and Carpenter, 1977), or if starch is being degraded by amylases.

All proteins show a general fall in digestibility when very severely heated at 7–20% moisture content for a long period. This seems to be explained by the formation of "unnatural" cross-linkages that resist, or greatly slow down their digestion in the mammalian gut. The cross-linking is accelerated by the presence of even low levels of sugars. It is known to be due in part to imide linkages forming between lysine and asparagine or glutamine units (Bjarnason and Carpenter, 1970), but the fall in digestibility may be considerably greater than the proportional fall in reactive lysine groups (e.g., Boctor and Harper, 1968).

Lastly, there are some foods in which the protein is of low digestibility even in the natural state. Sorghum grain seems a good example of this class. In strains of high tannin content digestibility is impaired—presumably because of the tannins having a non-specific inhibitory effect on trypsin and other digestive enzymes (Ford and Hewitt, 1979). As one might expect, analytical procedures applied to the food proteins themselves fail to detect the adverse biological effect of such non-protein factors as tannins.

SULFUR-CONTAINING AMINO ACIDS

Total Sulfur

It has been suggested that, in practice the sulfur-containing amino acids are always the first limiting in human diets, or that a score based on them is so close to the 'limiting' score that the difference is negligible (Miller and Naismith, 1958). The same workers also suggested that "total sulfur" content was an adequate measure of "cystine + methionine" so that one analysis was all that was necessary to obtain a useful prediction of the adequacy of the protein in a diet. This claim has not been buttressed by any further studies. In the particular case of peas and beans, which are generally limiting in methionine a significant proportion of the total sulfur may come from S-methylcysteine that has no nutritional value (Evans and Bandemer, 1967). Perhaps because of this, total S has been a useful predictor of protein quality in some studies (e.g., Porter *et al.* 1974) but not in others (e.g. Evans and Boulter, 1974).

Available Methionine

In view of the importance of methionine as a limiting factor in protein quality there have been many attempts to develop a specific chemical procedure for "available methionine". The concept depends, of course, on methionine residues in a protein reacting in some way to make them unavailable even when the protein as a whole is digested. It has been proposed that this could happen through oxidation at the sulfur atom to give the sulfoxide and that this is either unusable or is only released slowly during digestion.

With the Moore (1963) method of oxidizing proteins prior to hydrolysis so that both methionine and its sulfoxide are converted to the sulfone derivative, the extent to which methionine residues were already oxidized would go unnoticed. Since even ordinary procedures of acid hydrolysis result in some degree of oxidation of methionine residues, other methods have been developed to measure methionine residues in the intact protein of foods. Ellinger and Duncan (1976) have used cyanogen bromide which reacts with methionine residues to yield methylthiocyanate which can be determined rapidly by gas chromatography, thus providing a method which discriminates between oxidized and unoxidized residues. Two other procedures using dimethylsulfoxide are discussed by Lipton and Bodwell (1974).

Whether methionine residues form a more useful nutritional measure than the sum of "methionine plus its sulfoxide" now seems doubtful because the balance of the recent evidence indicates that although the sulfoxide can be formed during some proposed industrial procedures for the treatment of food proteins with hydrogen peroxide it has a high availability for rats (Table 14.2; also, reviewed by Cheftel, 1974). The sulphone, which all workers agree to be unavailable, does not seem to be formed in significant amounts under oxidizing conditions likely to be encountered in practice.

If it is confirmed that humans are like rats in their use of the sulfoxide, the only methods that would seem useful for determining available methionine are those depending on enzymic digestion, either as a preliminary to chemical analysis or as part of a microbiological assay (see Chapters 15, 16).

HYDROXYPROLINE IN MEAT

Meat provides a possible example of a material where a single amino acid analysis could serve as a useful general indicator of protein quality. The

TABLE 14.2.   THE INFLUENCE OF HYDROGEN PEROXIDE ON THE NUTRITIVE VALUE OF FISH PROTEIN (SJOBERG AND BOSTROM, 1977)

	H_2O_2 added, g/kg		
	0	20	40
methionine, g/kg protein	3.09	1.22	0.41
" sulfoxide "	0.41	2.12	2.48
" sulfone "	0.00	0.17	0.31
NPU test:			
N digestibility, %	0.923	0.918	0.906
Biol. value	0.948	0.863	0.854
NPU + 0.1% MET:			
N digestibility, %	0.932	0.937	0.921
Biol. value	0.970	0.961	0.940

major cause of variation in its protein quality is the proportion of connective tissue that it contains. The higher the proportion of connective tissue, the lower the level of all the essential amino acids including methionine (which is usually first limiting in rat assays of meat products) and lysine (Dvorak and Vognarova, 1969). Dvorak (1972) has shown that the level of hydroxyproline can serve as a good predictor of the proportion of connective tissue present and, conversely, as an indicator of both chemical score and actual NPU for rats (Table 14.3). This amino acid can be determined in protein hydrolysates without having to carry out a complete amino acid separation on an "Analyzer" column. The method used by Dvorak (1972) involved oxidation, distillation of 2-methyl pyrrole and a color reaction with an aldehyde (Serafini-Cessi and Cessi, 1964).

Dvorak's evaluation procedure would not serve to detect processing damage to meat products. Since lysine values are also well correlated in meats with those for methionine and other essential amino acids it would seem that "reactive lysine" would also serve a useful purpose here—a high value

TABLE 14.3.   ESTIMATED CHEMICAL SCORE AND NET PROTEIN UTILIZATION (NPU) OF ANIMAL TISSUES[1]

Test material	Hydroxyproline, (log g/16 g N)	'Estimated chemical score'	Protein, % in the test diet	NPU
Beef tenderloin	−0.657	85.6	8.9	89
Beef tenderloin	−0.402	81.8	8.1	84
Pork liver	−0.335	81.0	8.0	83
Beef round	−0.239	79.5	7.3	82
Beef liver	−0.130	77.8	5.4	77
Pork shoulder	0.000	75.3	5.5	79
Pork neck	0.130	70.4	8.4	77
Beef shoulder	0.230	70.0	7.1	75
Pork shoulder	0.253	67.3	8.2	75
Pork belly	0.403	65.3	5.8	68
Pork belly	0.403	65.3	5.1	60
Pork from the head	0.795	50.6	5.4	49

[1]From Dvorak (1972).

indicating both (a) that the raw material was of high quality *and* (b) that there had been no serious degree of processing damage. Some experiments carried out for a different purpose (Atkinson and Carpenter, 1970) have given results that are also quite encouraging in this respect (Table 14.4).

TABLE 14.4.   RELATION BETWEEN OVERALL PROTEIN QUALITY (RAT NPU) AND FDNB-REACTIVE LYSINE VALUE OF PROCESSED CARCASS TISSUES[1]

	N.P.U.	Reactive lysine, g/16gN
Freeze-dried beef fillet	74	8.4
Oven-dried beef fillet	78	7.2
Freeze-dried gut	75	6.5
Oven-dried gut	37	4.3
Autoclaved and oven-dried beef fillet	32	4.3
Heated gut and tendon mixes	28–34	4.0
Autoclaved and oven-dried gut	24	4.5
Freeze-dried tendon	24	4.7
Severely damaged gut .I.	17	3.4
Heated ossein	12	4.1
Severely damaged gut .II.	11	3.2
Ossein (bone protein)	8	4.3

[1]From Atkinson and Carpenter (1970).

LYSINE

Measurements of Reactive Lysine

As with methionine and other amino acids, estimates of available lysine may be made with *in vitro* enzymic procedures, followed by chemical analysis (e.g. Mauron *et al.* 1955) or microbiological assay (e.g., Riesen, *et al.* 1947) and it was work of this kind that first confirmed the extent to which lysine could become resistant to enzyme release from processed foods (Table 14.5). With the introduction of pronase, as a more powerful proteolytic agent than individual mammalian enzyme preparations, this approach promises to give both convenient and meaningful values for a wide range of foods (Rayner and Fox, 1978).

Even methods in this class require validation against biological assays. Hamad and Fields (1979) have published values for "available lysine" with a method that involves pre-treatment with papain followed by microbiological assay of the digest with *Pediococcus cerevisiae*. The common cereals all show a 4–15 fold increase in value with germination. The absolute values obtained for the ungerminated cereals range from 0.05–0.36 g/16g N, as compared with published total lysine values for the same cereals of approximately 3 g/16g N. The lysine of wheat is almost completely available for rats and pigs, and presumably for man in view of the agreement between chemical score and NPU in various experiments (Young *et al.* 1975). And, although germination can cause a small increase in the nutritional value of

TABLE 14.5.   RESULTS OF ANALYSES OR ASSAYS FOR LYSINE (g/16gN) IN COW'S MILK CONCENTRATED OR DRIED IN DIFFERENT WAYS (MOTTU AND MAURON, 1967)

Method preparation	Total after acid hydrolysis	FDNB-reactive[1]	Digestion *in vitro* and dialysis	Rat growth assay
Freeze-dried	8.3	8.4	$(8.3)^2$	8.4
Spray-dried	8.0	8.2	8.3	8.1
Evaporated	7.6	6.4	6.2	6.1
Roller-dried 2[3]	7.1	4.6	5.4	5.9
Roller-dried 4	6.8	3.8	4.5	4.0
Roller-dried 6	6.3	2.5	3.1	2.9
Roller-dried 8	6.1	1.9	2.3	2.0

[1] Determined by the Carpenter (1960) procedure.
[2] The actual values were all lower than given here because of the limitations of the procedure. The values listed were calculated *relative* to the freeze-dried sample, for which the true digestibility was assumed to be 100% (see Mauron *et al.*, 1955).
[3] The code numbers refer to samples prepared under increasingly severe conditions.

cereals, the effect is relatively small. The values reported by Hamad and Fields (1979) seem therefore to have no relevance to human or animal nutrition. The inability of the micro-organism to make effective use of the lysine in the untreated cereal may be explained by its being unable to rupture intact vegetable cells, or to digest uncooked starch.

Chemical Methods

These methods have all depended on the idea that it is the reaction of the ϵ-amino lysine groups in protein with carbonyl groups and other compounds that makes the availability of the lysine in processed food fall to a greater extent than that of other essential amino acids. The methods use a "tagging" compound to react with the test foods at ϵ-groups that have not already reacted with other components in the food itself. The amount of "tagged lysine" is then determined, usually by hydrolysing the treated food and measuring the amount of tagged lysine released.

The subject has been reviewed and discussed by a number of authors (Carpenter and Booth, 1973; Bodwell, 1976; Friedman, 1977). Although these are obvious exceptions to a generalisation that "reactive = available", the approach has been of considerable practical value in understanding and controlling processing damage to foods.

Fluoro-dinitrobenzene.—The classic use of a chemical reagent to determine the identity of free amino groups in a peptide chain was that of Sanger (1945) with 1-fluoro-2, 4-dinitrobenzene (FDNB). This reagent forms dinitrophenyl derivatives at each primary amino group in insulin (i.e., the terminal α-NH_2 groups of the two chains and the ϵ-NH_2 group of the lysine unit present). On acid-hydrolysis, the labelled amino acids were freed and the bright yellow DNP-amino acids could be separated by silica gel chromatography. This was an elegant procedure in the hands of an expert, and, with successive analysis of sub-units, it led to the classic elucidation of the structure of insulin. Lea and Hannan (1950a) then showed that FDNB no longer tagged the lysine units in casein that they had allowed to react with glucose and then re-purified by dialysis.

Carpenter and Ellinger (1955) had been attempting to devise a fairly simple quality control test for protein concentrates that could be used commercially, and they found that the FDNB procedure, even applied to foods and feedstuffs in quite a crude form, gave results showing better correlations with biological assays than any of the other tests available. Separation from any other DNP amino acids present was achieved without resorting to chromatography by using a 'chloroformate' blank (Carpenter, 1960), and was further slightly simplified (Booth, 1971). In these forms it has been used in many laboratories. An example of good agreement between FDNB values and biological values for milk powders damaged to different extents (Mottu and Mauron, 1967) is shown in Table 14.5.

The greatest problem with the procedure is that with samples containing a high level of starch or other polysaccharide, there can be some 20–30% loss of DNP-lysine during the acid-hydrolysis stage. This is apparently due to the sugars released by hydrolysis having a reductive, and thus decolorising, effect on some of the nitro groups in the DNP-lysine present. An excess of FDNB present, some of it by then hydrolysed to dinitrophenol, has some protective effect. The problem and attempts at circumventing it are discussed in detail elsewhere (Carpenter and Booth, 1973). Matheson (1968) has pointed out the positive source of error from humin color with high-carbohydrate materials in the Carpenter (1960) procedure which does not use chromatographic separation. Both Milner and Carpenter (1969) and Walz and Ford (1973) tried to correct for this with an additional blank when working with cereals. It seems the general conclusion that the procedure is best used in studies of the processing damage to a single material, rather in providing absolute "reactive" lysine values for different materials.

FDNB-reactive Lysine "by Difference".—Because of the problems just discussed, in obtaining quantitative recovery of DNP-lysine and of the increased distribution of automated "amino acid analyzers," Rao et al. (1963) suggested a different approach. On the assumption that the "total" lysine value of a damaged material came from (i) reactive lysine, *plus* (ii) some recovery of lysine from lysine with blocked ϵ-NH$_2$ groups, they did two total lysine analyses. The first used the test material in the ordinary way and the second used the test material pre-treated with FDNB. It was expected that in the second analysis only type "ii" lysine units would be recovered. Then from the difference between the two values one would get only the class "i" or reactive lysine. The advantage was that even though DNP-lysine might be partially reduced during the acid-hydrolysis stage, it did not break down to free lysine.

In practice the assumption seems to hold well with proteins severely heated, either with or without carbohydrates, but it does not do so for protein that has reacted with reducing sugars under mild conditions (Table 14.6). Under the latter conditions the major Maillard compound would be

TABLE 14.6. THE TOTAL- AND REACTIVE-LYSINE CONTENT (mg/g CRUDE PROTEIN) OF TEST MATERIALS[1]

(Values in parentheses are percentages of the values for the unheated samples)

	Total lysine	Total lysine after borohydride	FDNB-reactive lysine (difference method)	FDNB-reactive lysine (direct method)	TNBS-reactive lysine
Albumin mix + glucose					
Stored 10 d at 37°	86.3 (77)	81.1 (32)	84.7 (73)	81.6 (40)	53.6 (78)
Stored 30 d at 37°	(59)	(15)	(54)	(24)	(67)
α-Formyl-fructosyl-lysine[2]	(53)	(0)	(53)	(3)	(80)
Bovine plasma albumin Heated 27 h at 145°	130 (78)	128 (81)	127 (19)	122 (14)	113 (14)
Extracted chicken muscle Heated 27 h at 121°	96.5 (92)	87.5 (95)	93.3 (67)	89.4 (70)	78.2 (62)

[1] From Hurrell and Carpenter (1974).
[2] Values for this compound are expressed as percentages of its theoretical lysine content (i.e. 43.5 g/100g).

expected to be a desoxyketosyl derivative. Using a model compound, α-N-formyl-(ε-N-deoxyfructosyl)-lysine, Finot and Mauron (1972) were able to demonstrate that after treatment with FDNB there were several reaction products that broke down in complex ways after acid-hydrolysis yielding neither lysine nor DNP-lysine (or at any rate, very little). This has been discussed further by Hurrell and Carpenter (1980). The "difference" method does not therefore provide a sensitive measure of the type of damage occurring in stored milk powders. Thus, even though it has been adopted by A.O.A.C. (Couch, 1975), it must be used with caution.

Trinitrobenzene Sulphonic Acid (TNBS).—This reagent was introduced by Kakade and Liener (1969) as an alternative to FDNB. It has the advantage of being water-soluble. Further, it only appears to be necessary to acid-hydrolyse the tagged material long enough to break the protein down to soluble peptides. These give the same color per "tagged" lysine unit as TNP-lysine itself. Eklund (1976) recommends 90 minutes digestion at $100°$. It also seems that TNBS does not give a colored product with free arginine if this amino acid is present in free form, so that nothing equivalent to a "chloroformate blank" is required.

The first problem encountered is that TNP-lysine is even more easily reduced than is DNP-lysine during acid-digestion in the presence of carbohydrate (e.g. Ruderus and Kilberg, 1972; Posati et al. 1972). The second is the unexpected observation (Finot and Mauron, 1972; Hurrell and Carpenter, 1974) that TNBS will still react at the ε-N position in an early Maillard derivative of lysine, and that TNP-lysine remains after acid digestion. We have already seen that FDNB can also react with an early Maillard compound, but the difference is that the complex in that case does not break down to yield a significant proportion of tagged lysine, whereas the complex with TNBS apparently does do so (Table 14.6).

For these reasons TNBS cannot be used as an 'all-purpose' measure of damage. Datta and Datta (1976) advocate its use with a chromatographic separation stage for the evaluation of skim milk powders, but we would not expect it to be a sensitive detector of deterioration in this material. It can be used to good purpose in some conditions. Tella and Ashton (1978) have studied the problems arising with its application to potato waste.

Other Reagents.—O-methylisourea reacts with ε-lysine units to form a compound that breaks down on acid-hydrolysis to give homo-arginine. This can then be estimated on an ordinary ion-exchange aminogram (Mauron and Bujard, 1964) or converted to a volatile derivative and determined by GLC (Nair et al. 1978). It has proved sensitive to different kinds of heat damage (Hurrell and Carpenter, 1974) but cannot be recommended as a routine procedure because it is slow, and different materials require slightly different conditions for optimal reaction.

Methyl acrylate can also be used to react with ε-lysine groups forming the 'mono' and then the dicarboxyethyl derivative (Finley and Friedman, 1973; Friedman and Finley, 1975); so can ethyl vinyl sulfone to give predominantly bis-(ethyl sulfonyl) lysine that separates well on standard aminograms (Friedman, 1977). It is not yet known to what extent they would still react with ε-amino groups that have already entered into some other reaction.

It has been a common problem with all these methods of "tagging" unreacted ε-lysine groups that in very severely heated protein they do not reflect the full extent of the nutritional damage. An example of this is seen in Table 14.7. The usual explanation is that cross-linkages have developed which reduce the digestion of amino acid residues generally, including lysine units still with reactive ε-NH_2 groups.

TABLE 14.7.   NUTRITIONAL VALUE OF THE PROTEIN IN SEVERELY HEATED COD FILLETS IN RELATION TO THE VALUE OF UNHEATED MATERIAL (MILLER *ET AL.*, 1965).

Measure	'Heated' value as % of control[2]
Total lysine (g/16gN)	90
FDNB-reactive lysine	67
Lysine potency (assayed using chick growth)	52
Methionine potency (assayed using chick growth)	69
N.P.U. for rats	63
Nitrogen digestibility for rats	87

[2] The freeze-dried fillets were adjusted to 14% H_2O content and held at 116°C for 27 h.

Sodium Borohydride.—As discussed by Means and Feeney (1971), sodium borohydride can be used to reduce early Maillard-type compounds with lysine to a methylene linkage -CH_2-NH- that resists acid hydrolysis. The "total lysine" then recovered does not include any lysine units that were previously in an "unavailable" Maillard combination. This works well with protein-glucose and protein-gossypol reactions, but not for the damage occurring within severely heated proteins (Hurrell and Carpenter, 1974; Couch and Thomas, 1976).

Furosine.—This compound was first found in acid-hydrolysates of overheated milk powders (Erbersdobler and Zucker, 1966). Furosine can be easily detected on the short basic column of the amino acid analyzer where it appears after arginine. The peptide-bound lactulosyl-lysine in milk powders has been found on acid hydrolysis to give consistently 40% of its lysine content as lysine and 32% as furosine. This finding formed the basis of the furosine test for Maillard damage to milk powders (Bujard and Finot, 1978; Erbersdobler, 1979). A comparison with the results of other methods is shown in Table 14.8.

DYE-BINDING METHODS

Dye binding capacity (DBC) as a rapid indicator of the total protein content of samples of milk and of other foods of almost constant amino acid

TABLE 14.8. MEAN LYSINE VALUES (g/16gN) FOR THREE TYPES OF MILK PREPARATION AS DETERMINED BY DIFFERENT PROCEDURES (RECALCULATED FROM BUJARD AND FINOT, 1978).

Analytical procedure	Un-treated	Steri-lized	Roller-dried
'Total' (Aminogram of acid			
hydrolysate)	8.22	7.86	5.24
FDNB-direct procedure ($\times$ 1.09)	8.03	6.52	4.05
Homo-arginine (guanidination)	7.53	6.14	2.81
Total after reduction			
(borohydride)	7.73	6.88	2.69
(Furosine)	(0.0)	(0.33)	(1.58)
Total minus '1.23 $\times$ Furosine'	8.22	7.45	3.30
In vitro enzymic release	8.22[1]	6.93	3.25

[1]Relative values scaled up to equal the 'total' for untreated milk.

composition has been in use for many years (Ashworth, 1966; Udy, 1971). It has depended on the attraction between negatively charged azo dyes and the basic amino groups in the proteins of foods suspended in an acid medium. The loss of dye from solution, as measured after filtration, was used to estimate the protein in the sample, from an empirically calibrated graph.

With the development of interest in finding high-lysine cereals it was discovered that a high ratio of "DBC: Kjeldahl crude protein" in a sample, although in principle only an indicator of a high level of "histidine + arginine + lysine", in practice could be used to pick out high-lysine samples for more accurate analysis (e.g., Mossberg, 1968; Bhatty and Wu, 1975). Other workers also reported that the same ratio could serve in some circumstances as an indicator of protein quality in fish meals (Moran et al., 1963; Carpenter and Opstvedt, 1976). This was so although the method measured all basic groups and although the early Maillard compounds between ϵ-amino groups and reducing sugars were still basic enough to attract the azo dyes (Lea and Hannan, 1950).

D.C. Udy was probably the first to have the idea of doing two DBC determinations on a test sample, the first in the ordinary way and the second with the ϵ-amino groups blocked by some specific reaction so that only arginine and histidine would be determined. Then the difference between the two DBC values should be a more direct measure of "dye binding lysine" (DBL) specifically. Udy himself (private communication, 1979) attempted to use FDNB as the blocking agent, but found the procedure impracticable. Sandler and Warren (1974) used the same approach with ethyl chloroformate, and Jones and Lakin (1976) with TNBS, but neither procedure proved really practicable and is not being pursued.

We had been using propionic anhydride in earlier experiments as a simple way of blocking ϵ-amino groups specifically by acylation (Bjarnason and Carpenter, 1969; Varnish and Carpenter, 1975) and noticed that propionylation reduced the DBC of lactalbumin to the same extent as its FDNB—lysine value. After a long series of tests (Hurrell et al. 1979), we believe that propionic anhydride can serve as the reagent for a workable DBL procedure, and that it should be a more convenient replacement in quality control work for the older methods involving acid hydrolysis of "tagged" proteins. In particular, a sample can be carried through the procedure in $1-2$ hours.

Some of the results obtained so far are summarised in Table 14.9. Unfortunately we have so far only been able to compare our results with FDNB values rather than with biological assays for most materials. The ability of

TABLE 14.9.   DYE-BINDING VALUES (A) BEFORE AND (B) AFTER ALKYLATION AND THE DIFFERENCE (DBL) COMPARED WITH TOTAL AND FDNB-REACTIVE LYSINE IN MATERIALS HEATED TO VARYING EXTENTS. (HURRELL, LERMAN & CARPENTER, 1979)

Process applied	DBC m moles/16gN A - B	Lysine (g/16 g N) DBL	FDNB	Total
Dried chicken muscle:		(8.90)	(8.94)	(8.79)
None	120−59	100	100	100
4h-121°	109−58	84	88	95
8h- "	104−58	76	81	92
27h- "	97−57	66	69	93
Casein + 6% glucose:		(7.85)	(8.39)	(8.44)
None	109−55	100	100	100
5d-37°	109−62	86	82	89
2h-80°	108−68	75	69	82
Leaf concentrate:		(6.50)	(5.76)	(6.18)
Freeze-dried	101−56	100	100	100
Vac. at 90°	102−63	87	91	100
Hot air at 90°	101−66	79	84	92
Soy Flour:		(6.18)	(6.02)	(6.22)
None	102−60	100	100	100
From over-	103−66	86	91	93
heated	87−54	79	80	82
silo	67−48	49	47	61

the test to detect early Maillard damage despite the fact that these compounds are still positively charged and thus measured in the ordinary (or 'A') DBC reading seems due to the fact that they do not react with propionic anhydride, and thus, remain charged (and thus measured) in the second (or 'B') DBC determination. A group measured in both tests cancels out in the subtraction of one measure from the other. Thus, with skim milk powder, we see that the 'B' values rise with moderate heating. A hypothetical scheme is set out in Figure 14.1. There is now evidence that desoxyketosyl-lysine is still partly propionylated (to the extent of $30-40\%$) so that the DBL method does not fully reflect the extent of the early Maillard reaction (Hurrell, R.F.; unpublished results).

For protein foods containing little or no carbohydrate, dried chicken muscle serves as our example. Here it is the 'A' values that are lower in the damaged sample while the B values remain almost unchanged (Table 14.9).

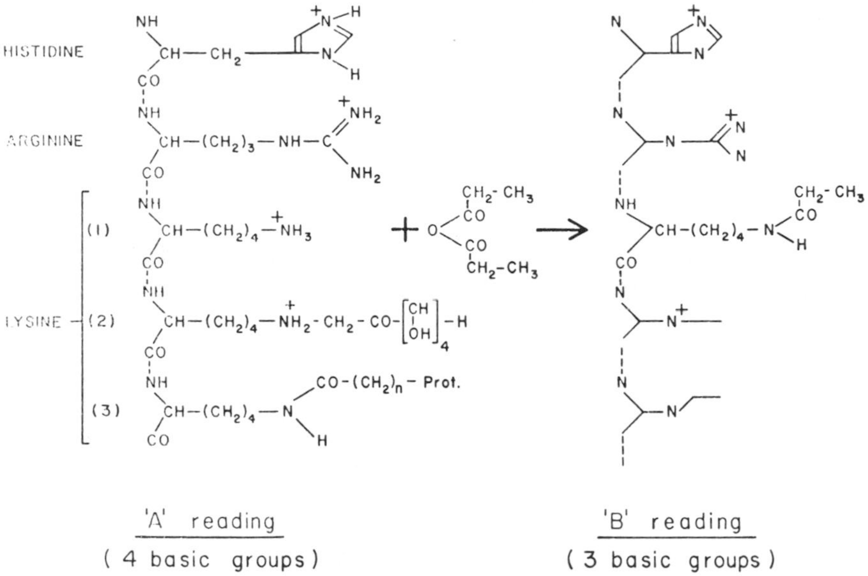

'A' reading

(4 basic groups)

'B' reading

(3 basic groups)

From Hurrell et al. (1979)

FIG.14.1.   POSTULATED SCHEME OF THE BASIC AMINO ACID UNITS IN PROCESSED FOOD PROTEINS AT ACID $_p$H AND THEIR REACTION WITH PROPIONIC ANHYDRIDE Lysine units are shown (1) unbound, (2) in a Maillard desoxy ketose compound with glucose (still basic) and (3) cross-linked to a glutamyl or aspartyl carboxyl group with a neutral imide link. Propionic anhydride is thought to react only with the first lysine unit, leaving the Maillard compound and arginine and histidine units still basic.

This would be consistent with ϵ-amino groups forming imide linkages with glutamine and asparagine units (cf. Bjarnason & Carpenter, 1970). Since such linkages are neutral they would not attract the dye, nor would they react with propionic anhydride.

There will almost certainly be interference from hydroxy-lysine and ornithine (as with the FDNB procedure) where these occur, nor is free lysine measured. Large quantities of free amino acids in a sample can interfere with the complete propionylation of ϵ-amino groups unless additional propionic anhydride is added. Fortunately, an excess of carbohydrate has had little effect in the tests carried out so far. Of course, it is always possible that a particular material may prove "impossible" for some reason. But the main remaining concern is an effect of concentration of dye after shaking for the standard time (we cannot really say "after equilibration") on the amount of dye bound. At least to standardize for this effect we are specifying fairly narrow limits within which readings are acceptable. This does, of course, mean that a greater weight of sample has to be taken for the B reading than for the A reading.

We understand that the procedure is being used in a number of laboratories without any special difficulties. Tame et al. (1978) worked at one-quarter scale for economy without any special problem but emphasized the need for careful sample preparation; a Tetsch centrifugal mill with a 0.2 mm screen proved satisfactory. Walker (1979) has applied it to leaf protein concentrates, and Almas and Bender (1980) to beans. Udy (private communication) has recommended an initial addition of 1 ml isopropanol to the test material, as a wetting agent. We also find that with the buffer used by Udy (1971) the results are less influenced by residual dye concentration (Hurrell, R.F. and Carpenter, K.J. unpublished results). The great advantage of the method is that it can be completed in 1–2 hours, does not involve a hydrolysis stage with its risk, nor automated equipment for ion-exchange chromatography.

GENERAL CONCLUSIONS

It is easy to get immersed in the details of analytical procedures, so that it becomes all the more difficult to stand back and consider for which real life problems these methods are going to be needed.

As we have noted previously, for the intensive animal feeding industry, the problem of providing 'least-cost' balanced diets is a clear-cut one. Protein is, in general, expensive so that requirements are met with minimal overage. Buyers of dried protein concentrates can work out, for example, how much cottonseed flours of different reactive lysine content are worth to them, and buy and re-formulate their mixes accordingly.

No such "rational" situation exists in human nutrition. The nearest approach is in breast milk-substitute foods where manufacturers and control agencies are concerned as to minimum levels of protein quality, especially in view of the protein typically coming from only one or two ingredients.

In Western countries there is, in general, no immediate problem of protein shortage and the wide range of foods contributing some protein to each individual's diet after infancy, makes the overall effect of damage to a single ingredient seem relatively trivial. Thus the milk used in the manufacture of chocolate bars is grossly damaged, and the products of the damaging reactions actually contribute to the flavor. But it is no worse than the gelatin of 'Jello' which inevitably has a zero chemical score.

Perhaps the realistic uses of control of protein quality for 'non-baby' foods is to ensure that foods traditionally of high protein quality do not lose their quality because of changes in processing methods; also that "substitutes" which are being introduced to the public as replacements for meat, i.e., textured vegetable protein isolates, should have minimum quality standards.

Again, one envisages two points of control: (i) at the manufacturing plant and (ii) at a laboratory representing the consumer. In the first instance the laboratory will have access to the raw materials and so be able to monitor changes due to processing. Provided that initial biological assays have confirmed that reactive lysine is a reasonably sensitive measure of damage that might take place, dye-binding would be the most rapid routine procedure and one not requiring special expertise.

In the case of milk powders, it has been found that even a simple ferricyanide reducing test can be used to give a good indication of the formation of lactulosyl-lysine (R.F. Hurrell, unpublished result). Tannenbaum (1974) has suggested that an experienced operator of a milk drying plant could be expected to control quality by inspection and simple physical properties of the products. This is obviously true for gross damage, but it seems very difficult to detect even as much as 15% damage to the lysine by this means.

Where time permits in a "consumer" laboratory and some of the samples can be given a 72-hour test, then digestion with pronase followed by column chromatographic analysis for lysine and methionine would seem promising. At the same time experience could be built up to determine how good the rapid dye binding-lysine procedure was at predicting which samples would and would not pass the longer test.

As with any routine quality control procedure the standard of work may slip without this being picked up unless good and bad 'coded' standards are regularly included for analysis.

REFERENCES

ALMAS, K. and BENDER, A.E.   1979.   Relation between storage, cooking time and available lysine in cow peas.   Proc. Nutr. Soc. *39*:115A.

ASHWORTH, U.S.   1966.   Determination of protein in dairy products by dye binding.   J. Dairy Sci. *49*:133.

ATKINSON, J. and CARPENTER, K.J.   1970.   Nutritive value of meat meals.   2. Influence of raw materials and processing on protein quality.   J. Sci. Fd. Agric. *21*:366.

BHATTY, R.S. and WU, K.W.   1975.   Lysine screening in barley with a modified Udy dye-binding method.   Can. J. Plant Sci. *55*:685.

BJARNASON, J. and CARPENTER, K.J.   1969.   Mechanisms of heat damage in proteins.   1. Models with acylated lysine units. Brit.   J. Nutr. *23*:859.

BJARNASON, J. and CARPENTER, K.J.   1970.   Mechanisms of heat damage in proteins.   2. Chemical changes in pure proteins.   Brit. J. Nutr. *24*:313.

BOCTOR, A.M. and HARPER, A.E.   1968.   Measurement of available lysine in heated and unheated foodstuffs by chemical and biological methods.   J. Nutr. *94*:289.

BODWELL, C.E.   1976.   Status of chemical methods to determine biologically available lysine in wheat proteins.   Proc. 9th Natl. Conf. on Wheat Utilization, Seattle:181.

BODWELL, C.E.   1977.   Problems associated with the development and application of rapid methods of assaying protein quality.   Nutr. Reports Int. *16*:163.

BOOTH, V.H.   1971.   Problems in the determination of FDNB-available lysine.   J. Sci. Food Agric. *22*:658.

BUJARD, E. and FINOT, P.A.   1978.   Measurement of the availability and of the blockage of lysine in industrial milk products.   Ann. Nutr. Aliment. *32*:291.

CARPENTER, K.J.   1960.   The estimation of the available lysine in animal-protein foods.   Biochem. J. *77*:604.

CARPENTER, K.J. and BOOTH, V.H.   1973.   Damage to lysine in food processing: its measurement and its significance.   Nutr. Abst. and Review *43*:423.

CARPENTER, K.J. and ELLINGER, G.M.   1955.   Estimation of 'Available Lysine' in protein concentrates.   Biochem. J. *61*:xi.

CARPENTER, K.J. and MARCH, B.E.   1961.   The availability of the lysine in groundnut biscuits used in the treatment of kwashiorkor.   2. Brit. J. Nutr. *15*:403.

CARPENTER, K.J. and OPSTVEDT, J.   1976.   Application of chemical and biological assay procedures for lysine to fish meals.   J. Agric. Fd. Chem. *24*:389.

CARPENTER, K.J. and WOODHAM, A.A.   1974.   Protein quality of feeding-stuffs.   6. Comparisons of the results of collaborative biological assays for amino acids with those of other methods.   Brit. J. Nutr. *32*:647.

CHEFTEL, J.C.   1977.   Chemical and nutritional modifications of food proteins due to processing and storage.   In: Food Proteins, J.R. Whitaker and S.R. Tannenbaum (Editors).   AVI Publishing Co., Westport, Conn., p. 401.

CLEGG, K.M.   1960.   The availability of the lysine in groundnut biscuits used in the treatment of kwashiorkor.   Brit. J. Nutr. *14*:325.

CLIFFORD, M.N., TANG, S.L. and EYO, A.A.   1979.   The development of analytical methods for investigating chemical changes during fish smoking.   Proc. Internat. Congr. Fish Sci. Technol., Aberdeen, Scotland. (Abstr.), p. 22.

COUCH, J.R.   1975.   Collaborative study of the determination of available lysine in proteins and feeds.   J. Assoc. Offic. Anal. Chem. *58*:599.

COUCH, J.R. and THOMAS, M.C.   1976.   A comparison of chemical methods for the determination of available lysine in various proteins.   J. Agric. Fd. Chem. *24*:943.

DATTA, S. and DATTA, S.C.   1976.   Thin layer chromatographic estimation of available lysine in dried milk powder.   J. Assoc. Off. Anal. Chem. *59*:1255.

DELLA MONICA, E.S., STROLLE, E.O. and McDOWELL, P.E.   1976.   A modified method for determining available lysine in protein recovered from heat-treated potato juice.   Anal. Biochem. *73*:274.

DVORAK, Z.   1972.   The use of hydroxyproline to predict the nutritional value of the protein in different animal tissues.   Brit. J. Nutr. *27*:475.

DVORAK, Z. and VOGNAROVA, I.   1965.   Available lysine in meat and meat products.   J. Sci. Fd. Agric. *16*:305.

DVORAK, Z. and VOGNAROVA, I.   1969.   The nutritive value of organ and tissue proteins from slaughter animals and its determination.   Nahrung. *13*:81.

EKLUND, A.   1976.   On the determination of available lysine in casein and rapeseed protein concentrates using TNBS as a reagent for free amino group of lysine.   Anal. Biochem. 70:434.

ELLINGER, G.M. and DUNCAN, A.   1976.   The determination of methionine in proteins by gas-liquid chromatography.   Biochem. J. 155:615.

ERBERSDOBLER, H.F.   1979.   Protein quality studies on the annual production of skim milk powder from a milk drying plant.   2. Damage to lysine during milk drying.   Milchwissenschaft. 34:325.

ERBERSDOBLER, H. and ZUCKER, M.   1966.   Lysine and available lysine in dried skim milk.   Milchwissenschaft. 21:564.

EVANS, I.M. and BOULTER, D.   1974.   Chemical methods suitable for screening for protein content and quality in cowpea (Vigna unguiculata) meals.   J. Sci. Fd. Agric. 25:311.

EVANS, R.J. and BANDEMER, S.L.   1967.   Nutritive value of legume seed proteins.   J. Agric. Fd. Chem. 15:439.

FINLEY, J. and FRIEDMAN, M.   1973.   Chemical methods for available lysine.   Cereal Chem. 50:101.

FINOT, P.A., BUJARD, E., MOTTU, F. and MAURON, J.   1977.   Availability of the true Schiff's bases of lysine: chemical evaluation of the Schiff's base between lysine and lactose in milk.   In: Protein Crosslinking B. Nutritional and Medical Consequences, M. Friedman (Editor).   Plenum Press, New York. p. 321.

FINOT, P.A. and MAURON, J.   1972.   Le blocage de la lysine par la reaction de Maillard.   II.   Properties chimiques des derives N-(desoxy-1-D-fructosyl-1) et N-(desoxy-1-D-lactulosyl-1) de la lysine.   (Blockage of lysine by the Maillard reaction N-(desoxy-1-D-fructosyl-1) and N-(desoxyl-1-D-lactulosyl-1) lysine).   Helv. Chim. Acta 55:1153.

FORD, J.E. and HEWITT, D.   1979.   Protein quality in cereals and pulses.   1. Application of microbiological and other in vitro procedures in the evaluation of rice, sorghum, barley and field beans.   Brit. J. Nutr. 41:341.

FRIEDMAN, M.   1977.   Effects of lysine modification on chemical, physical, nutritive and functional properties of proteins.   In: Food Proteins, J.R. Whitaker and S.R. Tannenbaum (Editors).   AVI Publishing Co., Westport, Conn., p. 446.

FRIEDMAN, M. and FINLEY, J.W.   1975.   Vinyl compounds as reagents for available lysine in proteins.   In: Protein Nutritional Quality of Foods and Feeds.   Part I. Assay Methods-Biological, Biochemical and Chemical, M. Friedman (Editor).   Marcel Dekker, New York. p. 503.

HAMAD, A.M. and FIELDS, M.L.   1979.   Evaluation of the protein quality and available lysine of germinated and fermented cereals.   J. Fd. Sci. 44:456.

HENRY, K.M. and KON, S.K.   1950.   Effects of reaction with glucose on the nutritive value of casein.   Biochim. Biophys. Acta 5:455.

HENRY, K.M., KON, S.K., LEA, C.H. and WHITE, J.C.D.   1948.   Deterioration on storage of dried skim milk.   J. Dairy Res. 15:292.

HURRELL, R.F. and CARPENTER, K.J.   1974.   Mechanisms of heat damage in protein.   4. The reactive lysine content of heat-damaged material as measured in different ways.   Brit. J. Nutr. *32*:589.

HURRELL, R.F. and CARPENTER, K.J.   1977.   Mechanisms of heat damage in proteins 8.   The role of sucrose in the susceptibility of protein foods to heat damage.   Brit. J. Nutr. *38*:285.

HURRELL, R.F. and CARPENTER, K.J.   1978.   Digestibility and lysine value of proteins heated with formaldehyde or glucose.   J. Agr. Fd. Chem. *26*:796.

HURRELL, R.F. and CARPENTER, K.J.   1980.   The estimation of available lysine in foodstuffs after Maillard reactions.   In "Maillard Reactions in Foods", Proceedings of Symposium, Uddevalla, Sweden.

HURRELL, R.F., LERMAN, P. and CARPENTER, K.J.   1979.   Reactive lysine in food-stuffs as measured by a rapid dye-binding procedure.   J. Food Sci. *44*:1221.

HUTCHINSON, J.B., MORAN, T. and PACE, J.   1960.   The quality of the protein in germ and milk breads as shown by the growth of weanling rats: the significance of the lysine content.   J. Sci. Fd. Agric. *11*:576.

JONES, G.P. and LAKIN, A.L.   1976.   Determination of lysine in barley by a modified dye-binding procedure.   Proc. Nutr. Soc. *35*:44A.

KAKADE, M.L. and LIENER, I.E.   1969.   Determination of available lysine in proteins.   Anal. Biochem. *27*:273.

LEA, C.H. and HANNAN, R.S.   1950.   Reaction between proteins and reducing sugars in the dry state.   III. Nature of the protein groups reacting.   Biochim. Biophys. Acta *5*:433.

LIENER, I.E.   1958.   Effect of heat on plant proteins.   In: Processed Plant Protein Foodstuffs, A.A. Altschul (Editor).   Academic Press, New York, p. 79.

LIPTON, S.H. and BODWELL, C.E.   1974.   Chemical approaches for estimating nutritionally available methionine.   In: Protein Nutritional Quality of Foods and Feeds, Part I.   Assay Methods—Biological, Biochemical, and Chemical, M. Friedman (Editor).   Marcel Dekker, New York, p. 569.

MATHESON, N.A.   1968.   Available lysine.   II. Determination of available lysine by dinitrophenylation.   J. Sci. Fd. Agric. *19*:496.

MAURON, J. and BUJARD, E.   1964.   Guanidination, an alternative approach to the determination of available lysine in foods.   Proc. 6th Internat. Congr. Nutr. (Edinburgh), p. 498 (Abstr.).

MAURON, J., MOTTU, F., BUJARD, E. and EGLI, R.H.   1955.   The availability of lysine, methionine and tryptophan in condensed milk and milk powder, *in vitro* digestion studies.   Arch. Biochem. Biophys. *59*:433.

MEANS, G.H. and FEENEY, R.E.   1971.   The Chemical Modification of Proteins.   Holden-Day, San Francisco.

MILLER, D.S. and NAISMITH, D.J.   1958.   A correlation between sulphur content and net dietary-protein value.   Nature, Lond. *182*:1786.

MILLER, E.L., CARPENTER, K.J. and MILNER, C.K.   1965.   Availability of sulphur amino acids in protein foods.   3. Chemical and nutritional changes in heated cod muscle.   Brit. J. Nutr. *19*:547.

MILNER, C.K. and CARPENTER, K.J.   1969.   Effect of wet heat-processing on the nutritive value of whole wheat protein.   Cereal Chem. *46*:425.

MOSSBERG, R.   1968.   Evaluation of protein quality by dye-binding; a tool in plant breeding.   FAO/IAEA Panel on New Approaches for Breeding for Plant Protein Improvement, Rostanga, Sweden.   International Atomic Energy Agency – Vienna, p. 151.

MOORE, S.   1963.   On the determination of cystine as cysteic acid.   J. Biol. Chem. *238*:235.

MORAN, E.T., JENSEN, L.S. and McGINNIS, J.   1963.   Dye binding by soybean and fish meal as an index of quality.   J. Nutr. *79*:239.

MOTTU, F. and MAURON, J.   1967.   The differential determination of lysine in heated milk.   II. Comparison of the *in vitro* methods with the biological evaluation.   J. Sci. Fd. Agric. *18*:57.

NAIR, B.M., LASER, A., BURVALL, A. and ASP, N.G.   1978.   Gas chromatographic determination of available lysine.   Food Chem. *3*:283.

PORTER, W.M., MANER, J.H., AXTELL, J.D. and KEIM, W.F.   1974.   Evaluation of the nutritive quality of grain legumes by an analysis of total sulfur.   Crop Sci. *14*:652.

POSATI, L.P., HEBINGER, V.H., De VILLBLISS, E.D. and PALLANSCH, M.J.   1972.   Factors affecting the determination of available lysine in whey with 2, 4, 6-trinitrobenzene sulfonic acid.   J. Dairy Sci. *55*:1660.

RAO, S.R., CARTER, F.L. and FRAMPTON, V.L.   1963.   Determination of available lysine in oilseed proteins.   Anal. Chem. *35*:1927.

RAYNER, C.J. and FOX, M.   1978.   Measurement of available lysine in processed beef muscle by various laboratory procedures.   J. Agric. Fd. Chem. *26*:494.

RIESEN, W.W., CLANDININ, D.R., ELVEHJEM, C.A. and CRAVENS, W.W.   1947.   Liberation of essential amino acids from raw, properly heated, and overheated soybean oil meal.   J. Biol. Chem. *167*:143.

RUDERUS, H. and KILBERG, R.   1972.   Estimation of available lysine in food proteins.   Rept. no. 33, Dept. of Applied Microbiology.   Karolinska Institute, Stockholm.

SANDLER, L. and WARREN, F.L.   1974.   Effect of ethyl chloroformate on the dye binding capacity of protein.   Anal. Chem. *46*:1870.

SANGER, F.   1945.   The free amino groups of insulin.   Biochem. J. *39*:507.

SERAFINI-CESSI, F. and CESSI, C.   1964.   An improved method for determination of hydroxyproline in protein hydrolyzates.   Anal. Biochem. *8*:527.

SJOBERG, L.B. and BOSTRAM, S.L.   1977.   Studies in rats on the nutritional value of hydrogen peroxide-treated fish protein and the utilization of oxidized sulphur amino acids.   Brit. J. Nutr. *38*:189.

TAME, M.J., NEEDHAM, D. and FISHER, C.   1978.   Protein Evaluation Group News Sheet, no. 4 A.P. Williams (Editor).   National Institute for Research in Dairying, Reading, England, p. 7.

TANNENBAUM, S.   1974.   Industrial Processing. In: Nutrients in Processed Foods: Proteins, P.L. White and D.C. Fletcher (Editors).   Publishing Sciences Group Inc., Acton, Mass., p. 131.

TELLA, A.F. and ASHTON, W.M.   1978.   Studies on 2,4,6-Trinitobenzene-sulphonic acid-reactive lysine using cowpeas, maize and DL-lysine.   J. Sci. Fd. Agric. 29:447.

UDY, D.C.   1971.   Improved dye method for estimating protein.   J. Amer. Oil Chem. Soc. 48:29A.

VARNISH, S.A. and CARPENTER, K.J.   1975.   Mechanisms of heat damage in proteins.   5. The nutritional values of heat-damaged and propionylated proteins as sources of lysine, methionine and tryptophan.   Brit. J. Nutr. 34:325.

WALKER, A.   1979.   The use of dye binding for the concurrent determination of protein and reactive (available) lysine in leaf protein concentrate.   Brit. J. Nutr. 42:445.

WALZ, O.P. and FORD, J.E.   1973.   The measurement of "available lysine" in protein foods: a comparison of chemical, biological and microbiological methods.   Z. Tierphysiol. Tierernahr Futtermitt. 30:304.

YOUNG, V.R., FAJARDO, L., MURRAY, E., RAND, W.N. and SCRIMSHAW, N.S.   1975.   Protein requirements of man: comparative nitrogen balance response within the submaintenance-to-maintenance range of intakes of wheat and beef proteins.   J. Nutr. 105:534.

DISCUSSION

DR. BODWELL: Dr. Carpenter, using one of your procedures for available lysine (the classic Carpenter procedure, I believe it was) some South Americans looked at a very limited number of samples, four or five, and compared available lysine values with digestibility as determined both in humans and rats. They found a high correlation between apparent digestibility and available lysine. With your classical methods, or with the dye binding method, has there been any comparative studies done relating digestibility, and *in vitro* chemical estimates of available lysine?

DR. CARPENTER: There has been no systematic work that I know of on the evaluation with humans of proteins judged from rat experiments to have been damaged by processing. There is just the one study in Uganda, mentioned in my paper, from which it seemed to come out that accidental lactose-protein damage had reduced the value of a protein supplement for children just recovering from kwashiorkor. With milk powder where the damage to reactive lysine can occur at very low temperatures, it seems that there can be virtually no fall in N digestibility for rats even when there is a great fall specifically in available lysine and in biological value. But where the damage is at higher temperatures, we do know of cases, particularly as Boctor and Harper showed and we found also, where if there's a very small amount of carbohydrate and very severe processing conditions, the fall in digestibility is considerably greater than we would predict from the fall in FDNB lysine. So we have to say that we cannot use the test to predict digestibility every time. However, it seems to apply to practically every situation that if there is no fall in reactive lysine there is no other kind of damage detectable.

DR. SATTERLEE: For the Amadori intermediate compounds you have shown us, what is the stability of those intermediate compounds during normal heat processing of foods?

DR. CARPENTER: I would like to pass that question to my former colleague, Dr. Richard Hurrell.

DR. HURRELL: The lactose-lysine in milk is the only form of sugar-lysine compound which is stable. This compound is formed under mild heating conditions and is often present right through storage. With further heat, it decomposes to the brown compounds which can react with other food components. Methionine and tryptophan could also react with the break-down products of the first formed sugar-lysine compound.

15

In Vitro Assays of Protein Quality Assays Utilizing Enzymatic Hydrolyses

Nina L. Marable and George Sanzone

INTRODUCTION

Although the amino acid composition of a protein is fundamentally related to its nutritional quality, protein quality cannot be reliably predicted from amino acid composition alone. Obviously bioavailability is one factor which must also be considered. Digestibility, absorbability, or utilizability (i.e., bioavailability) may be affected by characteristics of the protein itself including protein conformation, intramolecular bonding, and modification of amino acids (e.g. isomerization or oxidation). Bioavailability may also be affected by the matrix of the protein in the foodstuff, such as containment of the protein inside indigestible cell walls, the presence of inhibitors or toxic factors, and intermolecular bonding.

A list of some of these characteristics of protein foodstuffs which may affect protein quality is shown in Table 15.1. Amino acid composition measurements reflect primarily certain aspects of primary structure, and to a lesser extent are predictors of secondary structure. All other characteristics listed in Table 15.1 may significantly affect bioavailability. If the effects of these characteristics are to be assessed *in vitro*, amino acid composition measurements must be complemented with other assays.

Assays combining enzymatic hydrolyses with amino acid analysis represent one attempt to take all of these characteristics into account in an *in vitro* assay of protein quality. The use of enzymatic assays in the assessment of availability of individual amino acids or the estimation of protein nutritional quality in general has been reviewed several times (Sheffner, 1967; Mauron, 1970a, 1970b, 1973; Menden and Cremer, 1970; Stahmann and Woldegiorgis, 1975; Bodwell, 1977). These reviews cite many excellent references, most of which will not be repeated here.

The recent review by Bodwell (1977) tabulates some of the more familiar enzymatic assay techniques, including the pepsin digest residue (PDR)

TABLE 15.1.   "STRUCTURE" OF A PROTEIN FOODSTUFF: SOME PROTEIN FOOD-
STUFF CHARACTERISTICS WHICH MAY AFFECT PROTEIN QUALITY

A.   STRUCTURE OF THE PROTEIN

> *Primary Structure*
>
>> Amino acid composition: Total and relative amounts of
>> amino acids (may be modified by processing)
>> .Sequence
>
> *Secondary Structure*
>
>> Helix, β-conformation, other (may be modified by processing)
>> Intramolecular bonding (may be modified by processing)
>
> *Tertiary Structure* (may be modified by processing)
> *Quaternary Structure*

B.   MATRIX OF THE PROTEIN

> Cell Walls
> Inhibitors
> Intermolecular Bonding (may be modified by processing)

index of Sheffner *et al.* (1956), the pepsin pancreatin index (PPD) method of
Akeson and Stahmann (1964), Ford and Salter's gel filtration modification
(Ford and Salter, 1966; Ford, 1973), the pepsin pancreatin digest dialyzate
(PPDD) index of Mauron (1970a, 1973) and Satterlee's multi-enzyme digestion method for estimating a digestibility factor for use in predicting protein
efficiency ratio (PER) (Satterlee *et al.*, 1977). This review will concentrate
on the use of the PDR and PPDD methods of estimating protein nutritional
quality. The PPDD index will be considered from an experimental and
calculational point of view. Satterlee's method will not be considered since
it represents a different approach in that it measures a property proportional to the sum of digestibilities of all amino acids (see Chapter 18),
whereas the PDR and PPDD methods can measure both individual and additive digestibilities.

GENERAL DESCRIPTION OF THE PDR AND PPDD INDICES

As Bodwell (1977) has pointed out, both the PDR and PPDD index methods use similar experimental techniques and identical calculation methods:

1. Enzymatic digestion is carried out on the test protein and a reference
    (egg) using pepsin (Sheffner *et al.*, 1956) or pepsin and pancreatin
    (Mauron, 1970a, 1973).
2. The digested fraction or fractions and the total protein of the test
    substance and the reference are analyzed for amino acid composition.
3. Calculations in both methods are performed as described by Sheffner
    et al. (1956) to arrive at an estimated protein quality.

The experimental design in the PPDD technique is especially interesting.
According to Mauron (1970a, 1973), the proteins are first digested with

pepsin and then pancreatin. During the pancreatin digestion, the sample is contained in a dialysis bag which is suspended in buffer. At intervals during the pancreatin digestion, the dialyzing buffer is siphoned from the apparatus and replaced with fresh buffer. The siphoned fractions are pooled, concentrated, and analyzed for amino acid composition. Analysis of the concentration of each of the *separate* siphoned fractions would produce a series of numbers (A, B, C, D, Fig. 15.1) from which the average rates of digestion for each amino acid over the time periods between siphonings could be calculated. Similarly, compositing siphoned fractions and analyzing the composite produces an "average" composition ($\bar{X}$, Fig. 15.1) which corresponds, for each amino acid, to the composition at some fixed time ($t_{\bar{X}}$, Fig. 15.1) during the digestion. These composition numbers, $\bar{X}_i$ for each amino acid, are thus proportional to the average rates of digestion for each amino acid over the fixed time, $t_{\bar{X}}$.

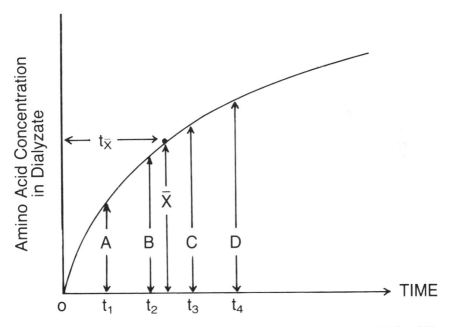

FIG. 15.1.   AMINO ACID CONCENTRATION CHANGE IN DIALYZATE DURING A PEP-SIN-PANCREATION DIGEST DIALYZATE EXPERIMENT:
A, B, C and D = concentration in dialyzate
t_1, t_2, t_3 and t_4 = siphoning times

USEFULNESS OF THE PDR AND PPDD INDICES

The PDR and PPDD experiments are necessarily more time-consuming than amino acid analysis of total protein alone. Sheffner (1967), Mauron (1970a, 1973), and Bodwell (1977) have addressed the question of whether or not the extra time and effort produce any additional information compared to amino acid analysis of the total protein alone.

Sheffner (1967) and Mauron (1970a, 1973) are relatively optimistic about the usefulness of their methods. Mauron (1970a) suggests that a principle advantage of the PPDD correlation is that it avoids the systematic bias of the high predictions which are characteristic of Essential Amino Acid Index (EAAI) calculations. Avoiding systematic bias (i.e. producing a PPDD index or any other index versus BV plot which has a zero intercept) is attractive because it suggests that the experiments and the equation used to calculate the index have considered all or most of the correct factors. In other words, it suggests that the experiments and the equation are somehow fundamentally correct. This suggested advantage of avoiding systematic bias is an advantage only for groups of proteins and not for individual proteins, unless the correlation also improves. If the correlation does not improve, the reliability of individual predictions does not improve. Mauron (1970a) points out that the correlation between the PPDD index and measured biological value (BV) is not better than the correlation of chemical score (Block and Mitchell, 1946−47) or Oser's essential amino acid index (EAAI) (Oser, 1951) with biological value.

Mauron (1970a, 1973) suggests that the PPDD index should be particularly useful for predicting protein quality of processed proteins. In fact, the PPDD predicted value for the soy isolate studied is 77 while the measured biological value is 46. The EAAI calculation estimates 82 (Mauron, 1973). Thus, the PPDD prediction is 67% high and the EAAI prediction is 78% high. Bodwell (1977) concludes on the basis of comparisons of PDR or PPDD index predictions with the results of both human and rat bioassays (Table 15.2) that these enzymatic assays do not result in improved esti-

TABLE 15.2.   COMPARISON OF ENZYMATIC ASSAY INDEX AND ESSENTIAL AMINO ACID INDEX PREDICTIONS WITH THE RESULTS OF RAT AND HUMAN BIOASSAYS[1]

Method And Protein Source	Index Value	Measured BV	EAAI	Percent Difference From Measured Value	
				Index	EAAI
PDR					
Egg Albumen	95	91[2]	96	3	5
		65[2]		46	48
Soy flour	69	66[2]	81	5	23
White flour	49	41[2]	64	20	56
PPDD					
Corn (cooked)[3]	68	50	74	36	48
White bread[3]	55	54	64	2	19

[1]From Bodwell (1977), pp. 94−95; Sheffner (1967), p. 162; and Mauron (1973), pp. 150−151.
[2]Literature values, obtained with human subjects; original references cited in Bodwell (1977).
[3]From Mauron (1973); rat assays, PPDD index, and EAAI were obtained on the same protein sources.

mates of protein quality for individual proteins compared to estimates derived from chemical scores or essential amino acid index values. The human assays were conducted separately from the enzymatic assays. Mauron (1973) has measured the PPDD index and biological value on the same protein sources but in many studies this is not done.

PROBLEMS WITH PDR AND PPDD METHODS

If these methods do not yield better predictions than amino acid analysis alone, there are three possible conclusions:

1. The theoretical basis for attempting to improve on simple amino acid analysis is unsound.
2. The experimental approaches are incorrect, or at least incomplete.
3. The results of handling the data during the final calculations are misleading.

If the first conclusion is rejected, then perhaps it is worthwhile to consider both the nature of the experiments and the calculations in more detail in an attempt to devise an improved prediction method.

Experimental Method

Basically it appears that the measurements made in the PPDD experiment are appropriate, since all of the factors listed in Table 15.1, including amino acid composition, are likely to influence rate of digestion. Some changes in experimental design could be considered. It is possible that average rates obtained over a different (shorter?) period of time might be more characteristic of the particular proteins. For example, initial rates might be more appropriate, although they also might be more difficult to obtain due to very low initial levels of amino acids in the dialyzate. It might be possible to simplify the experiments, since measurement of the composition of one siphoned fraction, taken at time $t_{\bar{x}}$, should be equivalent to the measurement Mauron describes. In addition, some investigators (Stahmann and Woldegiorgis 1975; Szmelcman and Guggenheim, 1967) have suggested that dialysis does not greatly influence the composition of the digest. If this is correct, the experiment could be greatly simplified by eliminating the dialysis as Stahmann and Woldegiorgis (1975) have done. On the other hand, the measurements suggested by Mauron may be entirely adequate, and the dialysis may affect the progress of digestion of some proteins. In any case, there does not seem to be anything basically wrong with the experiment.

Calculations

The calculations for the PPDD experiments are performed as described by Sheffner et al., (1956). Putting the various steps in Sheffner's description of the calculations together into one equation yields

(1) PDR or PPDD $=$

$$\left[\frac{\Sigma a_i}{\Sigma e_i}\left\{\frac{\Pi\dfrac{a_i}{\Sigma a_i}}{\Pi\dfrac{e_i}{\Sigma e_i}}\right\}^{1/n}\right]^{\frac{\Sigma e_i}{\Sigma E_i}}\left[\frac{\Sigma A_i-\Sigma a_i}{\Sigma E_i-\Sigma e_i}\left\{\frac{\Pi\dfrac{A_i-a_i}{\Sigma(A_i-a_i)}}{\Pi\dfrac{E_i-e_i}{\Sigma(E_i-e_i)}}\right\}^{1/n}\right]^{\frac{\Sigma(E_i-e_i)}{\Sigma E_i}}$$

In this equation a_i = amino acid composition for amino acid (i) of the digest or the composite of the digest—dialyzate fractions for the test protein,

e_i = amino acid composition for amino acid (i) of the digest or the composite of the digest—dialyzate fractions for egg,

A_i = total amino acid composition for amino acid (i) for the test protein, and

E_i = total acid composition for amino acid (i) for egg.

According to arguments just presented, the a_i's and e_i's are proportional to rates of digestion of the individual amino acids in the test protein or egg. The symbols Σ and π have the usual meanings of "sum" and "product". All sums and products include all the essential amino acids plus histidine, cystine, and tyrosine. Methionine and cystine and phenylalanine and tyrosine are considered together with some specific reservations. Therefore, there are nine terms in all sums and products, and thus the exponent $1/n = \dfrac{1}{9}$.

In spite of the limitations imposed by Sheffner that the ratios

$$\frac{a_i/\Sigma a_i}{e_i/\Sigma e_i} \quad \text{and} \quad \frac{(A_i-a_i)/\Sigma(A_i-a_i)}{(E_i-e_i)/\Sigma(E_i-e_i)}$$

should not be larger than 1 or less than 0.01 for a given amino acid, equation (1) will reduce to a simpler form. Therefore for purposes of simplicity and discussion, let

$$(2)\ \text{PPDD} \cong \left(\sqrt[n]{\prod_i^n \frac{a_i}{e_i}}\right)^{\frac{\Sigma e_i}{\Sigma E_i}} \left(\sqrt[n]{\prod_i^n \frac{A_i-a_i}{E_i-e_i}}\right)^{\frac{\Sigma E_i-\Sigma e_i}{\Sigma E_i}}$$

The first term in the equation is based on the composition of the digest of the test protein relative to the digest of egg. If the a_i's are proportional to rates of digestion for each amino acid as we have suggested, the first term is the geometric mean of individual amino acid digestion rates in the test protein relative to egg. The weighting exponent, $\Sigma e_i/\Sigma E_i$, is obviously based on the proportion of the amino acids of egg which appear in the digest. The second term is based on the composition of the undigested residue of the protein and again is a geometric mean with a weighting exponent derived from the composition of the undigested portion of the egg.

The general form of the equation, that is, the use of geometric means, is related to the form of the equation for calculating the essential amino acid index. Oser (1970) has pointed out that the basic rationale for the use of a geometric mean of amino acid levels relative to egg is the fact that "this is a probability concept". Therefore, according to Oser, a geometric mean (the root of a product of numbers) rather than an arithmetic mean is appropriate.

Although the general form of equations (1) and (2) may be reasonable, consideration of the details of the calculations reveals some difficulties. First, calculation of geometric means, by nature, has a leveling effect on the individual amino acid ratios which contribute to the means. For example, calculation of the a_i/e_i ratios from Mauron's data (Mauron, 1970a) for overheated roller-dried milk shows that while the individual ratios, (a_i/e_i), range from 0.33 for lysine to 1.14 for tryptophan, the geometric mean of the ratios, with none > 1 counted, is 0.66. Similarly, for soy the individual ratios range from 0.40 to 1.36 but the geometric mean is 0.90 (Table 15.3). It can be assumed that the crucial factor in determining protein quality is the timing of the availability of a proper mix of amino acids and/or peptides after digestion as Stahmann (Stahmann and Woldegiorgis, 1975) and many others have suggested. If this is true, a high rate of digestion for one amino

TABLE 15.3. RELATIVE RATES OF DIGESTION OF INDIVIDUAL AMINO ACIDS FOR SEVERAL PROTEIN SOURCES COMPARED TO EGG[1]

Amino Acid	Soya Meal	Over-heated roller-dried milk	Spray-dried milk	Microbial protein	Peanut meal	Fish meal
Lys	1.00	0.33	1.57	1.05	0.95	2.57
His	1.00	0.50	0.67	0.83	0.50	1.17
Thr	1.25	0.75	0.75	1.75	0.50	1.25
Val	1.25	0.67	0.58	1.00	0.50	1.17
Met	0.40	0.53	0.53	0.47	0.33	1.01
Cys	1.00	0.67	0.00	0.00	0.00	0.67
Met+Cys	(0.50)	(0.56)	(0.44)	(0.39)	(0.28)	(1.00)
Ile	1.36	0.64	0.45	1.09	0.64	1.18
Leu	0.95	0.98	0.84	0.79	0.67	1.00
Phe	1.03	0.80	0.70	0.67	0.90	0.93
Tyr	1.00	0.92	0.88	0.68	0.68	1.04
Phe+Tyr	(1.02)	(0.85)	(0.78)	(0.67)	(0.80)	(0.98)
Try	0.86	1.14	1.29	0.57	0.57	1.14
Geometric Mean[2]	0.90	0.66	0.70	0.77	0.57	1.00

[1]From the data of Mauron (1970a); rates of digestion are defined as the ratio of digest composition of amino acid (i) for test protein to digest composition of the same amino acid for egg: a_i/e_i.

[2]Calculated as $\sqrt[9]{\prod\limits_{i}^{n} \dfrac{a_i}{e_i}}$, with no ratio > 1 or < 0.01 and methionine and cystine and phenylalanine and tyrosine considered as pairs.

acid is not likely to compensate for a low rate for another amino acid in making a contribution to protein nutritional quality, and it would be incorrect to calculate the geometric or any other mean of individual digestion rates. Second, the weighting of the terms in equations (1) or (2) produced by the exponents

$$\frac{\Sigma e_i}{\Sigma E_i} \quad \text{and} \quad \frac{\Sigma(E_i - e_i)}{\Sigma E_i}$$

further dilutes the effect of variations in the composition of the digests (a_i's). This is especially true in the first term of equation (1) or (2) where the exponent calculated from Mauron's data (1970a) is 0.36. When the geometric means for over-heated roller dried milk and soy (0.66 and 0.90, respectively) are raised to the 0.36 power, the first terms in the simplified equation become 0.86 and 0.96. Thus, the PPDD experiment may reveal big differences between proteins in terms of individual a_i/e_i ratios. However the subsequent arithmetical manipulations dilute these differences substantially. Finally, the form of the equation, including the use of a term based on the digest and one based on the residue, is arbitrary, as Mauron (1970a) has pointed out. The choice of one set of weighting exponents based on the digestion of egg which is applied to all test proteins seems especially arbitrary. Therefore, careful consideration of the calculation of the PDR and PPDD indices suggests that the reason for the failure to obtain good predictions based on the enzymatic digestion experiments may be that we have taken a simple, good idea and a reasonably simple, good experiment and incorrectly manipulated the numbers.

ALTERNATIVE CALCULATIONS

Several equations which would make use of the data generated by the PPDD method (Mauron, 1970a) are fundamentally attractive as alternatives to a PDR—PPDD type calculation. For example, reconsideration of the protein foodstuff characteristics in Table 15.1 suggests that rate of digestion may reflect *all* of these characteristics, including to some extent, amino acid composition. If protein nutritional quality (PNQ) is determined by rate of digestion, and if mean rates are rejected based on previous arguments, then an appropriate general equation might be

$$(3) \qquad \text{PNQ} = \text{function of} \left(\frac{a_i}{e_i} \right)_L$$

where $(a_i/e_i)_L$ = the lowest or limiting rate for an individual amino acid relative to the rate for the same amino acid in egg. Stahmann (Stahmann and Woldegiorgis, 1975) has proposed the calculation of an Enzyme Score which is essentially $(a_i/e_i)_L$. However, they do not describe this as a rate term, and they suggest a linear relationship between protein quality and Enzyme Score.

In order to examine the possibility of a relationship between PNQ and $\left(\dfrac{a_i}{e_i}\right)_L$, data from Mauron (1970a) for six protein sources were used to calculate two values of an $(a_i/e_i)_L$ for each source as shown in Table 15.4.

TABLE 15.4.   LOWEST RATE OF DIGESTION FOR SEVERAL PROTEIN SOURCES[1]

Protein Source	$(a_i/e_i)_L$		Amino Acid Used	
	I	II	I	II
Soy	0.50	0.40	Met + Cys	Met
Over-heated milk	0.33	0.33	Lys	Lys
Spray-dried milk	0.44	0.45	Met + Cys	Ile
Microbial protein	0.39	0.47	Met + Cys	Met
Peanut meal	0.28	0.33	Met + Cys	Met
Fish meal	0.98	0.93	Phe + Tyr	Phe
		0.67?		Cys?

[1]Taken from Table 3. Columns labeled I result from considering methionine and cystine or phenylalanine and tyrosine together. Columns labeled II result from considering all amino acids in Table 3 separately. No ratios less than 0.01 were used. Lowest rate of digestion = $(a_i/e_i)_L$. Phenylalanine ratio used in subsequent calculations based on column II.

One value is based on considering methionine and cystine or phenylalanine and tyrosine together and the other is based on considering all essential amino acids plus cystine, tyrosine and histidine separately. The relationship between literature values for biological value for these six sources plus egg and the two $(a_i/e_i)_L$'s is shown in Figures 15.2 and 15.3. Instead of a linear relationship between BV and $(a_i/e_i)_L$, the data in Figures 2 and 3 suggest that the relationship may be something like

(4) $BV = 100 (1 - e^{-K(a_i/e_i)_L})$,
where $K \cong 2.9$.

It seems quite reasonable that the effect of increasing rate of digestion on BV or some other measure of protein quality might level off rather than continue to increase linearly. Biological values for each protein source were calculated from equation (4) based on the $(a_i/e_i)_L$'s in column I in Table 15.4, and compared to the literature values for BV and to the PPDD index (Table 15.5). The calculations from this exponential-type equation for these six protein sources plus egg seem at least as reliable as the PPDD index predictions. The experiments and the calculations are certainly much simpler. More important is the fact that the equation seems less arbitrary than the PDR—PPDD type equations. It should be pointed out that a better predictive equation might be obtained with more elaborate curve fitting tech-

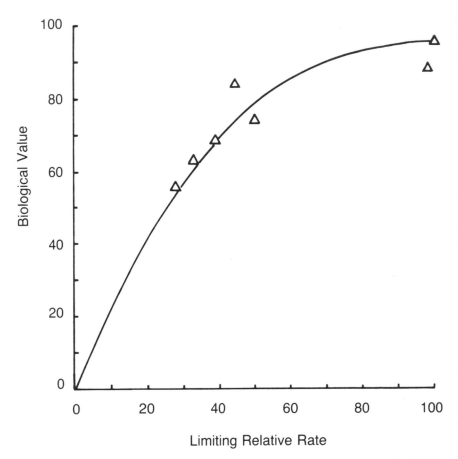

FIG. 15.2.   RELATIONSHIP BETWEEN LIMITING RELATIVE RATE AND BIOLOGICAL
VALUE
Literature values of biological value and data used to calculate relative rates were
taken from Mauron (1970a); limiting relative rates were calculated based on essential
amino acids with methionine and cystine and phenylalanine and tyrosine considered
as pairs (see Column I, Table 15.4).

niques than we used. The use of elaborate fitting techniques did not seem
worthwhile due to the small number of data points.

Certainly it cannot be demonstrated on the basis of digestibility data for
six proteins plus egg, especially when biological values determined on the
same sources are not available, that this relationship is either unique or
reliable for a wide variety of protein sources. The correlations may be
entirely fortuitous. In fact, we have tried with some success several other
relationships which we also consider to be fundamentally appropriate. For
example, the relationship between literature values for BV and $(a_i/e_i)_L \times$

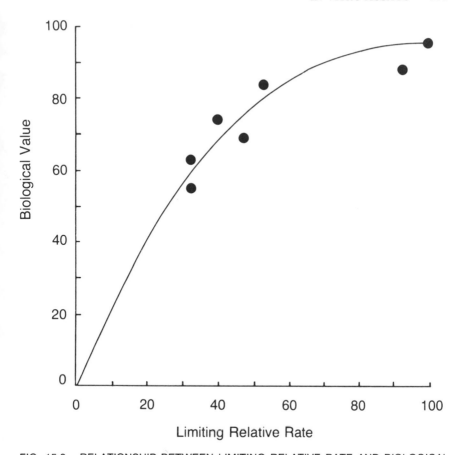

FIG. 15.3.   RELATIONSHIP BETWEEN LIMITING RELATIVE RATE AND BIOLOGICAL VALUE
Literature values of biological value and data used to calculate relative rates were taken from Mauron (1970a); limiting related rates were calculated based on all essential amino acids considered separately (see Column II, Table 15.4).

EAAI is shown in Figure 15.4. The $(a_i/e_i)_L$'s used in this calculation were from Column II in Table 15.4. The equation describing this relationship may also be exponential in form. Again, we did not attempt any serious curve fitting because of the small amount of data available. However, an equation of the form

(5) $BV = 100 \, (1 - e^{-0.038Q})$

where $Q = (a_i/e_i)_L \times EAAI$ estimates BV within 10% using Mauron's data and literature values of BV (Mauron, 1970a). Logically enough, the $(a_i/e_i)_L$ EAAI correlation seems to be better than with $(a_i/e_i)_L$ alone, since multipli-

TABLE 15.5.   CALCULATED BIOLOGICAL VALUE S(BV) AND PEPSIN PANCREATIN DI-GEST DIALYZATE (PPDD) INDEX COMPARED TO MEASURED BIOLOGICAL VALUE[3]

Protein Source	Calculated BV (1)	(2)	Measured BV[3]	PPDD Index[4]	Percent Difference from Measured BV		
					Calculated BV (1)	(2)	PPDD
Egg	94	98	96	100	2	2	4
Soy	77	72	74	79	4	3	7
Over-heated roller-dried milk	62	66	63	77	2	5	22
Spray-dried milk	72	79	84	86	14	6	<1
Microbial protein	68	70	68	64	0	3	6
Peanut meal	56	60	56	60	0	7	7
Fish meal	94	96	88	87	7	9	1

[1]Calculated from the equation $BV = 100 \left(1 - e^{-2.9} \left(\frac{a_i}{e_i}\right)_L\right)$, where $(a_i/e_i)_L$ is the lowest rate of digestion relative to egg taken from column I, Table 15.4

[2]Calculated from the equation $BV = 100 (1 - e^{-0.038Q})$ where $Q = (a_i/e_i)_L \times EAAI$; $\left(\frac{a_i}{e_i}\right)_L$ was from column II, Table 15.4

[3]Literature values taken from Mauron (1970a). Original references cited by Mauron.

[4]Taken from Mauron (1970a).

cation times EAAI sets limits on the effect of the lowest rate. In the case of microbial protein, for example, a relatively high $(a_i/e_i)_L$ is less effective because of a poor amino acid pattern in general. Unfortunately, this correlation requires determination of the amino acid composition of the total protein whereas the correlation with $(a_i/e_i)_L$ alone does not. This is probably a minor drawback, since in most cases, total amino acid composition would be determined anyway.

In spite of the fact that data from six protein sources plus egg are insufficient to determine whether one of these relationships is unique, the considerable reliability of the estimations strongly suggests that these and other similar relationships should be tested with data for more protein sources. Perhaps we have concentrated too long on linear one-to-one type relationships. In addition, we think that the relationships deserve testing because we believe that we are close to understanding them in terms of thermodynamic principles.

If the relationships are tested, biological value and/or net protein utilization in rats should be determined on the same sources. Some data of this type obviously exist even if they are not available in the literature. If one or more of the correlations discussed are supported, then validation tests should be extended to include human bioassays.

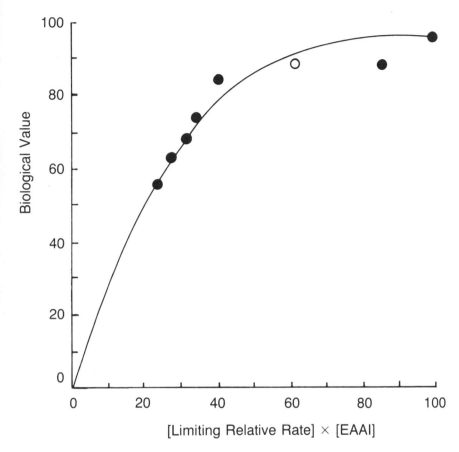

FIG. 15.4.   RELATIONSHIP BETWEEN (LIMITING RELATIVE RATE x EAAI) AND BIO-
LOGICAL VALUE
Literature values of biological value and EAAI values and data used to calculate
(Limiting Relative Rate x EAAI) were taken from Mauron (1970a); limiting relative rates
were calculated based on essential amino acids considered separately (see Column II,
Table 15.4).

SUMMARY

1. Existing enzymatic assay experiments and calculations do not result
   in better predictions of protein quality than amino acid analysis alone
   for either processed or unprocessed protein sources. In existing form,
   these assays have very little potential for use for regulatory purposes.
2. Rat bioassays and the PPDD method have been compared with at least
   28 of the same protein sources (Mauron, 1973). Poor correlation of the
   rat bioassays and the PPDD index probably accounts for the lack of
   attempts to validate this or other enzyme indices with human bioassays.

3. Poor correlations between enzyme indices and bioassays may be due to the choice of calculation method and/or choice of the equation describing the relationship between protein quality and the experimental measurement.

4. For six protein sources plus egg for which data is available, a relationship can be demonstrated between literature values of biological value and a limiting rate of digestion relative to egg. A similar relationship can be demonstrated between BV and a limiting rate × EAAI product. These and other fundamentally appropriate relationships, in our opinion, merit further testing, perhaps with existing data.

5. Subject to some validation of a new calculation method, the experimental design of the PPDD method probably could be simplified.

REFERENCES

AKESON, W.R. and STAHMANN, M.A. 1964. A pepsin pancreatin digest index of protein quality evaluation. J. Nutr. 83, 257–261.

BLOCK, R.J. and MITCHELL, H.H. 1946. The correlation of the amino-acid composition of proteins with their nutritive value. Nutr. Abstr. Rev. 16(2), 249–278.

BODWELL, C.E. 1977. Use of enzymatic assays for the determination of nutritional quality. *In* Nutritional Evaluation of Cereal Mutants, International Atomic Energy Agency, Vienna.

FORD, J.E. 1973. Some effects of processing on nutritive value. *In* Proteins in Human Nutrition, J.W.G. Porter and B.A. Rolls (Editors). Academic Press, New York.

FORD, J.E. and SALTER, D.N. 1966. Analysis of enzymically digested food proteins by Sephadex-gel filtration. Br. J. Nutr. 20, 843–860.

MAURON, J. 1970a. Nutritional evaluation of proteins by enzymatic methods. *In* Evaluation of Novel Protein Products, A.E. Bender, R. Kihlberg, B. Lofqvist and L. Munck (Editors). Werner-Gren Center International Symposium Series 14, Pergamon Press, Oxford.

MAURON, J. 1970b. Nutritional evaluation of proteins by enzymatic methods. *In* Improving Plant Proteins by Nuclear Techniques (Proc. Symp. Vienna, 1970). IAEA, Vienna.

MAURON, J. 1973. The analyses of food proteins, amino acid composition and nutritive value. *In* Proteins in Human Nutrition, J.W.G. Porter and B.A. Rolls (Editors). Academic Press, New York.

MENDEN, E. and CREMER, H. 1970. Laboratory methods for the evaluation of changes in protein quality. *In* Newer Methods of Nutritional Biochemistry, A.A. Albanese (Editor). Academic Press, New York.

OSER, B.L. 1951. Method for integrating essential amino acid content in the nutritional evaluation of protein. J. Am. Dietet. Assoc. 27, 396–402.

OSER, B.L. 1970. *In* Evaluation of Novel Protein Products, A.E. Bender, B. Lofqvist, R. Kihlberg and L. Munck (Editors). Pergamon Press, Oxford.

SATTERLEE, L.D., KENDRICK, J.G., and MILLER, G.A. 1977. Rapid assays for estimating protein quality. Food Tech. 31, 78–88.

SHEFFNER, A.L. 1967. *In vitro* protein evaluation. *In* Newer Methods of Nutritional Biochemistry, A.A. Albanese (Editor). Academic Press, New York.

SHEFFNER, A.L., ECKFELDT, G.A., and SPECTOR, H. 1956. The pepsin-digest-residue (PDR) amino acid index of net protein utilization. J. Nutr. 60, 105–120.

STAHMANN, M.A. and WOLDEGIORGIS, G. 1975. Enzymatic methods for protein quality determination. *In* Protein Nutritional Quality of Foods and Feeds, Vol. 1., M. Friedman (Editor). Marcel Dekker, Inc., New York.

SZMELCMAN, S. and GUGGENHEIM, K. 1967. Availability of amino acids in processed plant-protein foodstuffs. J. Sci. Fd. Agric. 18, 347–350.

DISCUSSION

DR. STAHMANN: Enzyme assays do have value but their greatest value is not to try to reproduce the same information that biological tests give. Biological tests involving rats or microorganisms tell only how much the rat or microorganism grew on the test protein compared to a standard protein. The enzyme assays may tell how much of each amino acid was released by the digestive enzymes. I think we can duplicate the action of the digestive enzymes in a test tube and obtain data which tells us not only how much of each amino acid is released but also which is the first and the second limiting amino acid and how the composition of the product should be changed to improve it. That, I think, has great merit.

When we started these studies, we had small amounts of leaf protein concentrates that looked good from their amino acid analysis, but we did not have enough for feeding. I was surprised and pleased at the high correlation that we obtained between the literature values from rats and the enzyme analyses of 12 proteins, including both very high and very low quality proteins. Our original method can be simplified. One problem is to separate the unhydrolyzed protein from the amino acids and peptides released by enzymes. This may be done by dialysis or by precipitation with picric acid as we first used or more simply by adding sulfosalicylic acid. I would urge that enzyme assays be used, not to try to obtain a single value for protein quality, but rather to look at the individual amino acids and see which ones are low or lacking. When these enzymes hydrolyze proteins in test tubes, the same limiting amino acids will be detected as when they're hydrolyzed in the gut.

I would urge that these assays be carried out, not with purified proteolytic enzymes as many have done, but rather with a mixture of the crude enzymes that assimilate those in the gut. That is, to use crude pepsin and crude pancreatin. We also added a crude intestinal peptidase. But that didn't seem to make very much difference. Proteins occur in association with carbohydrates and fats, hence it may be desirable to include the lipases and the carbohydrases as well as the proteolytic enzymes that are found in the crude enzyme preparations in such model systems.

I think enzyme assays would have a great value in studies of processing operations to detect their effects in protein quality. We have shown losses in protein quality that were inducted by oxidative enzymes. I would like to point out that in many foods, peroxidase, or polyphenol oxidase, will act at room temperature as soon as food is ground, to produce quinones which crosslink and precipitate proteins. They may oxidize lysine, cysteine and methionine. We can detect such losses in protein quality by the enzyme assay as well as losses from heat destruction or the reaction with carbohydrates. Such losses may not be detected by amino acid analyses of acid hydrolysates.

DR. MARABLE: Yes, perhaps I should have spent more time on the point that perhaps these enzyme methods can't be used to predict quality directly. I spent more time just looking at that aspect because I think it's fun, and maybe it makes somebody think of these things in a different way. But it certainly is true that if you look at the original data, not just at the calculation, you learn quite a bit about proteins. Even looking at sources, at the ratios of the digestibility for just these six proteins, the numbers make it apparent right away what some differences are in the proteins. The over-heated roller dried milk for example, for which the essential amino acid index and chemical score failed totally to predict protein quality is obviously digesting at a slow rate. On the other hand, chemical score for the fishmeal turns out to be something like 48 on the basis of tryptophan, but if you look at these rate ratios, the ratios are all quite high. And of course, the literature value for the biological value is high.

DR. HARPER: I'm impressed with your approach of using the relative rates of release. I wondered why you used the essential amino acid index as a corrective factor in this?

DR. MARABLE: It was already available and we felt that we needed some term to describe pattern. The limiting rates describe the availability of the pattern, but then you also have to have something that describes the pattern. My coauthor is a physical chemist and he has generated an enormous number of numbers, that I didn't have nerve enough to show you, that we can use as a sort of corrective factor in place of the EAAI (Essential Amino Acid Index). The EAAI was the first thing we tried and it seemed to be the right idea, so we've done some other things. If you take simply the sum of the net mole numbers for the total amino acid composition, that works as a correction factor.

DR. HARPER: I'd like to suggest that you use a simple chemical score which has a better biological basis than some of the other factors. If you did a calculation such as the classical chemical score and listed the amino acids and their scores compared to the standard, would you come up with a modified chemical score? You could see then if an amino acid that was in great excess was reduced in availability, say only by 20%. This might be trivial. But if one or two of those that were very low were reduced by 20%, this would lower the total chemical score. And I wonder if that approach mightn't apply to what you're doing—maybe you've already tried it?

DR. MARABLE: Well, I'm not sure that we tried exactly what you suggested, but I did look at the plot of chemical score and the plot of the limiting rate, and chemical score didn't look like the right correction factor. There are many things you could try as correction factors.

DR. SAMONDS: I'm a little concerned about your use of the word prediction. You really are using a curvilinear regression to describe the relationship between your calculated value and biological value. To say that you're predicting, using the same data set, really isn't true. What you're really doing is looking at the residual variance about your line.

DR. MARABLE: Oh, sure, and I apologize if I've used the term incorrectly. What I would like now is to have more data so that we could generate a better descriptive curve.

DR. SAMONDS: Prediction means using your best set of data to generate a line, then using other data to see how well it fits.

DR. MARABLE: Oh, sure, I quite agree.

DR. SATTERLEE: I do not feel that it is necessary to get a linear fit when relating chemical characteristics of a protein to its nutritional quality.

DR. MARABLE: I don't either. I didn't mean to give the impression that we wanted linear fits. I was trying to describe what I believe is the prevailing desire to get a linear fit. I quite agree with you. I really think we should stop looking for a linear fit.

16

Microbiological Methods for Protein Quality Assessment

J.E. Ford

Biological assay methods for measuring the nutritional value of proteins are laborious and expensive, and much too slow and imprecise to be suitable for routine use for evaluating large numbers of samples. Several microbiological methods—some with bacteria and others with the protozoan *Tetrahymena pyriformis*—have been recommended for this purpose, but none has yet been widely accepted, despite what might seem to be overwhelming advantages in simplicity and convenience. We must wonder why this is so. It is perhaps due partly to a reluctance of analysts engaged in quality control to introduce microbiological assays into their repertoire. But this cannot be an important factor. A greater problem has been to establish the practical usefulness of the tests for detecting other than gross differences in nutritional quality. These seem to be the central questions—the precision of the tests, and their accuracy as predictors of protein nutritional quality for the growing rat. What the rat bioassays tell us about the value of the protein in the human dietary is another question again, which happily does not concern me at the moment. I have been invited to comment on the usefulness of microbiological assays for measuring protein nutritional quality. In particular, I am asked to assess the present status of these assays in relation to the needs of the food industry and the regulatory agencies. I'll begin with *Tetrahymena*, whose potentialities have attracted so much attention in recent years.

ASSAYS WITH TETRAHYMENA

Our present interest in *Tetrahymena* stems from the observations that the organism is strongly proteolytic (Lawrie, 1937) and readily uses intact protein as the sole source of amino acids (Rockland and Dunn 1946), and requires much the same pattern of amino acids as the growing rat (Kidder and Dewey 1951). Dunn and Rockland (1947) measured the protein nutritional quality of lactalbumin and gelatin relative to that of a casein standard, and obtained values that agreed broadly with results of animal feeding tests, though they varied with the length of the incubation period and with the level of protein added to the test cultures. These authors encountered a problem that has persisted to the present day; namely, how best to assess growth of the protozoan in cultures containing particles of the test foodstuff in suspension.

Measurement of Growth Responses

Our aim in the assays is to measure the relative efficiencies with which N in the 'standard' and 'test' proteins is incorporated into protozoal cells. The direct approach would involve separating the organisms from undigested food particles and measuring their N content. It was attempted without success by Fernell and Rosen (1956), who employed differential centrifugation in sucrose solutions, and electromigration techniques.

If we assume that the average size and composition of the organisms does not vary between standard and test cultures, then the relative amounts of protozoal N will be proportional to the numbers of organisms. Most workers with *Tetrahymena* have accepted this assumption—sometimes with misgivings—and measured growth in their test cultures by microscopic counting in a hemocytometer. This is extremely tedious and it is a major source of variation in the assays.

Electronic cell counting with a Coulter counter offers a less laborious alternative, but the instrument does not discriminate between cells and undigested food particles, and difficulties may be encountered through blocking of the aperture with debris (Shorrock 1972). Teunisson (1971) used *Tetrahymena* in a study on peanut and cottonseed meal, and separated the cells from food residues by elutriation before counting them electronically. Commenting on this, Shorrock (1972) argued that if it is possible in this way to separate the cells from extraneous material, then it would generally be simpler to assess growth from the optical density of the cell suspension. More recently, Satterlee *et al.* (1979) described a procedure in which the test cultures are electronically counted twice, at 24 hr and 66 hr after inoculation. The 24 hr count was taken as a measure of the amount of interfering particulate material, and subtracted from the final count at 66 hr to give a corrected figure. This approach seems to embody an assumption that the numbers of food particles in the test cultures remain unchanged throughout the period of incubation.

Evancho *et al.* (1977) found that use of a Coulter counter instead of direct microscopic counting greatly increased speed and precision in their assays. With the aid of a 'channelizer' attachment to the counter they showed that cells increased both in numbers and in volume as the amount of available N in the medium increased. With a good protein source, maximum cell size was obtained with 0.3mg N/ml culture medium, whereas with a poor protein maximum size was not obtained even at 0.4mg N/ml. With the channelizer it is easily possible to measure the size distribution of the cells within selected limits, and so to obtain a corrected estimate of total cell yield. It seems therefore that this might be a worthwhile refinement of the cell-counting technique.

It is now generally accepted that enzymic predigestion of the test proteins is a necessary preliminary to the *Tetrahymena* assay (see overpage), and in consequence the test extracts are often relatively clear and free from material in suspension. Shorrock (1976) showed that for high protein feedstuffs (casein, fish meal, soya bean meal) growth could be accurately assessed from the optical densities of the test cultures. But with most foodstuffs this simple photometric determination would not be appropriate.

Besides these direct methods of assessing growth, several indirect methods have been used. Thus Dunn and Rockland (1947) measured acid production during 41 days incubation, a procedure that was dismissed by Anderson and Williams (1951) as being more a measure of oxygen deprivation than of cell growth. These latter authors developed a technique in which growth responses were measured after only $3-5$ days incubation, by determination of the red formazan pigment produced in the medium from 2, 3, 5-triphenyltetrazolium chloride (TPTC) under the action of the cells' dehydrogenase enzymes. Results for a selection of protein foods were not always similar to those found in rat tests, but good and poor quality materials were clearly differentiated.

Wang *et al.* (1979a) measured the relative nutritional value (RNV) for *Tetrahymena* of several mixed diets, comparing results calculated from direct microscopic counts, TPTC reduction, oxygen uptake, and ATP-bioluminescence. All the four sets of results were highly correlated with those obtained by rat PER test, and it was concluded that the various indirect measures of growth were satisfactory alternatives to counting under the microscope.

Tetrahymena cells contain a characteristic pentacyclic terpene, tetrahymanol, and Shepherd *et al.* (1977) showed that there was a linear relationship between the numbers of cells and the content of tetrahymanol in cultures. They described an assay procedure which involves solvent extraction of the tetrahymanol from the test cultures, saponification and determination of the sterol by GLC using an automatic injection system. Greater precision was obtained than with cell-counting techniques, but even so the authors concluded that the *Tetrahymena* assay is not ideal for routine work in quality-control laboratories. It was found that one worker could complete only eight assays in the working week.

A somewhat similar approach was adopted by Maciejewicz-Ryś and Antoniewicz (1978), who measured 2-aminoethylphosphonic acid (AEP) which is present at high concentration in *Tetrahymena* cells but generally absent from foodstuffs. Large differences were observed between the average dimensions of *Tetrahymena* cells grown with different protein sources, and AEP was preferred to cell count because it took these differences into account. The authors examined a varied selection of protein concentrates and food mixtures, comparing *Tetrahymena* ratings (RNV) and rat NPU values, and overall a rather poor correlation was found, whether the RNV values were derived from cell counts or AEP measurement. They concluded that the *Tetrahymena* assay in its present state of development has defects that make it unreliable for food mixtures and limit its usefulness in the evaluation of foodstuffs.

So much for the measurement of growth in the assays. Certainly with low protein materials it seems that we do not yet have the speed and precision that would be necessary if the procedure were to be suitable for screening large numbers of samples.

Besides this problem of measuring the growth responses, there is a more fundamental problem at the front end of the assay, in preparing the proteins for testing. It is a problem that is common to all the microbiological assay techniques.

Preparation of the Samples for Assay

Tetrahymena is strongly proteolytic, and early workers saw no need to specify enzymic pretreatment of the test samples as part of their assay procedures. So for example, Rosen and Fernell (1956) extracted their test materials successively with diethylether and ethylalcohol to remove non-specific growth stimulants, and then dried and ground them to pass a 72 mesh BS sieve. A sample of this ground material was then suspended in water for assay. Later work showed that many proteins were poorly utilized by *Tetrahymena* unless they were partially predigested before assay, and it is now accepted that enzymic pretreatment of the test samples is a prerequisite in the assay. But it has perhaps not yet been sufficiently recognised that the pretreatment may be critically important in determining the results. The assay procedures are highly empiric and, as with other *in vitro* tests for protein quality, their development must be guided by careful comparative testing with laboratory animals.

The present situation is that different workers have recommended quite different enzymic pretreatments. For example, Evancho et al. (1977) predigested with pepsin under carefully specified conditions, for 3 hr at 55°. In a study on 34 commercially prepared foods they found a high correlation (r = 0.90) between rat PER and calculated *Tetrahymena* PER values, and concluded optimistically that their assay technique can measure protein quality with much the same accuracy and precision as the rat assay, and shows great promise as a rapid screening procedure. They do, however, stress a need for further research on variables in the assay before the method is subjected to collaborative testing. Careful examination of the data given in this paper shows, surprisingly, that the *Tetrahymena* RNV values were almost as highly correlated (r = 0.84; P <0.001) with the protein content of the test samples. High protein content was associated with high RNV, and it seems that 70% of the variation in protein *quality* was somehow explained by the variation in protein *content*.

Frank et al. (1975) encountered problems in their use of pepsin, "due to ionic imbalances which adversely affected the growth of the organism" and used instead a combination of trypsin and bromelain. Scant details were given of their assay techniques, but results for ten foodstuffs agreed closely with rat PER values.

Shepherd *et al.* (1977) confirmed that enzymic predigestion was essential for all except soluble proteins, and described a procedure using Pronase which they applied in their *Tetrahymena* assays for available lysine, methionine and tryptophan in a varied selection of high protein feedstuffs. Good agreement was obtained between a chick bioassay and the *Tetrahymena* assay for available lysine, and results for available methionine and tryptophan were regarded as reasonable, though no biological assays were done. These workers recognised that too much predigestion might be as undesirable as too little, pointing out that ". . . if a protein source is subjected to very extensive enzymic degradation or too prolonged incubation with the microorganism, the availability of amino acids may well be greater than that determined by bioassay with chicks or rats, whereas insufficient pretreatment or an inadequate period of incubation may produce the opposite result".

Shorrock (1976) reported a similar study in which he used papain as the digesting enzyme. He found fair agreement between *Tetrahymena* and rat bioassay values for available lysine in fifteen protein-rich feedstuffs. The *Tetrahymena* values were closely correlated with values for available lysine obtained by the FDNB method of Carpenter (1960), and overall these two *in vitro* tests were equally successful as predictors of biological availability of lysine for the rat, whereas total lysine values showed no significant correlation with the rat bioassay results. *Tetrahymena* was also used to measure available methionine in the same fifteen samples, and the results were compared with those obtained by microbiological assay with *Streptococcus zymogenes* and by chick bioassay. The values obtained by the two microbiological methods were broadly similar and highly correlated (r = 0.96, P <0.001), as also were the chick assay values with both sets of microbiological assay values. Shorrock (1976) concludes his paper with a familiar warning: ". . .the findings indicate that the *Tetrahymena* assay can predict accurately the results of growth tests with rats, but it would be premature to recommend the present assay procedure for routine use. Further development is still needed, and in particular, the problem of finding a quicker alternative to cell counting remains to be solved".

Satterlee *et al.* (1979) subjected a varied selection of protein sources to more comprehensive *in vitro* digestion, first with a mixture of trypsin, chymotrypsin and peptidase and then with Pronase, before assaying with *Tetrahymena*. PER values were calculated by comparing growth responses with those to ANRC sodium caseinate, and were found to be in broad agreement with rat PER values.

There is a clear need for more systematic investigation of the influence of the enzymic predigestion on the results of *Tetrahymena* assays: the choice of procedures by different workers seems to have been altogether too arbitrary. There is general agreement that predigestion is necessary, but it remains to be seen whether different treatments give the same results. Shepherd *et al.* (1977) envisaged the possibility that too much might be as undesirable as too little, and there is no doubt that this would be so with some types of processed foodstuffs. Thus, Ford and Salter (1966) examined the effects of severe heat treatment of cod fillet on the protein nutritional quality, comparing the extent of enzymic release of amino acids *in vitro* with their biological availability as measured in rat growth assays. They found with the heated protein that much more of the methionine, isoleucine and lysine was susceptible to enzymic release *in vitro* than was biologically available to the rat. About 66% of the methionine, 82% of the isoleucine and 28% of the lysine were released *in vitro*, as against only 33, 37 and 12% measured in rat growth tests.

General Comments Concerning Tetrahymena Assays

All the *Tetrahymena* assay procedures are somewhat involved and demand a high level of technical competence, more so than conventional microbiological assays with bacteria. I think that none is yet ready for general adoption outside of the research laboratory, despite encouraging reports from various authors.

There are difficulties with the assay medium. Several of the procedures involve separate sterilization and aseptic addition of constituents of the medium, a complication that Shorrock (1976) found to be unnecessary in assays for available lysine and methionine. And problems may arise from non-specific stimulation or inhibition of growth by the test samples (cf Evans *et al.*, 1978, 1979; Evans, 1979 a, b, c; Satterlee *et al.* 1979). Several authors have advocated the removal of interfering substances by solvent extraction, but some would not be easily removed in this way. These problems of non-specific influences on growth of the test cultures, and the problems of measurement that arise from the presence of undigested food residues, are exacerbated by the low sensitivity of the test. The *Strep. zymogenes* assay for example is at least four times as sensitive.

Non-parallelism between the standard and test responses seems to be a common finding. It certainly is so in my experience, especially in RNV tests, where we are comparing responses to different proteins that may be limiting in different amino acids. Landers (1975) remarked on it and concluded that it was very important to choose an optimum nitrogen concentration for determining organism counts. This begs the question, how to decide on this optimum test level? Several workers have adopted the dubious expedient of testing at only one level of nitrogen, but it evades the problem.

Wang *et al.* (1979b) obtained *Tetrahymena* RNV values for a varied selection of protein sources, and compared single-point and four-point slope ratio assays. The correlation with rat assay results was better with the single-point assays, but even so these workers found no *significant* correlation (but see Wang *et al.* 1979a). They concluded that "precautions should be taken regarding the absolute nutritional requirements of test organisms . . .", and this is clearly of crucial importance. It is not enough that our assay microorganism should require the same amino acids as the growing

rat; it also needs to match the rat rather closely in its quantitative requirement for the different amino acids.

Some years ago I described an assay with *Strep. zymogenes* that graded
meat meals and fish meals very impressively in accordance with their rat
NPU values and also with their content of available lysine (FDNB lysine),
and this despite the absence of any requirement for lysine in the test
microorganism (Ford, 1960). Clearly, lysine was not the limiting amino acid
in the microbiological assays, and the meals were not improved when lysine
was added to them. The differences in protein nutritional quality reflected
differences in the availability of several or all of the amino acids, and not
only of the lysine. As long as I compared within groups of similar proteins
my results made very good sense and they revealed quite small differences
in protein quality. But in comparisons between different types of proteins
there were occasional marked anomalies: for example, the RNV values of
barleys and yellow sorghum were about the same as that of casein. The low
content of lysine in the cereals did not limit the growth of *Strep. zymogenes*,
and so the microbiological assay values were much too high in relation to
rat NPU values. Nevertheless, comparing a varied collection of protein
sources—casein, egg, skim milk powders, soya bean meals, food yeast, fish
meals, meat meals, groundnut meals, wheat gluten—there was a close
correlation between the microbiological values and rat NPU and NPR
(Ford, 1960). The results are shown in Table 16.1.

So here we have an assay with what might be seen as a fatal flaw—our
test bug has no requirement for lysine—and yet when we examined a
selection of proteins representing a wide spectrum of types and qualities we

TABLE 16.1.   RELATIVE VALUES OF DIFFERENT PROTEINS FOR *STREPTOCOCCUS ZYMOGENES* AND FOR THE RAT[1,2,3]

Test protein	RNV	NPU	NPR
Casein	99	84	3.92
Dried whole egg	94	95	4.66
Dried skim milk: 1	99	78	3.77
2	106	89	4.11
Dried buttermilk	107	83	4.21
Soybean meal: unsupplemented	72	70	3.37
+ 1% methionine	90	82	4.03
Soya protein isolate	56	63	2.71
Dried food yeast	80	69	3.10
Fish meal: 1	52	54	2.85
2	74	89	4.02
Meat meal: 1	39	40	0.98
2	50	54	2.40
3	19	43	1.70
Wheat gluten	54	54	2.79
Groundnut meal	54	58	2.72

[1]From Ford (1960); RNV (Relative Nutritive Value) values from microbiological assays;
NPU (Net Protein Utilization) and NPR (Net Protein Ratio) values from rat assays.
[2]NPU = 2.64 + 0.60 RNV; standard error of regression coefficient = ± 0.08; r = 0.90
(14 d.f.), P <0.001.
[3]NPR = 0.79 + 0.034 RNV; standard error of regression coefficient = ± 0.005; r = 0.89
(14 d.f.), P <0.001.

found a very fair correspondence between our microbiological and rat assay results. Of course it was partly fortuitous; if we had included mixed diets and cereals in the comparison the correlation might have been rather poor.

I wonder whether the same might be true for the *Tetrahymena* assays. *Tetrahymena* certainly requires lysine, and there is no obvious mismatch in its pattern of amino acid requirements. Many workers have shown a broad correspondence between *Tetrahymena* RNV's and rat PER's, but it is clear that the *Tetrahymena* assays are not *inherently* accurate. The results may be influenced by the conditions of enzymic pretreatment used to prepare the samples for test, by the length of the incubation period, the level of protein in the test cultures, and by several other factors.

There is a further important question, concerning the relative quantitative requirements for different essential amino acids. Rølle (1976) determined these requirements for *Tetrahymena* and compared them with the corresponding requirements in children. Table 16.2 shows her values, and also the NRC (1978) values for the growing rat. If we compare *Tetrahymena* and the growing rat, we see that *Tetrahymena* has a lower requirement for lysine and for methionine + cystine, and a higher requirement for isoleucine and threonine. It seems that with many animal protein foods the nutritional quality for *Tetrahymena* would be limited by isoleucine or threonine, rather than by methionine + cystine as in the rat.

TABLE 16.2. REQUIREMENT FOR ESSENTIAL AMINO ACIDS (g/100 g PROTEIN) IN TETRAHYMENA, GROWING RATS, INFANTS AND 10–12 YEAR OLD CHILDREN

Amino Acid	Tetrahymena[1]	Rat[2]	Infant[3]	Child[3] (10–12 years)
Isoleucine	6.6 (158)[4]	4.17 (100)[4]	3.50 (84)[4]	3.70 (89)[4]
Leucine	5.3–6.6 (86–106)	6.25 (100)	8.00 (128)	5.60 (90)
Lysine	4.0 (68)	5.83 (100)	5.25 (90)	7.50 (129)
Methionine + cystine	2.1–3.2 (42–64)	5.0 (100)	2.90 (58)	3.40 (68)
Phenylalanine + tyrosine		6.67 (100)	6.30 (94)	3.40 (51)
Threonine	6.6 (158)	4.17 (100)	4.40 (106)	4.40 (106)
Tryptophan	1.1 (88)	1.25 (100)	0.85 (68)	0.46 (37)
Valine	5.6 (112)	5.00 (100)	4.70 (94)	4.10 (82)

[1]Rølle (1975).
[2]NRC (1978).
[3]FAO/WHO (1973), assuming 2.0 and 0.8 g protein intake/kg body weight for infants and children, respectively.
[4]Figures in parenthesis are calculated on the basis that the rat requirement = 100.

Similarly when we compare the requirements of the rat with those of young children we find differences—for example with methionine + cystine and tryptophan. Perhaps these differences are more apparent than real; according to Irwin and Hegsted (1971) there is an urgent need for more reliable estimates of the requirements in children. But accepting these FAO/WHO requirement figures for the moment, I feel that we have in *Tetrahymena* a rather poor model for the growing rat, and the rat in turn seems to be an imperfect model for the growing child.

ASSAYS WITH BACTERIA

Several species of bacteria have been employed in assays for overall protein quality, mostly in procedures that involved extensive enzymic predigestion of the test samples. Thus for example, Horn *et al.* (1952, 1954) used *Leuconostoc mesenteroides* P-60 to grade a number of differently processed cottonseed meals. Samples were incubated for successive 24 hr periods with pepsin, trypsin and erepsin. The growth of the test organism with these enzymic digests was compared with that obtained with a digest of a reference standard meal, and 'indices of protein value' were computed which agreed fairly well with the findings in rat growth tests. The value of the method for grading other types of protein was not investigated. Halevy and Grossowicz (1953) assayed a small selection of proteins with a strain of *Streptococcus faecalis*, which like *L. mesenteroides* has little ability to utilize intact protein. They hydrolysed the test proteins for 48h with pancreatin and determined the quantities of the hydrolysates needed to promote half-maximum growth in their cultures. Only about 40% of the proteins was hydrolysed during the enzymic digestion, but the results of the tests were considered to be broadly comparable with those of rat-growth assays. Ford (1960) used a related organism, *Strep. zymogenes*, which is powerfully proteolytic and grows quickly with an adequate intact protein as the main source of nitrogen. Nevertheless, enzymic predigestion of the test samples proved necessary. It speeded growth in the assays and improved the linearity and reproducibility of the dose-response curves. Protein quality ratings for a variety of foodstuffs correlated closely with rat assay values and it seemed that, at least for the limited range of food proteins examined, results obtained with *Strep. zymogenes* would support predictions of NPU values for the rat, and would indicate reliably any diversity of quality within groups of similar materials.

Another strongly proteolytic bacterium that has been recommended for use in protein quality testing is *Clostridium perfringens* (welchii). It requires nine of the essential amino acids, is fast growing and, under assay conditions specified by Boyd *et al.* (1948), it is virtually immune from bacterial contamination. Solberg *et al.* (1979) employed it in a rapid (24 hr) protein quality assay procedure. The test samples were ground or mashed and suspended in the culture medium, which was then inoculated and incubated anaerobically at 46°. Growth was measured manometrically, and expressed relative to that obtained with a standard mixture of amino acids. Results for an assortment of 25 foods correlated only moderately well (r = 0.74) with PER values, but the authors concluded that the method shows promise and warrants further development. Ford (1964) took a less optimistic view. He used *Clostridium perfringens* for the measurement of 'total' and 'available' methionine in animal protein feedstuffs, under conventional test conditions similar to those used in assays with *Streptococcus zymogenes* on the same test samples. The two organisms graded the samples similarly as sources of available methionine, though results with *C. perfringens* were unaccountably variable from assay to assay. Ford (1964) concluded that any possible advantage in the use of *C. perfringens* would be outweighed by the potential danger of the organism as a pathogen. It has been increasingly implicated in cases of food poisoning in recent years, and it would seem unwise to encourage its widespread use in the food industry.

MICROBIOLOGICAL ASSAY OF AVAILABLE AMINO ACIDS

In Britain, in 1955, the Agricultural Research Council (ARC) formed a Protein Quality Group to undertake collaborative work on the development of laboratory procedures for the assessment of protein quality in animal feedstuffs. Boyne *et al.* (1961) gave a comprehensive progress report on the work of the Group to 1960. Thirteen laboratories took part in a comparative study of chemical, microbiological and biological assay procedures as applied in the evaluation of 130 samples representing seven types of protein concentrate. Useful correlations were found between protein values for rats and chicks and those for *Tetrahymena, Strep. zymogenes* and *Strep. faecalis*, and it was evident that the microbiological methods had considerable potential. A Working Panel was formed to promote their development and to undertake collaborative testing. Exploratory studies (Boyne *et al.*, 1967) led to the conclusion that microbiological tests of overall protein quality might prove useful as grading tests, as they are well capable of distinguishing 'good' from 'poor' samples of the same type of material. But such grading tests are of limited value and, in practice, overall protein quality, however determined, is probably of greater academic than practical interest. Certainly for the compounder of animal feedstuffs, the practical problem is generally to evaluate proteins as sources of individual amino acids so that they can be used most economically in mixed diets. And much the same considerations must apply in assessing the value of protein foods in the human dietary. The Panel therefore decided that attention should be concentrated initially on the measurement of available amino acids, and specifically on testing the *Strep. zymogenes* procedure as applied in the assay of available methionine (Ford, 1962). The procedure is simple and conventional, except that it involves a pretreatment of the test samples with papain, which has to be carefully standardized. Miller *et al.* (1965) examined a range of meat meals and fish meals of widely varying nutritional quality and found that their *Strep. zymogenes* assay values for available methionine were highly correlated and quantitatively similar to those from chick growth assays (Fig. 16.1). The same finding emerged from the ARC Working Panel report (Boyne *et al.*, 1975) which stated: "...We may fairly conclude that for fish meals and meat meals and probably for other classes of protein-rich foodstuffs, the *Streptococcus zymogenes* assay procedure is capable of grading samples in a similar order to that obtained using chick or rat assay procedures." This is not necessarily true for all protein sources; for example, plant proteins whose nutritional quality for animals may be influenced by toxic constituents. Clearly, no one *in vitro* test will serve as the single predictor of protein quality. But to the extent that we are interested in monitoring differences in the quality of proteins as sources of

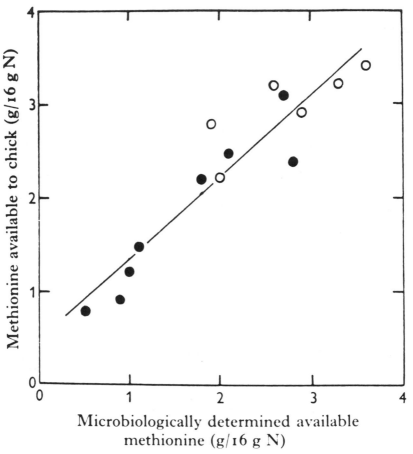

From Miller et al. (1965)

FIG. 16.1. RELATIONSHIP BETWEEN VALUES OBTAINED BIOLOGICALLY (CHICK) AND THOSE OBTAINED MICROBIOLOGICALLY *(STREPTOCOCCUS ZYMOGENES)* FOR AVAILABLE METHIONINE IN PROTEIN CONCENTRATES

The response metameter used for calculating the results of the chick assays was food conversion efficiency. The micróbiological values were obtained by assay with *Strep. zymogenes* after predigestion with 0.36% crude papain. O, laboratory preparations of cod muscle; ●, commercial meat, fish and whale-meat meals. The least squares regression line, 'chick value = 0.49 + 0.87 (microbiological value)', is also shown.

methionine, the *Strep. zymogenes* assay may be safer than any alternative chemical test.

Shorrock (1976) reported that *Strep. zymogenes* and *Tetrahymena* gave essentially the same values for available methionine, but concluded that the *Strep. zymogenes* method offered several advantages. It was quicker (48 hr vs 96 hr) and more sensitive, and more easily applied to samples rich in carbohydrates. *Tetrahymena* offered the advantage that it can be used to measure both lysine and methionine in the same test digest.

Precision of The Assays

We recognize that methods for quality control of manufactured foods must be quick and accurate, and reasonably precise and reproducible both within and between laboratories. Table 16.3 gives some data on the preci-

TABLE 16.3.   VARIATION WITHIN AND BETWEEN ASSAYS FOR METHIONINE AND LYSINE IN FISH MEALS

	Mean	No. of samples	Within assays SD	Within assays Resolution[1]	Between assays SD	Between assays Resolution[1]
Methionine assays[2] (Strep. zymogenes):						
'total'	3.01	42	0.12	0.33	0.15	0.42
'available'	2.41	42	0.12	0.33	0.14	0.38
Available lysine assays[3]	6.49	6	0.17	0.47	0.22	0.61

[1]Smallest difference between samples that may be resolved within an assay. Calculated as t (P = 0.05)$\sqrt{2}\times$SD
[2]For the *Strep. zymogenes* assay, four sets of papain digests and acid hydrolysates were prepared for each sample, and each was assayed twice. For available lysine each of the six fishmeals was assayed six times, on six separate occasions.
[3]Measured using a modification of Carpenter's FDNB method (Booth, 1971).

sion of the *Strep. zymogenes* assay as applied in the measurement of total and available methionine in 42 fish meals. Four independent sets of papain digests and acid hydrolysates were prepared for each sample, and each was assayed twice. The mean SD within an assay was 0.12 (3−4%), which means that we would not resolve differences of less than 8−10% between samples within an assay. The variation between assays was greater, but it could be reduced by the inclusion of a reference sample within each assay to provide a correction factor, and of course by further replication of the assay.

We obtained smaller variation in the assays for available lysine, but this was probably beginner's luck. We used a modification of Carpenter's method (Booth, 1971) for which a CV of 3.8% has been reported. So it seems that none of these tests would resolve differences between samples of less than about 8−10% within one assay on duplicate samples. But this is quite as good as we could expect in practice from any other method of amino acid assay, including the newer automated amino acid analysers (Knipfel *et al.* 1971). Certainly with methionine, the figure of ± 3% that is commonly given for the reproducibility of these assays is far too optimistic.

Shorrock (1972) reported on the reproducibility of *Tetrahymena* assays. Ten high protein materials were assayed repeatedly for available lysine in a series of independent tests. His findings are shown in Table 16.4. The coefficients of variation ranged from 7.2% for a hydrocarbon yeast to 14.4% for a poor quality fish meal.

TABLE 16.4. REPRODUCIBILITY OF *TETRAHYMENA* ASSAYS FOR AVAILABLE LYSINE. VARIATION BETWEEN ASSAYS [1]

Test Material	Number of assays	Mean lysine content, g/16gN	Standard Error	Coefficient of variation, %
Fish meal:				
FM 101	16	5.44	0.12	9.0
FM 102	8	3.94	0.20	14.4
FM 113	6	6.48	0.18	6.9
Whale meal:				
WM 1	5	2.95	0.10	7.9
WM 3	5	4.47	0.21	10.5
WM 7	8	2.10	0.08	11.2
WM 9	5	7.15	0.34	11.5
Hydrocarbon yeast:				
HY 101	9	5.97	0.18	9.3
HY 104	9	7.13	0.17	7.2
Casein	14	7.40	0.17	8.7

[1]Data from Shorrock, 1972.

Evaluation of Cereals and Mixed Diets

The *Strep. zymogenes* assay for available methionine has been extensively tested with fish meals and several other protein-rich foodstuffs (cf Ford, 1962; Ford and Salter, 1966; Miller *et al.* 1965; Carpenter and Woodham, 1974; Boyne *et al.* 1975), but there is as yet little information for cereals and cereal-based mixed diets. Papain predigestion as specified in the standard assay procedure (Boyne *et al.* 1975) may be inappropriate, and the fineness of grinding of the test samples may strongly influence the results. Thus in rice, the endosperm protein is present mainly as discrete $1-4\mu m$ particles in the interstices between the compound starch granules, and in a coarse flour many of these particles might be inaccessible to the digesting enzyme. And the protein is especially rich in glutelins, which are poorly soluble except at high pH and are better digested by Pronase or other alkaline protease. A further difficulty in extending the microbiological tests to cereals is the lack of precise comparative information from tests with animals. The biological assessment of protein quality is difficult in materials of low protein content, and particularly so if the protein is of poor quality and limiting in lysine, when the biological value may be strongly influenced by the level of protein in the test diet (cf Hegsted and Juliano, 1974).

Table 16.5 shows the effects of finer milling, and of pretreatment with different enzymes, on results obtained for available methionine in wheat, barley and rice. Values for total methionine are given for comparison. With all three cereals, results obtained with Pronase were higher than with papain, though the differences were small. Finer grinding of the test samples gave generally higher results, though again the increase if any was small. For further tests it was decided arbitrarily to adopt Pronase digestion

TABLE 16.5.   INFLUENCE OF FINENESS OF GRINDING OF THE TEST SAMPLE AND OF DIFFERENT ENZYMIC PRETREATMENTS ON THE AVAILABLE METHIONINE IN WHEAT, BARLEY AND RICE AS MEASURED WITH *STREPTOCOCCUS ZYMOGENES*[1]

Test sample	Fineness (mesh size)	Papain	Pepsin	Pronase	Total
		Methionine, g/kg protein			
		Available			
Wheat	40	13.8	14.6	15.6	16.4
	80	14.8	14.8	15.3	
Barley	40	12.9	14.3	16.9	17.9
	80	13.9	16.9	16.6	
Rice	40	19.9	21.7	21.7	24.9
	80	21.3	22.9	24.9	

[1]For a description of the enzymic pretreatments, see Ford and Hewitt (1979a).

and, whenever possible, to ball-mill the samples to pass an 80 mesh sieve. Table 16.6 shows results obtained in this way for available methionine in nine samples of rice flour. The values are means of four independent assays. The coefficient of variation (based on the 'rice x assay' interaction mean square in the analysis of variance) was 0.028 and the least significant difference (P = 0.05) was 1.0. Comparing these values with the 'total' values, the availability of methionine appears to be uniformly high, as might be expected from the high values for true digestibility of the N reported by Eggum and Juliano (1973).

TABLE 16.6.   TOTAL AND AVAILABLE METHIONINE IN RICE FLOURS, AND TRUE DIGESTIBILITY OF THE NITROGEN

Test material	N (g/kg)	Total[1]	Available	Available Total	Digestibility of N[2]
		Methionine (g/kg protein)			
Intan	9.55	32.8	30.3	0.92	1.001
IR 8a	12.30	23.8	26.6	0.92	0.962
IR 8b	16.32	23.2	23.4	1.01	0.954
IR 22a	12.61	22.0	28.9	1.31	—
IR 22b	16.00	23.4	25.7	1.10	0.985
IR 480-5-9	18.88	25.4	24.7	0.97	0.945
IR 1103	18.56	24.5	24.8	1.01	0.959
BP 176-1	24.32	17.1	18.0	1.05	0.944
IR 480-5-9 (Brown)	17.44	26.0	25.2	0.97	0.908

[1]From Bressani *et al.* (1971).
[2]From Eggum and Juliano (1973).

   In other food grains, part of the protein may be poorly digestible, or become indigestible during processing through interaction with polyphenols which are widely present in higher plants. This complicates the interpretation of amino acid analysis in terms of protein nutritional quality. The problem is starkly illustrated in sorghums, which exhibit wide differences in digestibility that reflect corresponding differences in tannin content.

Table 16.7 shows this wide variation in protein nutritional value and availability of methionine in sorghum protein in relation to the tannin content.

TABLE 16.7.   PROTEIN NUTRITIONAL VALUE AND AVAILABILITY OF METHIONINE IN SEEDS OF TEN VARIETIES OF SORGHUM, IN RELATION TO THEIR TANNIN CONTENT

Variety	Tannin content (g/kg)	RNV[1]	Methionine, g/kg protein Total	Available	Available Total
De Kalb E57	4.9	93	16.6	15.8	0.95
TE 66	5.8	88	18.5	15.5	0.84
ORO 7	6.9	98	18.4	17.7	0.96
RS 617	16.0	48	18.5	8.5	0.46
AKS 663	17.8	58	17.0	9.7	0.57
Shoobird	19.8	49	18.1	9.2	0.51
Savannah	20.6	50	17.7	9.9	0.56
Ga 615	23.4	39	17.4	7.4	0.43
AKS 618	23.7	31	16.3	6.3	0.39
BR 76	26.0	37	17.5	6.6	0.38
LSD (P = 0.05)	2.6	5.1	1.3	1.4	—

[1]RNV = Relative Nutritive Value as measured with *Strep. zymogenes*; casein = 100.

Nelson *et al.* (1975) showed with these same sorghum varieties that amino acid digestibilities in chickens were correlated with the tannin content (r = 0.82; P <0.01). Ford and Hewitt (1979b) examined a selection of sorghums and field beans representing high and low tannin varieties, and of barleys. Results of bioassays with rats and chickens were closely consistent with those from microbiological tests with *Strep. zymogenes*. From this preliminary work it seems that the *Strep. zymogenes* assay for available methionine gives sensible results for cereals and cereal-based mixed diets, as well as for protein-rich concentrates.

GENERAL

No single amino acid will serve as a reliable indicator of protein quality, though in some classes of manufactured foods we may expect that differences in protein nutritional quality would reflect corresponding differences in the biological availability of several or all of the constituent amino acids. Thus, in twelve whale meat meals of closely similar 'total' amino acid composition there were large differences in the nutritional quality; *Strep. zymogenes* values for available methionine, tryptophan, leucine and arginine were closely correlated with each other and with rat assay (NPU) and FDNB-lysine values (Ford, 1962). Similarly with eight fish meals, chosen to represent a range of source materials and nutritional quality in commercial products, values for available lysine and methionine measured by chick bioassay and by *in vitro* methods were all closely correlated (Table 16.8; Ford and Hewitt, unpublished data). But different amino acids are not always equally susceptible to processing damage. In a study on the effects of

TABLE 16.8. AVAILABLE METHIONINE AND LYSINE (g/16gN) IN FISH MEALS

Sample	Available methionine			Available lysine	
	Chick growth test	Strep.[1] zymogenes assay	Tetrahymena[1] assay	Chick growth test	Dye-binding lysine assay[2]
A	3.24	3.07	3.05	7.66	7.38
B	2.53	2.79	3.02	7.15	6.25
C	1.49	1.51	1.67	2.97	3.84
D	3.00	2.80	2.72	7.62	6.53
E	2.65	2.65	2.70	7.80	6.76
F	2.93	2.76	2.81	8.84	6.81
G	2.24	2.16	2.38	5.12	5.24
H	2.20	2.23	2.20	6.41	5.90
r	0.95 (P <0.001)			0.95 (P <0.001)	

[1]The test samples were predigested with pronase (Ford and Hewitt, 1979a) and the same digests used in both assays.
[2]Hurrell and Carpenter (1975; 1976).

severe heat treatment of dried cod fillet, Ford and Salter (1966) found a marked differential effect of the treatment in retarding *in vitro* the enzymic release of several amino acids, and feeding tests with rats also showed marked differences in the extent to which lysine, methionine and isoleucine were damaged (Table 16.9). Heating reduced the biological availability of all three amino acids, but apparently more so that of lysine than of methionine and isoleucine. A likely reason for this greater loss of lysine was the presence of ribose, derived from nucleic acids. The freeze-dried cod fillet contained 335mg ribose/100g, but after heating for 18 hr at 140° the level had fallen to 142mg, and on enzymic digestion this strongly-heated material gave rise to unavailable peptides that were strongly yellow-brown in color (Ford, 1973).

The Maillard reaction is of major importance in food technology and it may cause rapid deterioration in some foods—notably dairy products—during processing and storage, with disproportionate loss of available lysine from lysine-rich proteins (Frangne and Adrian, 1972a,b). There is a growing awareness that 'unavailable peptides', and premelanoidins representing the soluble early products of the Maillard reaction, may be positively antinutritive and distort the normal processes of protein breakdown and absorption in the small intestine (Buraczewski *et al.* 1967; Shorrock and Ford, 1978; for a review, see Adrian, 1974). Thus, the nutritionally effective amino acid composition of damaged proteins may well be different from that indicated by the amino acid analysis. It has been said in this connection that the advent of the more accurate methods of 'total' amino acid analysis offers only a more refined means of obtaining the wrong answer, though with greater precision than hitherto. This is putting it much too strongly, but clearly we should be cautious in interpreting amino acid analyses in terms of nutritional quality. With some manufactured foods we need to take account of the biological availability of the amino acids, and our present problem is how best to do this. Microbiological assays have a lot going for them. I have been especially interested in the use of

TABLE 16.9. INFLUENCE OF SEVERE HEAT TREATMENT OF COD MUSCLE PROTEIN ON THE AVAILABILITY OF LYSINE, METHIONINE AND ISOLEUCINE

Heat treatment of freeze-dried cod fillet	True digestibility of nitrogen for the rat, %	Available lysine		Available methionine		Available isoleucine	
		Rat growth assay	DNFB-test[1]	Rat growth assay	Strep. zymogenes assay	Rat growth assay	Strep. zymogenes assay
None	99.9	10.9	9.0	3.64	3.11	5.57	5.20
18 hr/135°	88.9	4.3	5.3	2.97	2.40	4.44	4.46
18 hr/145°	56.1	1.3	3.0	1.22	1.65	2.06	3.40

[1]Determined with dinitrofluorobenzene by the method of Carpenter (1960).

Strep. zymogenes for measuring available methionine, and I think that this assay is now at a promising stage for collaborative testing. It is straightforward and well founded, and it is simple as microbiological assays go. My experience with *Tetrahymena* over the years has been on the whole discouraging, though it may well prove useful for the assay of methionine + cystine (see below), and of available lysine in some foods—for example high tannin grain, severely overheated protein concentrates, some dairy products treated by the high temperature—short time process—for which the simpler and more precise dye-binding method (Hurrell and Carpenter, 1975, 1976; Hurrell *et al.* 1979) gives misleadingly high results. But apart from this I think that *Tetrahymena* has no indispensable advantages over bacteria whose use is more familiar to analysts, and which allow of greater speed and precision of measurement.

Sheffner *et al.* (1956) described an enzymatic method for protein quality assessment, the Pepsin Digest-Residue Amino Acid Index (PDR Index), which combines the pattern of essential amino acids released by *in vitro* pepsin digestion with the amino acid pattern of the remainder of the protein, to produce an integrated index. Values obtained for a selection of protein sources were impressively similar to NPU values for the growing rat and for adult man. But the amount of work involved was considerable, and so Sheffner (1976) introduced a much simpler procedure which involves the measurement of only three amino acids—lysine, methionine and tryptophan—". . .since it is the availability of these amino acids which is generally reduced during food processing." Whether or not this is strictly true, the simplified procedure gave much the same values as the original PDR-Index for the selected protein sources. McLaughlan *et al.* (1959) had earlier concluded that for many common foods, deficient in lysine or methionine + cystine, it was sufficient to determine only these three amino acids to obtain a fair measure of the nutritive value of the protein. Perhaps some such simplified index of protein nutritional quality, based on the measurement of only two or three *available* amino acids, would meet our present needs. Available lysine (by dye-binding) and available methionine (*Strep. zymogenes* assay) can be determined with reasonable accuracy and precision in a wide range of foodstuffs and it might prove possible to combine the values into a measure of protein quality that would be adequate for most primary screening purposes. A problem here is how best to take into account the cystine component of the requirement for total sulfur-containing amino acids, since cystine may be exceptionally labile to severe heat processing. With *Tetrahymena* there is in principle no difficulty in assaying for methionine + cystine; it entails only the deletion of both cystine and methionine from the culture medium. Cystine spares methionine in *Tetrahymena* as in higher animals. But there may be differences between *Tetrahymena* and the growing rat in the extent of this sparing action, as there are between the rat and other animal species.

This problem may prove of more academic than practical importance; the food industry and regulatory agencies could—indeed must—learn to live with some form of quick and precise quality grading test that is less than perfect in the nutritional sense. Of course, it might not be sufficient to measure only available methionine and lysine; perhaps some other essential amino acids should be short listed as being commonly limiting in practical diets. The *Strep. zymogenes* assay is versatile in this respect, and it can be extended to the measurement of tryptophan, isoleucine, leucine, valine, arginine and histidine, though there is a need for comparative biological and microbiological testing on a variety of food proteins to check the adequacy of the enzymic predigestion procedure as applied with these amino acids. Thus, results for available tryptophan in fish meals that had been predigested with papain were too low, though they were closely correlated with chick assay values (Carpenter and Woodham, 1974). In this case, substitution of Pronase for papain gives higher and more acceptable values (Ford, unpublished results).

Like other *in vitro* tests for available amino acids, the microbiological methods are empiric and their development must be guided by the results of parallel tests on laboratory animals. The distinctions we make between 'available' and 'unavailable' are arbitrary and relate to the particular conditions of the tests: the results are valid only in the practical sense that they predict accurately the results of biological tests, and these might vary with the species of animal used and with several other factors not directly related to the intrinsic availability of the amino acids. We do not avoid this problem by trying to measure 'overall' protein quality, as in the various microbiological assays for relative nutritional value (although, as predictors of protein quality for humans, our microbiological tests might give the wrong answer much faster and cheaper than do rat assays!). There is a place for such tests, as for example in ranking protein quality within categories of foodstuffs, and for routine in-house quality control. But we cannot yet safely apply them for grading dissimilar foods and food mixtures, and indeed I think that to attempt this is mistaken and has little relevance to practical nutrition. I am sure that microbiological assay methods will have an important role in protein quality testing, but the emphasis should be placed squarely on measurement of the biological availability of individual amino acids. We should be more concerned with meeting amino acid requirements, and less with protein quality.

REFERENCES

ADRIAN, J.   1974.   Nutritional and physiological consequences of the Maillard reaction. *In* World Review of Nutrition and Dietetics, G.H. Bourne (Editor), Vol. 19. S. Karger, Basel, p. 71.

ANDERSON, M.E. and WILLIAMS, H.H.   1951.   Microbiological evaluation of protein quality: 1. A colorimetric method for the determination of the growth of Tetrahymena gelii W. in protein suspensions.   J. Nutr. *44* : 335.

BOOTH, V.H.   1971.   Problems in the determination of FDNB—available lysine.   J. Sci. Fd. Agric. *22* : 658.

BOYD, M.J., LOGAN, M.A. and TYTELL, A.A.   1948.   A microbiological procedure for the assay of amino acids with Clostridium perfringens (Welchii) BP 6K. J. Biol. Chem. *174* : 1027.

BOYNE, A.W., FORD, J.E., HEWITT, D. and SHRIMPTON, D.H. 1975. Protein quality of feeding-stuffs. 7. Collaborative studies on the microbiological assay of available amino acids. Br. J. Nutr. *34* : 153.

BRESSANI, R., ELIAS, L.G. and JULIANO, B.O. 1971. Evaluation of the protein quality of milled rices differing in protein content. J. Agric. Fd. Chem. *19* : 1028

BURACZEWSKI, S., BURACZEWSKA, L. and FORD, J.E. 1967. The influence of heating of fish proteins on the course of their digestion. Acta. Biochim. Polonica *14* : 121

CARPENTER, K.J. 1960. The estimation of available lysine in animal protein foods. Biochem. J. *77* : 602.

CARPENTER, K.J. and WOODHAM, A.A. 1974. Protein quality of feeding-stuffs. 6. Comparisons of results of collaborative biological assays for amino acids with those of other methods. Br. J. Nutr. *32* : 647.

DUNN, M.S. and ROCKLAND, L.B. 1947. Biological value of proteins determined with Tetrahymena geleii H. Proc. Soc. Exp. Biol. Med. *64* : 377

EGGUM, B.O. and JULIANO, B.O. 1973. Nitrogen balance in rats fed rices differing in protein content. J. Sci. Fd. Agric. *24* : 921

EVANCHO, G.M., HURT, H.D., DEVLIN, P.A., LANDERS, R.E. and ASHTON, D.H. 1977. Comparison of Tetrahymena pyriformis W and rat bioassays for the determination of protein quality. J. Fd. Sci. *42* : 444.

EVANS, E. 1979A. Alterations in composition of nutrient media and growth of Tetrahymena pyriformis. I. Nucleic Acid Bases. Nutr. Rep. Int. *20* : 125.

EVANS, E. 1979B. Alterations in compositions of nutrient media and growth of Tetrahymena pyriformis. II. Vitamins. Nutr. Rep. Int. *20* : 131.

EVANS, E. 1979C. Alterations in composition of nutrient media and growth of Tetrahymena pyriformis. III. Minerals. Nutr. Rep. Int. *20* : 137.

EVANS, E., KHOUW, B.T., LIKUSKI, H.J. and WITTY, R. 1978. Evaluation of factors affecting the use of Tetrahymena pyriformis W to determine the relative nutritive value of foods. Can. Inst. Food Sci. Technol. J. *11* : 82

EVANS, E., KHOUW, B.T., LIKUSKI, H.T. and WITTY, R. 1979. Effects of altering carbohydrate and butter on the growth of Tetrahymena pyriformis and on the determination of the relative nutritive value of foods. Can. Inst. Food Sci. Technol. J. *12* : 36

FAO/WHO (Food and Agriculture Organization/World Health Organization). 1973. Energy and Protein Requirements. Report of a Joint FAO/WHO ad hoc expert committee. Wld. Hlth. Org. Techn. Rep. Ser., No. 522. World Health Organization, Geneva.

FERNELL, W.R. and ROSEN, G.D. 1956. Microbiological evaluation of protein quality with Tetrahymena pyriformis W. 1. Characteristics of growth of the organisms and determination of relative nutritive values of intact proteins. Br. J. Nutr. *10* : 143.

FORD, J.E.   1960.   A microbiological method for assessing the nutritional value of proteins.   Br. J. Nutr. *14* : 485.

FORD, J.E.   1962.   A microbiological method for assessing the nutritional value of proteins. 2. Measurement of 'available' methionine, leucine, isoleucine, arginine, histidine, tryptophan and valine.   Br. J. Nutr. *16* : 409.

FORD, J.E.   1964.   A microbiological method for assessing the nutritional value of proteins. 3. Further studies on the measurement of available aminoacids.   Br. J. Nutr. *18* : 449.

FORD, J.E.   1973.   Some effects of processing on nutritive value. *In* Proteins in Human Nutrition, J.W.G. Porter and B.A. Rolls (Editors).   Academic Press, London, p. 515.

FORD, J.E. and HEWITT, D.   1979a.   Protein quality in cereals and pulses. 1. Application of microbiological and other in vitro methods in the evaluation of rice (Oryza sativa L), sorghum (sorghum vulgare Pers.), barley and field beans (Vicia faba L).   Br. J. Nutr. *41* : 341.

FORD, J.E. and HEWITT, D.   1979b.   Protein quality in cereals and pulses. 3. Bioassays with rats and chickens on sorghum (Sorghum vulgare Pers.), barley and field beans (Vicia faba). Influence of polyethylene glycol.

FORD, J.E. and SALTER, D.N.   1966.   Analysis of enzymically digested food proteins by Sephadex gel filtration.   Br. J. Nutr. *20* : 843.

FRANGNE, R. and ADRIAN, J.   1972a.   La réaction de Maillard. 6. Réactivité de diverses protéines purifieés.   Ann. Nutrit. *26* : 97.

FRANGNE, R. and ADRIAN, J.   1972b.   La réaction de Maillard. 7. Mise en évidence d'une fraction labile de lysine dans les protéines.   Ann. Nutrit. *26* : 107.

FRANK, O., BAKER, H., HUTNER, S.H., RUSOFF, I.I. and MORCK, R.A.   1975.   Evaluation of protein quality with the phagotrophic protozoan Tetrahymena.   *In* Protein Nutritional Quality of Foods and Feeds, Part 1. Assay Methods—Biological, Biochemical and Chemical, M. Friedman (Editor), Marcel Dekker Inc., New York, p 203.

HALEVY, S. and GROSSOWICZ, N.   1953.   A microbiological approach to nutritional evaluation of proteins.   Proc. Soc. Exptl. Biol. Med. *82* : 567.

HEGSTED, D.M. and JULIANO, B.O.   1974.   Difficulties in assessing the nutritional quality of rice protein.   J. Nutr. *104* : 772.

HORN, M.J., BLUM, A.E. and WOMACK, M.   1954.   Availability of amino acids to microorganisms. II. A rapid microbial method of determining protein value.   J. Nutr. *52* : 375.

HORN, M.J., BLUM, A.E., WOMACK, M. and GERSDORFF, C.E.F.   1952.   Nutritional evaluation of food proteins by measuring availability of amino acids to microorganisms. 1. Cottonseed proteins.   J. Nutr. *48* : 231.

HURRELL, R.F. and CARPENTER, K.J.   1975.   The use of three dye-binding procedures for the assessment of heat damage to food proteins.   Br. J. Nutr. *33* : 101.

HURRELL, R.F. and CARPENTER, K.J.   1976.   An approach to the rapid measurement of reactive lysine in foods by dye binding.   Proc. Nutr. Soc. *35* : 23A.

HURRELL, R.F., LERMAN, P. and CARPENTER, K.J.   1979.   Reactive lysine in foodstuffs as measured by a rapid dye-binding procedure.   J. Food Sci. *44* : 1221.

IRWIN, M.I. and HEGSTED, M.   1971.   A conspectus of research on protein requirements of man.   J. Nutr. *101* : 539.

KNIPFEL, J.E., AITKEN, J.R., HILL, D.C., McDONALD, B.E. and OWEN, B.D.   1971.   Amino acid composition of food proteins: inter- and intralaboratory variation.   J. Assoc. Off. Anal. Chem. *54* : 777.

LANDERS, R.E.   1975.   Relationship between protein efficiency ratio of foods and relative nutritive value measured by Tetrahymena pyriformis in bioassay techniques.   *In* Protein Nutritional Quality of Foods and Feeds, Part 1. Assay Methods—Biological, Biochemical and Chemical, M. Friedman (Editor).   Marcel Dekker Inc., New York, p. 185.

LAWRIE, N.R.   1937.   Studies in the metabolism of protozoa. III. Some properties of a proteolytic extract obtained from Glaucoma pyriformis.   Biochem. J. *31*-1: 789.

MACIEJEWICZ-RYS, J. and ANTONIEWICZ, A.M.   1978.   2-Aminoethylphosphonic acid as an indicator of Tetrahymena pyriformis W growth in protein-quality evaluation assay.   Br. J. Nutr. *40* : 83.

McLAUGHLAN, J.M., ROGERS, C.G., CHAPMAN, D.G. and CAMPBELL, J.A.   1959.   Evaluation of protein in foods. 4. A simplified chemical score.   Canad. J. Biochem. Physiol. *37* : 1293.

MILLER, E.L., CARPENTER, K.J., MORGAN, C.B. and BOYNE, A.W.   1965.   Availability of sulphur amino acids in protein foods. 2. Assessment of available methionine by chick and microbiological assays.   Br. J. Nutr. *19* : 249.

NELSON, T.S., STEPHENSON, E.L., BURGOS, A., FLOYD, J. and YORK, J.O.   1975.   Effect of tannin content and dry matter digestion on energy utilization and average amino acid availability of hybrid sorghum grains.   Poult. Sci. *54* : 1620.

NRC (NATIONAL RESEARCH COUNCIL).   1978.

ROCKLAND, L.B. and DUNN, M.S.   1946.   The microbiological determination of Tryptophan in unhydrolysed casein with Tetrahymena geleii H.   Arch. Biochem. *11* : 541.

RØLLE, G.   1975.   Research on the quantitative requirement of essential amino acids by Tetrahymena pyriformis W.   Acta Agric. Scand. *25* : 17.

RØLLE, G.   1976.   Research concerning the usefulness of Tetrahymena pyriformis W as a Test Organism for the Evaluation of Protein Quality.   Acta Agric. Scand. *26* : 282.

ROSEN, G.D. and FERNELL, W.R.   1956.   Microbiological evaluation of protein quality with Tetrahymena pyriformis W. 2. Relative nutritive values of proteins in feedstuffs.   Br. J. Nutr. *10* : 156.

SATTERLEE, L.D., MARSHALL, H.F. and TENNYSON, J.M.   1979.   Measuring protein quality.   J. Am. Oil Chem. Soc. *56* : 103.

SHEFFNER, A.L.   1967.   *In vitro* protein evaluation.   *In* Newer Methods of Nutritional Biochemistry, Vol. III, A.A. Albanese (Editor).   Acad. Press. N.Y., p. 125.

SHEFFNER, A.L., ECKFELDT, G.A. and SPECTOR, H.   1956.   The pepsin digest residue (PDR) amino acid index of net protein utilization.   J. Nutr. *60* : 105.

SHORROCK, C.   1972.   Studies on the biological availability of amino acids in feedstuffs.   PhD thesis, University of Reading, England.

SHORROCK, C.   1976.   An improved procedure for the assay of available lysine and methionine in feedstuffs using Tetrahymena pyriformis W.   Br. J. Nutr. *35* : 333.

SHORROCK, C. and FORD, J.E.   1973.   An improved procedure for the determination of available lysine and methionine with Tetrahymena.   *In* Proteins in Human Nutrition, J.W.G. Porter and B.A. Rolls (Editors).   Academic Press, London, p. 207.

SHORROCK, C. and FORD, J.E.   1978.   Metabolism of heat damaged proteins in the rat. Inhibition of amino acid uptake by unavailable peptides isolated from enzymic digests of heat damaged cod fillet.   Br. J. Nutr. *40* : 185.

SOLBERG, M., BERKOWITZ, K.A., BLASCHEK, H.P. and CURRAN, J.M.   1979.   Rapid evaluation of protein quality of foods using Clostridium perfringens.   J. Fd Science *44* : 1335.

STOTT, J.A. and SMITH, H.   1966.   Microbiological assay of protein quality with Tetrahymena pyriformis W. 4. Measurement of available lysine, methionine, arginine and histidine.   Br. J. Nutr. *20* : 663.

TEUNISSON, D.J.   1971.   Elutriation and Coulter counts of Tetrahymena pyriformis grown in peanut and cottonseed meal media.   Appl. Microbiol. *21* : 878.

WANG, Y.Y.D., MILLER, J. and BEUCHAT, L.R.   1979a.   Comparison of four techniques for measuring growth of Tetrahymena pyriformis W with rat PER bioassays in assessing protein quality.   J. Fd Science *44* : 540.

WANG, Y.Y.D., MILLER, J. and BEUCHAT, L.R.   1979b.   Comparison of Tetrahymena pyriformis W, Aspergillus flavus and rat bioassays for evaluating protein quality of selected commercially prepared food products.   J. Fd Science *44* : 1390.

DISCUSSION

DR. STEINKE: I thought your presentation was excellent. I have one question. Most of the products that you have evaluated have been fairly purified or simple materials. How applicable would these procedures be to, say, a mixed food system such as a piece of pie or beef stew where you might have any variety of spices and salts and other things that might affect growth of the organisms?

DR. FORD: I think that we might occasionally meet technical problems, but I have little experience with such mixed foods. One advantage of the assay with lactic acid bacteria is that we can measure growth indirectly by titrating the lactic acid production in the assay. This can be done quite precisely and it is a quick and simple alternative to measurement of optical density.

DR. HUTNER: I regret that I take exception to practically every conclusion that Dr. Ford has drawn. Dr. Ford seems not to have seen a paper in a 1978 issue of *Nutrition Reports International*. My colleague here, Dr. Rusoff of Nabisco, furnished us with 300 samples of foodstuffs from the production lines of Nabisco. The techniques we used were far advanced over our joint 1975 paper. We used a simple one-step digestion with bromelain activated with mercaptosuccinic acid. The standardization was simple. All we had to do was centrifuge after awhile and discard the residue. That gave a clear solution to assay. We applied traditional microbiological methods— slope ratio—with such not very high technology instruments as a Bausch and Lomb Spectronic 20, a Coleman Junior spectrometer, or Fred Kavanaugh's gadget from Elanco.

I think Dr. Rusoff will corroborate my assertion that whenever *Tetrahymena* varied from the rat, then we'd get the rat data from him. (I assure you we had no mole, British or American style, in the Nabisco lab to break the code for us). It turned out that there was something wrong with the rats each time.

We have experience in chemotherapy. It's a chancy affair: one deals here with vagaries of mice, rats, and people because we work with immunosuppressive protozoa and immunosuppressive drugs. If a stranger walks into a rat or mouse room there's a very good chance of having a whole experiment wiped out by intercurrent infection. There is no problem, however, with aseptic *Tetrahymena* cultures.

The trouble seems to be that workers don't accept *Tetrahymena* on its own terms. It's a specialized bacterial eater. Apparently all one has to do is present it with food in soluble form. We also don't seem to know how digestion really does place. Much digestion occurs in the brush-border cells of the intestine and since these enzymes are cathepsin-type, like bromelain and papain, it's hard to get them out. Therefore, they're unsuitable for doctoral dissertations and as you know, most research is done by graduate students while their mentors serve on committees on how to do research.

I agree with Dr. Ford that assays should be done on an individual amino acid basis but I would like to point out a problem that has never been addressed systematically: the gap between stock diets and those based on "crystalline amino acids". We're attacking that problem. I think we found out what some of the unknown stimulating factors are. The limiting factor often seems to be ion-transport compounds, such as siderophores and the like for iron.

I hope I'm not condescending in mentioning that one worker whom I respect deeply took violent exception to the way we select sample size by means of an empirical factor based on nitrogen content. If one's cynical and uncharitable, it's a fudge factor. Why do we have to make that empirical adjustment? It has to do probably with a whole melange of stimulating factors. But it works!

My colleague Dr. Baker, who does a great deal of clinical assays with *Tetrahymena*, calculated for my benefit that if one really wished to automate to the standards of the pharmaceutical industry, and once one had written off the capital costs, one assay costs 12 cents. If one uses conventional microbiological assay techniques it climbs steeply to 25 cents a test.

The literature shows that *Tetrahymena*, at least all that are thermophilic, have mating behavior. The ancient W is non-mating. Therefore, the geneticists are anxious for high precision defined media for *T. thermophila,* i.e., one based on crystalline amino acids—just what is needed for amino acid assays. We're checking our *Tetrahymena* results against another protozoan, which is as remote from *Tetrahymena* as we are phylogenetically, and it's crithidia. The lysine assays are checking nicely. We're just about to try our digestion procedure on it. As for the time factor that our FDA friends mentioned, I think if one were to use an instrument like a biophotometer there would be no trick at all to monitor onstream production for lysine.

DR. FORD: Thank you, Dr. Hutner. I'm sorry I overlooked your paper. I was trying to give some sort of assessment of the literature and I can see that I've overlooked an important paper. I think it is fair for me to say that in my experience (and I've been running a microbiological assay sweat shop for the past thirty years), we do find *Tetrahymena* a difficult bug to use. The assays tend to be nonlinear and poorly reproducible, and they are easily contaminated. Perhaps this is a reflection on our rather sloppy bacteriological techniques but I think they are better than would be found in many quality control laboratories. I shall be very keen to try out your improved procedure, Dr. Hutner. I am not concerned to denigrate *Tetrahymena*. I am simply arguing that conventional bacterial assays might be quicker and simpler, and more familiar to most of us. It may be that *Tetrahymena* will prove useful for measuring available lysine: I would very much like to have a microbiological assay for this purpose. But I think *Tetrahymena* responds to lysinoalanine and perhaps also to lysine bound in premelanoidins.

DR. BODWELL: Dr. Ford, you have given some data correlating amounts of available methionine or available lysine with NPU and, in one of your slides, with digestibility. I have two questions. Firstly, do you think microbiological assays have potential for measuring digestibility of foods for humans? Either specific proteins or mixed foods or both? And then, secondly, is the U.K. moving toward amino acids, single amino acids, as a basis for nutritional labeling of any sort?

DR. FORD: I can't answer that second question. Perhaps Dr. Carpenter can. As for digestibility, the answers we get from *in vitro* methods may be strongly influenced by the conditions of enzyme treatment used. This is as true for the microbiological tests as for the purely enzymic procedures. All these methods need to be calibrated against biological assays, and carefully standardized. This raises the question, whether the microbiological assays have any important advantages. I think that they have. They offer a higher degree of biological specificity. Also, it is possible in a single test to assay 30–40 samples for available methionine, for example. And if you are interested only in methionine and not all the other amino acids, the microbiological method is arguably the simplest and most effective way to measure it.

DR. BODWELL: Of course, what we're really interested in is amino acid bioavailability, not digestibility.

DR. FORD: Yes.

DR. SATTERLEE: I think Dr. Steinke was bringing up a point that deals with the use of food additives. We have seen in studies with *Tetrahymena* that a food additive can create artifacts by inhibiting the growth of the organism. Additives from antioxidants to spices, erythrobates and sorbates, all commonly found in foods, will have some effect on the organism. I think *Tetrahymena* has very good use as an in-house method for determining protein quality, but caution must be exercised when used on foods directly off the grocers' shelf.

DR. ROTRUCK: This is a question that Dr. Carpenter touched on earlier and that's regarding the potential oxidation of methionine and cysteine to unavailable forms like methionine sulfone and cysteic acid. I wonder if anybody's thought about or is concerned about the potential for amino acid analysis to overestimate available methionine and cysteine because of the oxidation of methionine to methionine sulfone or of cysteine to cysteic acid. I see some potential for this to happen but I don't know whether it can really happen in diets. The basic question boils down to whether methionine sulfone or cysteic acid occurs in diet samples on a routine basis.

DR. CARPENTER: The only examples I've seen of high levels of methionine sulfoxide were materials that had been treated with hydrogen peroxide. One table in my paper contains an example in which two-thirds of the methionine had been converted to the sulfoxide and yet the biological value of the protein was very little reduced. That was the basis for my saying that I thought it didn't matter that the sulfoxide was included with the methionine. It would be a bigger error not to measure the sulfoxide and just to measure the methionine. Methionine sulfone is agreed to be biologically inactive but little was formed under those severe conditions of treatment with hydrogen peroxide.

DR. PELLETT: A small comment in relation to methionine sulfone and methionine sulfoxide. There are, of course, many procedures for identifying these separately. We found, however, in our part of the collaborative assay that, except for soy isolate, we got almost identical answers for cystine and methionine whether we used the performic oxidation or the ordinary acid hydrolysate. There was a difference with soy isolate, however. It may well be that there was some methionine sulfone already present, thus 'true' methionine could be overestimated by the peroxidation procedure and the ordinary hydrolysis could give more accurate values.

DR. HOPKINS: One would get the implication from hearing the talks today that perhaps we should be determining available lysine and available methionine in all foods, from pizza to cheeses, et cetera. Is this the case or do we need amino acid availability assays to measure specific ingredients with which we might have problems, such as ingredients that have been treated with peroxide and this type of thing? And then could we use a more generalized procedure in areas where we think there is no problem?

DR. FORD: As I see it, a problem in our total protein quality value tests is to know what to do with the results when we've got them. We can obtain much the same relative nutritive values for proteins of quite different amino acid composition and limiting in different amino acids. If we are interested in assessing the value of the combination of foods that goes together to make a complete diet, we cannot compute it from the individual PER or RNV values.

DR. EDWARDS: Richard Link at Argonne Laboratories has reported that one of the mechanisms for metabolism of methionine is its breakdown to 5-methylthioadenosine and alpha butyric acid and he has shown that this mechanism exists in yeasts. In our own laboratories, we have demonstrated that in the rat, an alternative metabolic cycle involves the production of 5-methylthioadenosine and alpha butyric acid rather than the complete degradation of methionine. We have also demonstrated, using ^{35}S labeled 5-methylthioadenosine and nonlabeled alpha butyric acid, that the rat can produce methionine from those two metabolites. Similarly using nonlabeled 5-methylthioadenosine and ^{14}C labeled alpha butyric acid, the rat produces labeled ^{14}C methionine. Now, this mechanism appears to occur to a greater extent at lower levels of protein intake. I'm wondering, if such a mechanism for methionine metabolism occurs, whether or not false values from assays might be obtained at some levels of the protein source. I'm wondering if anyone has observed any anomalies at, say, lower levels of the protein source which cannot be explained in the methionine assays.

I don't know the extent to which other organisms beside yeast are able to perform this change, but there are alternate sources of sulfur. If the organism was breaking down the methionine to utilize its sulfur and then had available alpha butyric acid and other sources of sulfur, the organism could make itself more methionine. The assay result might not be correct. Has anyone run across anything in methionine assays that is unexplainable when other methods yield different methionine results?

DR. HUTNER: A protozoan which is the best B_{12} assay organism is *Ochromonous malhamensis* or, as it's called today, *Poteriochromonas malhamensis*. (The academic mind is wonderful!) It goes through these, the whole methylthioadenosine pathway. The literature on *Ochromonas* is very good.

17

Estimating Protein Digestibility for Humans From Rat Assays

S.J. Ritchey and L.J. Taper

Using data from the literature, estimates of apparent digestibility by human subjects were derived for several sources of protein. Mean apparent digestibilities range from 70.7 to 87.8 percent for adults and from 64.0 to 90.4 percent for growing children for different sources. For each source of protein, considerable variability was found, but this was expected because of differences which existed between studies. Data on apparent digestibility, protein efficiency ratio (PER) and net protein utilization were collated from rat assays of the same proteins for purposes of comparing with information from human studies. Apparent digestibilities in the human were not predicted well from PER, NPU, or apparent digestibility data from animal assays.

Interest in animal assays of protein quality and utilization goes back to the 1870s and the era of Voit who observed that gelatin would not support growth. Osborne and Mendel in the period from 1911 to 1920 initiated the development of quantitative methods of evaluating protein. In association with Ferry (1919), they developed the protein efficiency ratio (PER) technique with the objective of expressing in some numerical sense the growth-promoting properties of proteins from different sources. During the same general time frame, Thomas (1909) suggested the biological value method for protein quality assessment in adult human nutrition, but in a much later publication, Thomas (1954) admits that he lost interest because of doubt that nitrogen content provided an adequate expression of protein. In the 1920s, Mitchell (1924) adapted the Thomas method to evaluate protein utilization in both the growing and adult states.

One of the problems associated with the development of methodology is that the work with methodology becomes an end in itself and investigators lose perspective of the primary goal. This probably has happened in the area of protein evaluation; investigators became so enthralled with the method—development, problems, and improvements—that the need for transfer to the human was lost in the shuffle. We must remind ourselves at frequent intervals that animal assays which do not reflect human utilization rather well are of little or no value.

A group similar to this conference reviewed the range of questions relevant to proteins in processed foods (White and Fletcher 1974). This group recommended the accumulation of much more data from human studies in order to provide an improved basis for correlating data from animal and human studies.

Our assignment was to provide some notion about the usefulness of animal assays in estimating or predicting the digestibility of proteins by the human. Thus, we have avoided any discussion of protein quality and amino acid composition, although these parameters are obviously important in the totality of protein utilization. Further, we have attempted to utilize only those assay methods which include the digestibility factor. By definition, measures such as protein efficiency ratio (PER), net protein utilization (NPU), and net protein ratio (NPR) include digestibility. In each of these assays, the denominator includes protein or nitrogen consumed and reflects use of protein as a part of the total intake. Digestibility data from rat assays provide the most direct comparison for predicting digestibility in the human and have been considered.

The growing rat has been the most frequently utilized assay model, but there is disagreement about the validity of data from this animal in estimating human needs. Hegsted (1971) questioned the validity of this approach because a much larger percentage of the dietary protein is used for maintenance in the growing child and the adult human than by the rapidly growing rat. However, after reviewing the literature, Campbell and McLaughlan (1971) suggested that assays of protein quality which use the growing rat are a useful predictor of protein quality for humans. Mitchell (1954) compared the biological values of a group of proteins in growing rats, adult rats and adult man. Correlations were higher between the growing rat and adult man (0.915) than between the growing rat and the adult rat (0.658) or between the adult rat and adult man (0.670).

The FAO/WHO report on protein and energy requirements (1973) provides comparative data for NPU for children and rats. A major variable in these data from children is the percentage of energy from protein; this ranges from 2−3 to 18−21%; the values from rats were obtained in studies in which the protein represented 10% of the dietary energy. We ignored the energy variable for the children, averaged the NPU values for the energy levels from 2 to 10 percent, and plotted these values against corresponding NPU values for rats (Fig. 17.1). This relationship (r = 0.90) suggests that NPU values from the rat predict the utilization of protein by humans.

In a review of protein evaluation using rats, Dreyer (1973) presented data to indicate a linear relationship between intake and absorption for three proteins. He demonstrated also that the digestibility of combinations of egg and haricot bean followed the same linear relationship. Thus, Dreyer concluded that digestibility is a constant characteristic and may be determined in absolute terms using established biometrical procedures. In a recent

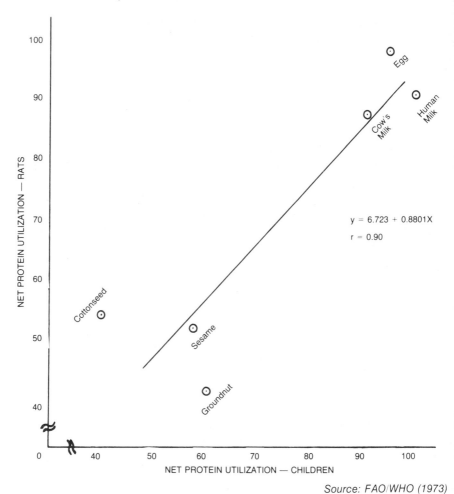

Source: FAO/WHO (1973)

FIG. 17.1.   RELATIONSHIP BETWEEN NPU IN RATS AND CHILDREN

report, Bodwell and co-workers (1980) compared digestibility data from human and rat assays. They reported correlations between animal and human assays in the range of 0.369 to 0.976 for different protein sources. The higher correlations were found for plant proteins and the lower correlations were found when all proteins (plant and animal) were included in the calculations.

As background for this paper, we extracted from the literature data about digestibility of proteins by humans. Information was obtained from 77 different publications which reported digestibility or which provided data from which digestibility could be calculated. Details of this literature survey will be presented in another publication (Ritchey *et al.* 1980). Apparent digestibilities of proteins from different sources of proteins are presented for adults (Table 17.1) and for children (Table 17.2). In preparing these summaries, factors such as level of protein, variation in product preparation, energy intake, and major discrepancies in age were ignored. We recognized that these factors and others may affect digestibility, but we recognized also that the inclusion of all factors would prevent the development of any summary.

Mean apparent digestibilities range from 70.7 to 87.8 percent for the various protein sources for adults (Table 17.1) and from 64.0 to 90.4 percent for growing infants and children (Table 17.2). Within each source of protein, there are considerable variations from study to study. However, it seems valid to conclude that humans utilize most proteins reasonably well and that apparent digestibilities can be anticipated to range from 70 percent upwards (also, see Chapter 11).

Data were extracted from the literature to permit comparisons of apparent digestibility, PER, and NPU values from animal assays with digestibility data from human studies. Means and ranges for PER and NPU were

TABLE 17.1. APPARENT DIGESTIBILITIES BY ADULTS OF PROTEINS FROM DIFFERENT SOURCES

Source of Protein	Apparent Digestibility Mean (%)	Range (%)	No. of Studies
Amino acid mixtures	82.4	46.7–90.5	7
Beef	81.9	81.4–82.7	3
Corn, white	70.7	63.2–75.9	6
opaque-2	71.9	65.5–76.5	7
Cottonseed	87.8	84.1–92.7	3
Egg	81.1	67.4–92.5	26
Fish	87.2	83.5–90.8	2
Rice	77.6	66.8–82.5	8
Soy	79.9	79.2–81.6	6
Wheat	81.6	52.1–92.7	9

TABLE 17.2. APPARENT DIGESTIBILITIES BY INFANTS AND CHILDREN OF PROTEINS FROM DIFFERENT SOURCES

Source of Protein	Age of Subjects	Apparent Digestibility Mean (%)	Range (%)	No. of Studies
Amino acid mixtures	Infants	90.4	89.7–91.0	2
Beef	15 yrs.	82.0	—	1
Cottonseed	<2 yrs.	72.4	68.0–79.3	3
Fish	2–12 yrs.	70.9	65.0–75.7	3
Milk, cow's	Infants to 11 yrs.	85.1	82.1–92.9	7
Milk, human	Infants	83.5	—	1
Rice	8–9 yrs.	64.0	62.9–66.1	6
Soy	<3 yrs.	73.4	—	9
Wheat	2–5 yrs.	76.6	70.9–87.3	4

calculated (Table 17.3) ignoring the influencing variables, although these assays tend to follow standard conditions. Apparent digestibility data from human and rat assays for several sources of protein are compared in Table 17.4. These data represent mean values from several studies.

Relationships between data from animal assays and apparent digestibilities for human subjects were examined by calculating correlations. Those correlations were generally poor, ranging from about 0.1 to 0.6 (Table 17.5). Attempts to plot these data suggested a nonlinear relationship. However, a more precise comparison may reveal a better relationship than the comparison of average values for sources of proteins. Digestibility is only one of the factors involved in the utilization of dietary protein, thus assays

TABLE 17.3. PROTEIN EFFICIENCY RATIOS (PER) AND NET PROTEIN UTILIZATION (NPU) OF PROTEINS FROM DIFFERENT SOURCES AS DERIVED FROM ANIMAL ASSAYS

Source of Protein	Mean	PER Range	No. of Studies	Mean	NPU Range	No. of Studies
Beef	2.40	2.12–2.82	4	62	—	1
Cottonseed	2.25	—	1	56	53–59	2
Egg	3.92	—	1	94	—	1
Fish	3.48	3.42–3.55	2	72	65–80	2
Milk (cow's)	2.98	2.86–3.09	2	82	—	1
Rice	2.18	—	1	57	—	1
Soy	1.99	0.80–3.38	13	42	27–61	8
Wheat	1.55	0.60–1.53	3	49	39–67	3

TABLE 17.4. COMPARISON OF APPARENT DIGESTIBILITY OF DIFFERENT SOURCES OF PROTEIN BY CHILDREN, ADULT HUMANS AND RATS

Source of Protein	Apparent Digestibility (%)[1] Children	Adult	Rats
Beef	82.0	81.9	89.0
Casein-lactalbumin	—	84.2	82.4
Egg	—	81.1	90.0
Fish	—	87.2	90.3
Milk	85.1	—	87.6
Peanut	—	81.5	89.0
Soy	73.4	79.9	83.0
Wheat	76.6	81.6	89.6

[1]Mean values calculated from several studies.

TABLE 17.5. CORRELATIONS BETWEEN APPARENT DIGESTIBILITY IN HUMANS AND ANIMAL ASSAYS

Apparent Digestibility (%)	Data From Animal Assays % dig	PER	NPU
Children	0.54	0.09	0.32
Adults	0.17	0.56	0.56

based on growth and/or protein retention may not provide a reasonable basis for predicting digestibility in the human. Correlations calculated from these data, and as reported by Bodwell *et al.* (1980), do not give much assurance of using animal assays as a predictor of digestibility in humans in a general fashion. Careful selection of protein sources may indicate positive relationships between digestibility data from human and rats. However, we do not always have the option of careful selection and matching of data.

At the present time, the nutrition literature appears to have considerable data on digestibility from human studies. These data may be more useful in predicting the response from humans to a given protein or mixture of proteins than animal assays. If we are interested in predicting digestibility in humans from rat assays, studies should be designed to provide solid data in this specific area. There have been a few studies (Bodwell *et al.*, 1980), but additional experiments appear warranted.

REFERENCES

BENDER, A.E. 1969. Newer Methods of Assessing Protein Quality. Chem. and Ind. *27*:904.

BODWELL, C.E., SATTERLEE, L.D. and HACKLER, L.R. 1980. Protein digestibility of the same protein preparations by human and rat assays and by in vitro enzymatic digestion methods. Am. J. Clin. Nutrition *33*:677.

BRESSANI, R., VITERI, F., ELIAS, L., DE ZAHI, S., ALVARADO, J. and ODELL, A. 1967. Protein quality of a soybean protein textured food in experimental animals and children. J. Nutr. *93*:349.

BRICKER, M., MITCHELL, H. and KINSMAN, G. 1945. The protein requirement of adult humans. J. Nutrition *30*:269.

CAMPBELL, J.A. and McLAUGHLAN, J.M. 1971. Applicability of animal assays to humans. *In* Proceedings SOS/70, Third International Congress of Food Science and Technology. Inst. Food Technologists, Chicago, IL.

CHICK, H., BOAS-FIXSEN, M., HUTCHISON, J. and JACKSON, H. 1935. The biological value of proteins. VII. The influence of variation in the level of protein in the diet and of heating the protein on its biological value. Biochem. J. *29*:1712.

DE GROOT, A.P. 1968. The influence of dehydration of foods on the digestibility and the biological value of the protein. Food Technol. *339*:103.

DREYER, J.J. 1973. Protein evaluation with rats. *In* Proteins in Human Nutrition. J.W.G. Porter and B.A. Rolls (Editors). Academic Press, New York.

FAO/WHO (Food and Agriculture Organization/World Health Organization). 1973. Energy and Protein Requirements. Wld. Hlth. Org. Techn. Rep. Ser., No. 522. World Health Organization, Geneva.

FOOD AND AGRICULTURE ORGANIZATION. 1957. Committee on Protein Requirements. FAO Nutritional Studies No. 16. World Health Organization, Geneva.

GREAVES, E., MORGAN, A. and LOVEN, M.   1938.   The effect of amino acid supplements and of variations in temperature and duration of heating on the biological value of heated casein.   J. Nutrition *16*:115.

HEGSTED, D.M.   1971.   Nutritional research on the value of amino acid fortification—experimental studies in animals.   *In* Amino Acid Fortification of Protein Foods. N.S. Scrimshaw and A.M. Altschul (Editors).   MIT Press, Cambridge, Mass.

HSU, H.W., VAVAK, D.L., SATTERLEE, L.D. and MILLER, G.A.   1977.   A multienzyme-automatic recording technique for in vitro protein digestibility. J. Food Sci. *42*:1269.

JUNEJA, P.K., KAWATRA, B.L. and BAJAJ, B.   1980.   Nutritive value of tritical and the effects of its supplementation to wheat and bengal gram *(Cicer arietenium)* flour.   J. Food Sci. *45*:328.

MILLER, D.S. and PAYNE, P.R.   1969.   Assessment of protein requirements by nitrogen balance.   Proc. Nutrition Society *28*:225.

MITCHELL, H.H.   1924.   A method for determining the biological value of protein.   J. Biol. Chem. *58*:873.

MITCHELL, H.H.   1948.   The biological utilization of proteins and protein requirements.   *In* Protein and Amino Acids in Nutrition. M. Sahyun (Editor). Reinhold Pub. Corp., New York.

MITCHELL, H.H.   1954.   The dependence of the biological value of proteins and amino acid interrelationships.   *In* Die Bewertung der Futterstoffe und andere Probleme der Tierernäbrung, Vol. V. K. Nehring (Editor). Abhandl, Deutsche Akad. der Landuirtscheftswiss., Berlin.

MITCHELL, H. and BEADLES, J.   1950.   Biological values of six partially purified proteins for the adult albino rat.   J. Nutr. *40*:25.

MITCHELL, H. and BLOCK, R.   1946.   Some relationships between the amino acid contents of proteins and their nutritive values for the rat.   J. Biol. Chem. *163*:599.

MURLIN, J., EDWARDS, L. and HAWLEY, E.   1944.   Biological values and true digestibilities of some food proteins determined on human subjects.   J. Biol. Chem. *156*:785.

NJAA, L.R., UTNE, F., and BRAEKKAN, O.R.   1968.   Protein value of cod and coalfish and some products for the young rat.   Fiskeridirektoratets Skrifter, ser. Teknol. Undersok. *4*:4.

OSBORNE, T.B., MENDEL, L.B. and FERRY, E.L.   1919.   A method of expressing numerically the growth-promoting value of proteins.   J. Biol. Chem. *37*:223.

PLATT, B. and MILLER, D.   1961.   The protein value of human breast milk. Proc. Nutr. Soc. *20*:8.

RITCHEY, S.J., SHEEHAN, P., McNEIL, D., and HACKLER, L.R.   1980. Protein digestibility in the human.   Am. J. Clin. Nutrition (paper submitted).

SAENZ DE BURUAGA, F.M.   1968.   E studio comparativo de algunas technicas quimicas y biologicas de control de la calidad nutritiva de las proteinas. An. Bromtol. *20*:49.

SATTERLEE, L.D., KENDRICK, J.G. and MILLER, G.A.   1977.   Rapid in vitro assays for estimating protein quality.   Food Technol. *31*:73.

THOMAS, K.   1909.   Über die biolegische Wertigkeit der Stickstoffsubstanzen in verschiedenen Nahrungsmitteln. Beiträge zur Frage dem physiologischem Stickstoffminimum.   Arch. Anat. Physiol., Physiol Abt. *1909*:219.

THOMAS, K.   1954.   Fifty years of biochemistry in Germany. Prefactory chapter.   Ann. Rev. Biochem. *23*:1.

WHITE, P.L. and FLETCHER, D.C.   1974.   Nutrients in Processed Foods—Proteins.   Publishing Sciences Group, Inc., Acton, Mass.

YOUNG, E., VASQUEZ, M., and SANCHEZ, F.   1979.   Texturization of sunflower/soy flour mixtures: Chemical and nutritive evaluation.   J. Food Sci. *44*:1714.

DISCUSSION

DR. SATTERLEE: Dr. Ritchey, since you were working with apparent digestibility, as was possibly shown yesterday, the level of protein in the diet can have quite an impact on the apparent digestibility value obtained. Do you think this is one of the variables leading to poor correlations between the rat and human digestibility data? Are some of the human digestibility data from studies where there were very low protein intakes?

DR. RITCHEY: I'm sure it is. The level of protein, the level of energy, variation in product development, perhaps discrepancies or differences in age, all of these variables influence that comparison.

DR. SATTERLEE: I was interested to note that digestibility in the young rat does not relate to that of the adult rat, but does predict that of man fairly well. Do you have any explanation?

DR. RITCHEY: I think it's the luck of probability, actually. I really cannot explain that. Those were Mitchell's calculations in 1954. I have a feeling it's just happenstance.

DR. EGGUM: We heard something about this yesterday from Dr. Hopkins. I have a lot of data on rats which are all true digestibility data. Just looking at the true digestibility values given by Dr. Hopkins yesterday and comparing them with my own rat data, I think we are very, very close. Anyway, I'm going home and do some calculations on this.

The question was raised as to whether or not digestibility in the rat changes with age. We have measurements from 60 gram rats up to 500 gram rats. Values are very constant, both for protein and energy, the whole way through from 60 grams to 500 grams.

I think it would be very good to have more comparative studies on digestibility between rat and man. We have values between rats and pigs and the pig is supposed to be even closer to man. Digestibility coefficients, obtained on rats and pigs, had a correlation coefficient of 0.97 for energy and 0.93 for protein.

DR. HOPKINS: Dr. Eggum, you and I discussed yesterday a little bit about the possible influence of coprophagy on measurement of digestibility. Do you want to say a word on this?

DR. EGGUM: Well, this question is often raised but I don't think this is a problem on average, good quality diets. It might become a problem on very poor diets. We have monitored rats for 24 hours across several days and it's very, very rare that we observe coprophagy. We can always look at the stools because if you have this problem, the stools are broken into pieces. The stools are usually intact. It's very seldom that we can see this problem other than on very poor quality diets.

DR. HARPER: I was wondering if some of the problem with the adult rats isn't that apparent digestibility values are used and that the levels of protein for many of these tests are only of the order of 4%. Fecal nitrogen excretion of rats fed a protein-free diet which is normally used to correct for true digestibility would represent a value nearly as high as the amount excreted when the protein was fed. In fact, Mitchell used rats fed 4% of high quality protein to correct for true digestibility as the fecal nitrogen was essentially the same as for those fed a protein-free diet.

DR. SARWAR: My question is for Dr. Eggum. I just wanted to know about the method of calculating protein digestibility in his laboratory. Is it a total collection method, or is an indicator used for calculating the digestibility in rats?

DR. EGGUM: This is a total collection. We have a preliminary period for four days and a balance period for five days in which we make total collections, both for energy studies and for protein studies.

If I can come back to Dr. Harper's question regarding digestibility, I don't think this is a problem with younger versus older rats. We feed the rats at the 9.4% protein level, both the young rats and the old rats. We find the digestibility coefficients to be very constant all the way through. The same is the case in pigs from six weeks up to 90 kilograms. Digestibility coefficients are very constant and independent of age in this interval.

DR. HOPKINS: I think Dr. Harper's point was that, if you use apparent digestibilities, then there might be a bigger difference between young and old rats whereas as you're pointing out, if you use true digestibility, it wouldn't really make a difference.

DR. RITCHEY: The point Dr. Harper also was making is that as the level of protein intake decreases, the fecal nitrogen represents an increasing percentage of the intake, and may result in a dramatic alteration in the apparent digestibility figure.

DR. EGGUM: I think I mentioned this yesterday. I believe that the difference between true and apparent digestibility is dependent on dry matter consumption. When the level of dry matter fed is controlled, as is done in our studies, a constant factor can be used to go from apparent to true digestibility.

18

Estimating Apparent Protein Digestibility from *In Vitro* Assays

L.D. Satterlee, J.G. Kendrick, D.K. Jewell and W.D. Brown

In the past five years, concern has been expressed by scientists from the food industry, food regulatory agencies and academia on the need for rapid assays that will allow for the close surveillance of protein nutritional quality in foods and food ingredients during processing. Anderson (1978) stated that the food processing industry is in dire need of a rapid assay that would replace the slow PER bioassay, by predicting protein nutritional quality that correlates well with data obtained from the PER bioassay. Anonymous #1 (1977) stated that any rapid assay, in addition to being fast and accurate, must be applicable to a wide range of foods, including processed cereal grain products, meats, milk products, products from oilseeds, potato products and a host of other products of commerce. Anonymous #2 and Bodwell (1977a) also stated that any rapid assay, to be of use, must be applicable to a wide variety of complex foods.

One difficulty that has arisen in the development of rapid chemical or enzymic methods for measuring protein nutritional quality is that a method may work well for one class of foods (i.e., meat products) and poorly on other classes of foods (i.e., cereal or complex foods). Among the limiting factors causing an assay to be able to predict protein nutritional quality accurately for one class of foods but not for other classes are:

1. Limited data base—only one class, or very similar classes, of food proteins were used in the development of the assay.
2. Unique physical and/or chemical properties—certain physical/chemical characteristics may prevent the accurate estimation of nutritional quality in certain classes of foods. Examples are foods that contain proteolytic enzyme inhibitors, oxidized sulfur amino acids, or intact cellulosic cell walls. All three of these factors will affect eventual amino acid availability and subsequent estimates of protein nutritional quality, especially by rapid assays that are not sensitive to the presence of such factors.

316

Three basic parameters that dictate the nutritional quality of a protein are 1) the quantities of the essential amino acids (EAA) that are present in the protein, 2) the digestibility of the amino acids that have been released from the protein via digestion and 3) the bioavailability of the EAA released during protein digestion. If a rapid assay measures either directly or indirectly the three basic parameters known to be important in protein nutritional quality, and is also built around a wide variety of foods, then the two above listed limiting factors of a rapid *in vitro* assay can be minimized.

Bioassays, such as the rat based PER or the human based nitrogen balance assay, measure the three parameters in aggregate, even though they do not yield specific data on each parameter.

If a rapid assay is to be able to predict protein nutritional quality successfully, it must measure the above mentioned parameters, either directly or indirectly. For example, the Chemical Score technique for assaying protein nutritional quality does not take into account the digestibility of the protein or the bioavailability of the amino acids in a protein, and therefore does not accurately predict the nutritional quality of foods where processing has affected protein digestibility and amino acid bioavailability (Anonymous #1, 1977). Steinke (1979) discussed the use of amino acid analysis of a protein and then compared the EAA profile to the EAA requirements of humans, as a means of rapidly predicting protein quality based upon human requirements. But Steinke (1979) stated that consideration must also be given to the digestibility of a protein and the availability of its amino acids, if the rapid assay was to be successful.

Thus far the discussions concerning the development of rapid assays for protein nutritional quality have emphasized the need for the assay to be applicable to a wide range of foods, and to be based on the three aforementioned parameters, known to affect protein nutritional quality.

PROTEIN QUALITY ASSAYS NOT INCORPORATING A DIGESTIBILITY COMPONENT

Alsmeyer *et al.* (1974), in an effort to develop a rapid assay for predicting PER, developed three regression equations using the amino acid profiles obtained from a small number (<12) of beef muscle and connective tissue samples. Each of the regression equations developed utilized the quantities of two to four amino acids (both essential and nonessential amino acids) to predict the PER of a protein. Table 18.1 lists all three regression equations

TABLE 18.1.   REGRESSION EQUATIONS FOR PREDICTING PER.[1]

Equation 1.
$$PER = -0.684 + 0.456 \text{ (leucine)} - 0.047 \text{ (proline)}.$$

Equation 2.
$$PER = -0.468 + 0.454 \text{ (leucine)} - 0.105 \text{ (tyrosine)}.$$

Equation 3.
$$PER = -1.816 + 0.435 \text{ (methionine)} + 0.780 \text{ (leucine)} + 0.211 \text{ (histidine)} - 0.944 \text{ (tyrosine)}.$$

[1]From Alsmeyer et al (1974).

developed by Alsmeyer *et al.* (1974). All three equations worked well in estimating the PER for beef products, but were unable to accurately estimate the PER for meat-vegetable combinations, food products containing beans, vegetables, meat and dairy combinations and pasta and dairy combinations. The authors felt that the inability of their regression equations to accurately predict PER for foods other than meats, was due to 1) the protein digestibility differences, known to exist between classes of foods, was not taken into consideration by either of the three equations, and 2) the limited base of foods upon which the equations were based, led to equations which were limited to only that narrow class of foods (i.e., meats). Happich *et al.* (1975) further tested the equations of Alsmeyer *et al.* (1974) on meat, meat-soy, soy and whey based products. When used alone, none of the regression equations were able to accurately predict the PER of the meat-soy combinations, soy or whey products. Happich *et al.* stated that when protein digestibility was used to adjust the amino acid profiles of the soy products, the predicted PER was in close agreement with the rat based PER for 4 or 5 (depending on whether equation 1 or equation 2 was used) of the 7 samples.

Lee *et al.* (1978) developed a rapid assay for predicting the PER of meats which contain varying levels of connective tissue. The assay is based on the amino acid composition of the meat sample, with special emphasis on the hydroxyproline content as an indicator of collagen content. The assay equation, given below, works well on meat products.

$$PER = -0.02290 \text{ (collagen content)} + 3.1528$$
where collagen content is expressed
as mg collagen/g sample.

The authors did caution any potential users of the equation in that it is to be used only on products which are predominantly meat, since it has been shown to yield inaccurate predictions of PER for non-meat and combination meat foods.

Schuster and Bodwell (1979) used regression analysis to compare the EAA content (individual EAA as well as selected combinations of EAA) to the nutritional quality of seven proteins, as measured by human and rat bioassay. The authors found that the correlations between the quantities of selected EAA and nutritional quality were high, and were not improved when protein digestibility data was included in the regression analyses. Schuster and Bodwell (1979) do not list the protein digestibility for the seven proteins included in their study, but data obtained from Dr. Bodwell, supported from data obtained in our own laboratory, indicates that all of the proteins included in the study had high apparent protein digestibilities, within the narrow range from 82 to 87% digestibility (mean values). Given the small variance in the digestibilities among the sources used in their study, the regression data of Schuster and Bodwell can be placed in a similar category with the regression models of Alsmeyer *et al.*(1974) and Lee *et al.* (1978). Thus, we conclude that assays derived for a class of foods which possess a high and consistent level of protein digestibility, need not incorporate digestibility data into their predictive equations. But caution must then be exercised in the use of these equations on other foods that possess lower protein digestibilities. In these situations the predicted nutritional quality will be significantly higher than that measured in an animal or human bioassay.

In general, it can be stated that rapid assays for measuring protein nutritional quality need not include a parameter for measuring protein digestibility, if the assay is to be specifically used for only one class of foods (i.e., meat products) or foods possessing similar digestibilities (i.e., isolated plant proteins and animal protein products). But it should always be noted that these assays are then limited to these unique classes and/or types of foods, and thus will have limited value in widespread quality control and regulatory applications.

IN VITRO ASSAYS FOR ESTIMATING PROTEIN DIGESTIBILITY—THEIR VARIOUS USES

In the preceding section, we discussed the necessity of utilizing protein digestibility and/or availability of data when constructing models that are to estimate the nutritional quality of proteins in a wide variety of foods. In this section, the various uses of data that are obtained from *in vitro* assays for protein digestibility will be discussed.

In Vitro Assays for Determining Protein Digestibility

Possibly one of the oldest assays for estimating protein digestibility is that of Akeson and Stahmann (1964) which utilizes the enzymes, pepsin and pancreatin, and was originally intended for use in estimating protein quality. Saunders *et al.* (1973) used the pepsin pancreatin digestion (PPD) assay of Akeson and Stahmann (1964) on various alfalfa proteins and found a correlation of 0.88 between the *in vitro* digestibility data and digestibility data obtained from rat bioassays. Buchanan (1969) found that a papain-trypsin digestion of leaf protein also yielded *in vitro* protein digestibilities that were highly correlated with their respective rat values. Ford and Salter (1966) utilized a protease preparation from *Streptomyces griseus* to predict protein digestibility for a variety of food ingredients. Maga *et al.* (1973) used trypsin to successfully estimate the apparent digestibility of selected food proteins.

Hsu *et al.* (1977) described a 10 minute multienzyme-automated assay that could successfully predict (r = 0.90) rat apparent protein digestibility. The assay utilized the pancreatic enzymes, trypsin and chymotrypsin, and the intestinal enzyme, amino-peptidase. The Hsu *et al.* assay was found to have an excellent relationship to *in vitro* digestion in the rat, as noted by the *in vitro* assay's sensitivity to: 1) trypsin inhibitors, 2) the effects of processing on food proteins and 3) its insensitivity towards fats and inherent buffering salts in the 23 foods tested. Even though the assay of Hsu *et al.*

(1977) was built around 23 foods and food ingredient samples, over two-thirds of the samples were proteins of plant origin and the only animal proteins used were those of dairy origin. Therefore, when used on meat or egg protein based foods, the Hsu *et al.* assay gave low estimates of apparent protein digestibility. Satterlee *et al.* (1979) modified the assay of Hsu *et al.* (1977) by: 1) increasing the sample base to over 50 foods and food ingredients (including meat and egg based foods), and 2) by adding an additional enzyme (protease from *Streptomyces griseus*). The modified 20 minute assay, as described by Satterlee *et al.* (1979), was able to accurately estimate apparent protein digestibility on a vast range of foods.

In Vitro Protein Digestion Assays, for Use in Determining Amino Acid Availability

If properly designed, *in vitro* assays that are used to estimate protein digestibility can also be used to estimate the availability of amino acids. Riesen *et al.* (1947) used a digestion assay to release the EAA from raw and heated soy products, in order to determine ideal heat processing times. Mottu and Mauron (1967) measured lysine availability from *in vitro* enzyme assays, the Carpenter chemical assay and by *in vivo* feeding trials, and found a high correlation between all three assays. Stahmann and Woldegiorgis (1975) discussed the PPD assay, which can also be used to determine availability of all amino acids in a protein and indicated that the PPD assay was more sensitive in detecting early losses of lysine in heated foods than the Carpenter assay for available lysine, as described by Booth (1971). Floridi and Fidanza (1975) described an *in vitro* digestion assay which continuously ultrafilters the enzyme digest, termed the enzyme-ultrafiltrate-digest (EUD). This assay yields, in the ultrafiltrate, the available amino acids from the protein under investigation.

In Vitro Protein Digestion Assays, for Use in Determining Protein Quality

The PPD assay, as described by Akeson and Stahmann (1964) and Stahmann and Woldegiorgis (1975), involves an initial pepsin-pancreatin digest of a protein, followed by an amino acid analysis of the hydrolyzate. The PPD index derived from this assay is essentially a chemical score of the protein, but the amino acids used in chemical scoring are those of the hydrolyzate and not those of the unhydrolyzed protein. The PPD index correlates well with *in vivo* estimates of protein nutritional quality, since it takes into consideration the quantity and bioavailability of all of the EAA in the protein.

Satterlee *et al.* (1977) and Hsu *et al.* (1978) used *in vitro* protein digestibility data, along with either amino acid profile or *Tetrahymena* growth data on various food proteins, to accurately predict the PER of these food proteins. Since both of the above mentioned studies utilized the *in vitro* protein digestibility assay of Hsu *et al.* (1977), now known to underestimate digestibility of egg and meat proteins, Satterlee *et al.* (1979) modified the three enzyme assay by the addition of a fourth enzyme, a protease for *Streptomyces griseus*, as discussed previously (p. 319). Satterlee *et al.* then used data from the modified digestion assay in combination with either amino acid profile or *Tetrahymena* growth data, in the estimation of PER, on over 50 foods and food ingredients. Bodwell (1979b) compared protein nutritional quality data from the rat (PER) and human (relative nutritive value) to data obtained from the C-PER assay of Satterlee *et al.* (1979), which utilized amino acid profiles and *in vitro* protein digestibility data to predict protein quality, expressed as a computed PER or (C-PER). In evaluating the data of Bodwell (1979b), we conclude that the C-PER and PER data compared favorably on all food proteins tested via the human bioassay, except where both tended to underestimate the nutritive value of very low quality proteins, such as wheat gluten.

Mariani and Spadoni (1979) utilized the EUD assay developed by Floridi and Fidanza (1975), described earlier, to estimate protein nutritional quality. Protein quality, as estimated by the EUD *in vitro* assay and the RPV (relative protein value) rat assay, had a correlation of 0.808. The authors noted that EUD data obtained on animal proteins was more highly correlated with the respective rat data, than with EUD data obtained on plant proteins.

Jewell *et al.* (1980) demonstrated that from amino acid profiles alone, the protein nutritional quality of a protein could be accurately estimated. To do this, Jewell and coworkers developed a computer model that used discriminant analysis. The model could accurately predict apparent protein digestibility from data on the concentrations of lysine, leucine, aspartic acid, proline, cystine and ammonia contained in the protein under study. The accuracy of the digestibility data, as predicted from amino acid profiles, was 2.2 times better than that obtained with the four-enzyme assay of Satterlee *et al.* (1979). The improved accuracy in determining digestibility also led to a more accurate prediction of PER (1.5 fold increase in accuracy), as is evidenced by comparing the C-PER of Satterlee *et al.* (1979) and the DC-PER (discriminant based C-PER) of Jewell *et al.* (1980). Table 18.2 lists the discriminant equations used to categorize each food protein, and the subsequent equations used to calculate protein digestibility for each protein. All three discriminant function values are computed to determine which *in vitro* digestibility equation to use. The appropriate equation for a particular protein is that associated with the group with the highest discriminant function value. Table 18.3 is taken from the work of Jewell *et al.* (1980) and compares the apparent protein digestibilities for 42 food proteins, as determined by rat bioassay, four enzyme *in vitro* assay and as computed from the amino acid profile. Table 18.4 lists the protein quality (PER) for 65 different foods as determined by rat bioassay, the C-PER *in vitro* assay (Satterlee *et al.*, 1979), and the DC-PER assay, which utilizes only the amino acid profile to compute digestibility and PER (Jewell *et al.*, 1980). As can be noted in Table 18.3, the standard error of estimation (Sx) for digestibility is acceptable for the four enzyme assay (5.43), but much

TABLE 18.2. DISCRIMINANT FUNCTIONS AND *IN VITRO* PROTEIN DIGESTIBILITY EQUATIONS USING THE CONCENTRATIONS (g/100 g PROTEIN) OF SPECIFIC AMINO ACIDS. JEWELL ET AL (1980).

Group 1	
Discriminant Function	= −265.5083 − 10.80226(LYS) + 16.30705(LEU) + 20.24899(ASP) + 17.54825(PRO) + 22.82687(CYS) + 14.91474(AMM)
In Vitro Digestibility (Percent)	= 146.8403 + 1.41157(LYS) − 1.70511(LEU) − 3.27409(ASP) − 1.71325(PRO) − 5.77153(CYS) − 2.82013(AMM)
Group 2	
Discriminant Function	= −208.3152 − 6.9285(LYS) + 12.96636(LEU) + 17.84711(ASP) + 15.27969(PRO) + 20.7666(CYS) + 14.32163(AMM)
In Vitro Digestibility (Percent)	= 90.594 + 1.35495(LYS) − 0.48207(LEU) − 0.29468(ASP) − 0.44271(PRO) − 1.76862(CYS) − 0.79387(AMM)
Group 3	
Discriminant Function	= −151.0272 − 3.30725(LYS) + 9.95592(LEU) + 14.57734(ASP) + 12.48677(PRO) + 16.54636(CYS) + 12.57811(AMM)
In Vitro Digestibility (Percent)	= 115.6262 − .06479(LYS) + 1.34446(LEU) − 2.42408(ASP) − 2.23058(PRO) + 1.57095(CYS) − 0.9924(AMM)

improved for the assay which utilizes the amino acid profile (2.47). Table 18.4 illustrates how the improvement in predicting digestibility by the discriminant model improves its (DC-PER) ability to more accurately estimate protein quality, when compared to the C-PER assay which utilizes the four enzyme *in vitro* assay.

Since the assay of Jewell *et al.* (1980) has been developed, subsequent testing has shown that it works excellently on all food proteins, except those encased in a tough cell wall (i.e., single cell protein), or those which have been improperly heat treated and still possess significant quantities of proteolytic enzyme inhibitors. These types of foods or food ingredients are not common, but do exist and could easily be detected by any of the common enzyme based *in vitro* assays for protein digestibility.

A COMPARISON OF PROTEIN DIGESTIBILITY, AS DETERMINED BY HUMAN AND RAT BIOASSAY AND *IN VITRO* ASSAYS

Within the past five years there have been several studies published where protein nutritional quality data obtained from rat and human bioassays have been directly compared (Bodwell, 1977b; 1979a, 1979b; Bodwell *et al.* 1980; and Schuster and Bodwell, 1980). These comparisons show that both similarities and differences in protein quality estimation between man and the rat do exist. The question that arises is, are the differences due to

TABLE 18.3.   COMPARISON OF *IN VIVO* AND *IN VITRO* PROTEIN DIGESTIBILITIES FOR VARIOUS FOODS AND FOOD INGREDIENTS. JEWELL ET AL (1980).

Food/Ingredient Sample	Apparent Protein Digestibility (%) Via:		
	Rat Based *In Vivo*[1] Assay	Satterlee et al (1979) *In Vitro* Assay	Discriminant Amino Acid Composition Based *In Vitro* Assay
Corn milo grain	72.00	72.89	73.87
Cornmeal	68.69	75.82	69.80
Hard red winter white flour	81.94	86.42	83.94
Durum wheat flour	82.76	85.29	78.64
Shredded Wheat breakfast cereal	74.77	81.68	77.87
Wheat distillers protein concentrate	82.00	80.33	83.06
Oat breakfast cereal	76.40	81.91	74.99
Corn distillers protein concentrate	79.20	84.39	77.82
Whole corn	78.20	71.76	74.63
Fortified cookie product	67.70	79.88	71.76
Hard red winter wheat grain	77.20	80.11	80.87
Fortified pasta product	74.80	83.26	72.69
High protein flour	77.52	69.05	78.63
Soy isolate	87.06	89.13	87.68
Soy flour	79.14	83.71	79.41
Fortified pasta ingredients	78.20	82.36	77.59
Macaroni & cheese dinner	87.49	83.04	84.94
Breakfast bars	84.18	84.84	83.77
Fortified bread ingredients	84.40	77.17	80.36
Turkey pot pie	89.52	80.56	83.26
Chicken dumpling dinner	83.00	83.26	85.53
Cottonseed meal	85.05	83.04	86.02
Beef noodle dinner	81.00	78.53	81.24
Steak analogue product	79.10	85.52	83.71
Steak analogue ingredients	81.90	85.52	83.87
Sausage analogue ingredients	81.80	83.49	82.69
Fortified breakfast cereal	86.00	84.39	83.63
Egg white	91.81	86.20	88.58
Sausage analogue product	81.80	80.11	82.58
Pizza #1 ingredient	86.90	79.88	83.79
Pizza #1 product	85.40	79.88	84.13
Pizza #2 ingredient	82.50	81.23	85.86
Pizza #2 product	81.00	81.91	83.21
Wheat protein concentrate	89.90	84.39	88.95
Lactalbumin	88.96	84.62	90.20
MDM (neutral)	90.61	84.84	91.95
Lean beef	91.72	85.07	93.73
Fish fillets	90.64	79.65	88.65
MDM (acid)	92.36	88.23	90.85
Beef round	91.39	84.62	89.74
NFDM	84.67	81.91	86.34
Whole egg, boiled	91.12	88.00	90.97
S_x		5.43	2.47

[1]Assumes zero variance.

differences in man's and/or the rat's ability to digest the protein, or is it in their ability to utilize the released amino acids, or both? Secondly, do the rapid *in vitro* assays for determining apparent protein digestibility yield data which correlates with data from human bioassays, especially since almost all *in vitro* assays were developed to predict digestibility as measured by rat bioassay?

TABLE 18.4. A COMPARISON OF PER, C-PER AND DC-PER FOR 65 DIFFERENT FOODS AND FOOD INGREDIENTS. JEWELL ET AL (1980).

Sample	PER[1]	C-PER	DC-PER
ANRC casein	2.5	2.7	2.6
Beef noodle dinner	1.8	2.4	2.1
Beef round	2.8	2.4	2.5
Breakfast bars	1.8	2.2	2.0
Chicken dumpling dinner	2.1	1.9	1.9
Corn distillers protein concentrate	1.3	2.2	1.4
Corn meal	0.7	1.5	0.9
Corn: milo grain mixture	0.7	0.6	0.5
Cottage cheese	2.3	2.6	2.1
Cottonseed meal	2.3	2.0	1.9
Dried skim milk	2.7	2.7	2.9
Durum wheat flour	0.9	0.9	0.6
Egg white	2.5	2.5	2.5
Fortified bread ingredients	2.6	2.4	2.4
Fortified breakfast cereal	2.2	2.4	2.0
Fortified cookie product	1.4	1.8	1.2
Fortified pasta #2 ingredients	1.7	2.1	1.6
Fortified pasta #2 product	1.6	2.2	1.6
Fortified snack #1 ingredients	1.7	2.0	1.6
Fortified snack #1 product	1.4	1.9	1.3
Fortified snack #2 ingredients	2.2	2.2	2.0
Fortified snack #2 product	2.0	2.1	1.9
Ground meat	2.3	2.2	2.3
Ground meat + textured soy protein	2.0	2.1	2.2
Hard red winter wheat grain	1.4	2.0	1.4
Hard red winter white flour	0.7	1.0	0.7
High protein flour	1.2	1.5	1.4
Lactalbumin	2.4	2.5	2.6
Lean beef	2.5	2.3	2.4
Macaroni & cheese dinner	1.8	1.9	1.9
MDM (acid)	2.6	2.9	3.0
MDM (neutral)	2.5	2.9	3.0
NFDM	2.8	2.5	2.5
Oat breakfast cereal	1.2	1.9	1.3
Oat cereal	1.2	1.8	1.4
Peanut flour (cooked)	0.8	2.1	1.6
Pizza #1 ingredients	2.0	2.5	2.1
Pizza #1 product	2.2	2.4	2.1
Pizza #2 ingredients	2.1	2.6	2.1
Pizza #2 product	2.2	2.4	2.1
Potato protein	1.7	2.4	2.3
Sausage analogue ingredients	1.7	2.5	2.0
Sausage analogue product	2.0	2.3	2.0
Shredded Wheat breakfast cereal	1.1	1.9	1.4
Soy concentrate	2.0	2.5	2.1
Soy flour (20PDI)	1.6	2.2	2.1
Soy flour (70PDI)	1.6	1.4	1.5
Soy isolate	1.3	1.3	1.5
Soy isolate #1	1.7	2.5	2.1
Soy isolate #2	1.9	2.3	1.9
Soy isolate #3 (cooked)	1.4	2.1	1.8
Spray dried egg	2.9	2.9	2.7
Steak analogue ingredients	1.9	2.8	2.2
Steak analogue product	2.2	2.7	2.1
Textured soy protein #1	2.1	2.3	2.1
Textured soy protein #2	1.9	2.1	1.9
Tuna	2.3	2.9	2.6
Turkey pot pie	2.4	2.4	2.2
Wheat cereal	0.3	0.0	0.4
Wheat distillers protein concentrate	1.3	1.5	1.0

TABLE 18.4.   (Cont'd.)

Sample	PER[1]	C-PER	DC-PER
Wheat gluten	0.3	1.2	0.7
Wheat protein concentrate	2.1	1.7	2.1
Whey	2.9	2.8	2.8
Whole corn	1.4	1.7	1.2
Whole egg, boiled	3.2	2.9	3.0
Sx		0.36	0.24

[1]Assumes zero variance.

Marshall et al. (1979) compared the apparent protein digestibilities of 33 plant, animal and combination plant-animal protein based foods and food ingredients. This comparison was made using in vitro digestion assays which utilized human pancreatic fluid, rat pancreatin and porcine pancreatin. After carefully analyzing the rate and extent of digestion by each enzyme source on all 33 foods, Marshall et al. concluded that in vitro digestion is identical across all three pancreatins tested, but food class was a factor that affected rate of digestion. Figures 18.1 and 18.2 illustrate how all three pancreatins respond to food class (i.e., animal, plant, combination) in a similar fashion, with animal proteins always undergoing digestion at a slower rate, and to a lesser extent during the 20 minute assay period. The combination plant and animal protein foods and the plant based foods were digested in a similar manner by all three pancreatins. Marshall et al. (1979) concluded that their data derived from studies on pancreatins would dictate that each food class should have a separate equation for determining in vitro protein digestibility.

Bodwell et al. (1980) used six food proteins (three of animal origin, three of plant origin) to compare protein digestion as measured by human and rat bioassays and the four enzyme in vitro assay of Satterlee et al. (1979). The proteins were fed to the human subjects at a level of 0.4 g protein/kg body weight/day, a level well below that consumed in normal American diets. The proteins were fed to the rats at a 10% level in the diet. Correlations between human and rat apparent and true digestibility data were 0.375 and 0.455, respectively. The relatively poor correlation could be attributed to the low level of protein fed and to a sample (tuna) that we believe was an outlier which biased the data. Bodwell et al. stated that if the data for tuna was eliminated from their data comparisons, then true digestibility, as measured in humans and rats, was highly correlated (r = 0.964). When Bodwell et al. compared in vivo protein digestibility data for all six proteins against the data obtained by the four enzyme in vitro assay, they found, as had been shown by Marshall et al. (1979), that strong correlations (r > 0.90) exist when foods were compared by class (i.e., animal or plant protein foods), instead of being compared as a group.

Rich (1978) and Rich et al. (1980) also compared protein digestibility between humans, rats and the four enzyme assay of Satterlee et al. (1979). Rich fed six food proteins (three legume based proteins and three non-legume proteins) to human subjects at a level of 1 g protein/kg body weight/day; and to rats at a 10% level in their diet. Rich found that human and rat apparent protein digestibilities for these six foods were highly

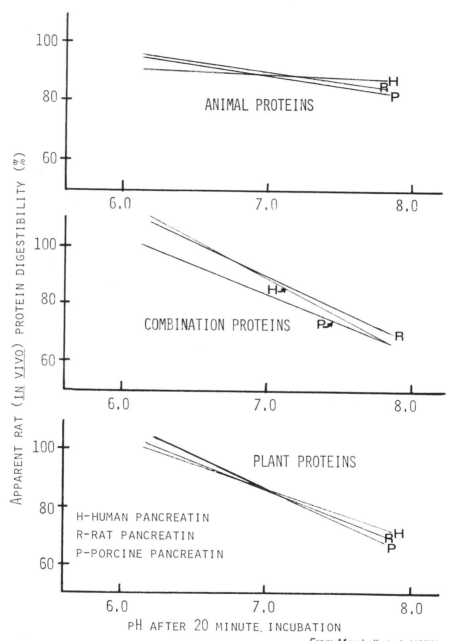

From Marshall et al. (1979)

FIG. 18.1.   REGRESSION LINES FOR DIGESTION OF DIFFERENT PROTEIN CATEGORIES WITH RAT, HUMAN AND PORCINE PANCREATINS

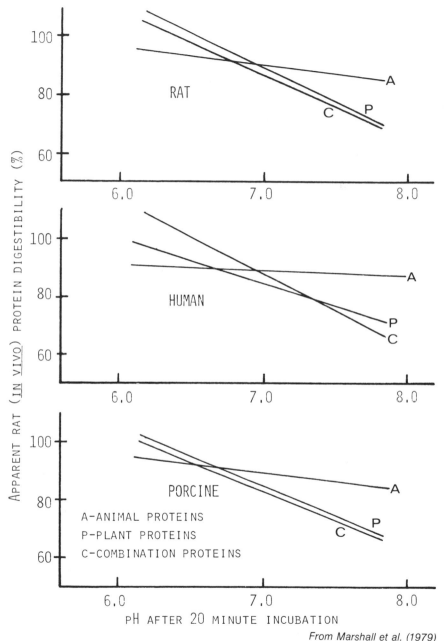

From Marshall et al. (1979)

FIG. 18.2.   REGRESSION LINES FOR DIGESTION OF THE SAME PROTEIN CATEGORIES WITH RAT, HUMAN AND PORCINE PANCREATINS

correlated (r = 0.957). The four enzyme *in vitro* digestibility data correlated very well with human apparent digestibility data (r = 0.859), and marginally well with rat apparent protein digestibility data (r = 0.781). The initial objective of the research reported by Rich (1978) and Rich *et al*. (1980) was to see if legume proteins and nonlegume proteins were digested differently by humans, rats and *in vitro* enzyme systems. Since casein was the only animal protein in the study, no plant versus animal class comparisons could be performed. Analysis of the data did show that *in vivo* apparent protein digestibility data for both legume and nonlegume proteins, as measured by human and rat bioassays, was nearly identical (correlation = 0.997). *In vitro* apparent protein digestibility data correlated best with human and rat data (r = 0.877 and 0.840) for the nonlegume foods, and not as well (r = 0.811 and 0.761, respectively) for the legume based foods (soy products and dry edible beans). Table 18.5 lists the various correlation coefficients and their levels of significance from the study by Rich *et al*. (1980).

TABLE 18.5.   CORRELATION COEFFICIENTS AND (LEVEL OF SIGNIFICANCE) COMPARING PROTEIN DIGESTIBILITIES FOR ALL SIX FOODS WHEN ANALYZED AS A GROUP OR SEPARATELY AS LEGUME AND NONLEGUME BASED FOODS. RICH ET AL (1980).

	Rat *In Vivo* Apparent Digestibility	Four Enzyme *In Vitro* Digestibility
For All Study Foods		
Human *in vivo* apparent digestibility	0.957 (99%)	0.859 (95%)
Rat *in vivo* apparent digestibility		0.781 (90%)
For Nonlegume Foods		
Human *in vivo* apparent digestibility	0.997 (99%)	0.877 (95%)
Rat *in vivo* apparent digestibility		0.840 (95%)
For Legume Foods		
Human *in vivo* apparent digestibility	0.997 (99%)	0.811 (95%)
Rat *in vivo* apparent digestibility		0.761 (90%)

The data that is provided by the studies of Rich (1978), Marshall *et al*. (1979), Bodwell *et al*. (1980) and Rich *et al*. (1980) indicates that both true and apparent protein digestibilities, as measured by human and rat bioassays, seem to be similar, and independent of food class. This generalization does not hold true when data from enzyme based *in vitro* assays are compared to *in vivo* digestibility data from humans and rats. The *in vitro* assays do moderately well on all foods, but can estimate human or rat protein digestibility much more accurately if specific equations, which are related to food class, are used. The necessity of having to know food class prior to determining the *in vitro* protein digestibility and eventually protein quality of a food sample, is in direct contradiction to guidelines that are necessary for the ideal properties of a rapid assay for protein nutritional quality. The knowledge needed to classify an unknown food sample arriving for analysis is often not available nor should it be for quality control or regulatory applications. Also, most formulated foods are combinations of plant and animal proteins, and once finely ground, the original components lose all identity.

Yet foods can be classified, not by the scientific investigator or laboratory technician, but by the model described by Jewell *et al.* (1980), once the amino acid profile of the sample is known. The discriminant based model of Jewell *et al.* utilizes information on the lysine, leucine, aspartic acid, proline, cystine and ammonia concentrations to classify proteins into one of three groups. As can be seen in Table 18.6, the model places food and food ingredients into classes that resemble, for the most part, classification as plant, combination plant and animal and animal protein foods. Therefore, in the Jewell *et al.* model, proteins are being classified automatically, but without knowledge of, or bias by, the technician in charge. Once a sample is classified by the model, the protein digestibility is determined using equa-

TABLE 18.6. DISCRIMINANT GROUPS FOR *IN VITRO* PROTEIN DIGESTIBILITY.[1]

Group 1

Corn DPC	Fortified pasta #2 product
Corn meal	Oat cereal
Corn:milo grain	Shredded wheat breakfast cereal
Durum wheat flour	Wheat DPC
Fortified cookie product	White wheat flour
Fortified pasta #2 ingredients	Whole corn grain

Group 2

Beef & noodle dinner	Pizza #1 product
Breakfast bars	Pizza #2 ingredients
Chicken & dumpling dinner	Pizza #2 product
Egg white	Sausage analogue ingredients
Fortified bread ingredients	Sausage analogue product
Fortified breakfast cereal	Soy isolate
HPF	Steak analogue ingredients
Lactalbumin	Steak analogue product
Macaroni & cheese dinner	Turkey pot pie
Pizza #1 ingredients	Whole wheat grain

Group 3

Beef round	MDM (neutral)
Cotton seed meal	NFDM
Fish fillets	Soy flour (20 PDI)
Lean beef	Whole egg, boiled
MDM (acid)	WPC

[1]From Jewell et al (1980).

tions appropriate for each group of foods (see Table 18.2 for the digestibility equations for each classification). This unique approach has greatly improved the accuracy of the *in vitro* digestion assay. Digestibility data from the Jewell *et al.* assay, when compared to *in vivo* rat data for 42 foods, had a correlation of 0.92 and a standard error of the estimate of 2.47.

ANIMAL VS. PLANT PROTEINS—IS THERE A CHEMICAL BASIS FOR THEIR DIFFERING RATES OF *IN VITRO* DIGESTION?

Data obtained from both enzyme based (Marshall *et al.* 1979 and Bodwell *et al.*, 1980) and amino acid profile based (Jewell *et al.* 1980) *in vitro* assays for protein digestibility indicate that real differences exist between the two broad classes of food protein (animal and plant proteins) in their rate and extent of digestion by the various *in vitro* assays.

Recent data obtained in our laboratories characterized the hydrolyzates obtained from the four enzyme *in vitro* protein digestibility assay of Satterlee *et al.* (1979). This data was collected on nine food proteins, six of which were also used in human bioassays for true and apparent protein digestibility and human serum amino acid analyses.

If a food protein sample is assayed for percent protein digestibility using the four enzyme *in vitro* assay, at the end of the 20 minute hydrolysis period the hydrolysis mixture (insoluble protein + hydrolyzed soluble protein + enzymes) can be immediately frozen and freeze dried. The resulting powder can later be rehydrated at pH 2 (the low pH will prevent further digestion by the proteolytic enzymes) and centrifuged to remove the *insoluble protein fraction*. The soluble protein, peptides and amino acids can be further separated using a Sephadex G-50 column into intermediate molecular weight (average Mol. wt. ~ 13,700 daltons) and small molecular weight (average Mol. wt. ~ 1,400 daltons) components.

The sequence that should be observed when a food (processed foods such as those used in this study have little, if any, soluble protein) is digested by proteases and peptidases is as illustrated below.

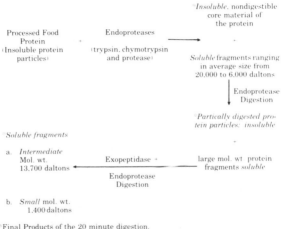

*Final Products of the 20 minute digestion.

Then, a protein that is highly digestible, should have primarily small mol. wt. ($\sim$ 1,400 dalton) peptides with very little insoluble core material after the 20 minute digestion. A protein which has a low digestibility should have a relatively large insoluble core fraction and only small amounts of soluble intermediate and small molecular weight peptides after the 20 minute digestion period. Figure 18.3 summarizes the quantities of the three end products (insoluble protein, intermediate and small molecular weight peptides) from the *in vitro* digestion of nine proteins possessing a wide range of *in vivo* rat digestibilities.

Casein, a protein that is highly digestible, follows the pattern expected for that class of proteins. Corn bread, whose sole protein ingredient is corn endosperm protein, is of intermediate to low digestibility and again (as is seen in Figure 18.3) this protein yields a pattern that was expected for that class of protein. The other seven proteins act anywhere from close to what would be expected, to completely the opposite of what would be expected, knowing in advance their *in vivo* digestibilities. For example, beef which is as digestible *in vivo* as casein, has much more insoluble protein and much less soluble small mol. wt. peptides, than would be expected. Pizza, nearly equal to casein and beef in *in vivo* digestibility, possesses a very large insoluble protein component and a proportionally smaller small mol. wt. peptide fraction. This is just the opposite of what would be expected, knowing its high degree of *in vivo* digestibility.

Certain plant proteins also act in an unusual fashion. For example, soy flour (SF), which has an *in vivo* digestibility similar to that of white wheat flour (WWF) and slightly below that of textured soy protein (TSP), is more readily converted into soluble peptides than either the WWF or TSP.

The general relationship of the *in vivo* protein digestibility to the amount of insoluble core remaining after a 20 minute *in vitro* protease digest is from fair to good at best (r = 0.67). Yet the relationship of the amount of insoluble core remaining to the amount of small mol. wt. peptides produced during the 20 minute *in vitro* hydrolysis is excellent (r = 0.97), thus indicating that the general digestion scheme, as diagrammed above, is what is occurring in the *in vitro* assay. Secondly, the discrepancies in the amounts of these fractions present after the 20 minute hydrolysis, from what would be expected knowing the *in vivo* digestibility of the sample, are probably an artifact of conditions that are present in the rat's digestive system and missing in the *in vitro* protease system. These conditions could be:

1. *Time for digestion*, known to be 20 minutes in the *in vitro* system, but undetermined in the rat.
2. *Specific sequence of pH alterations and attack by numerous enzymes in vivo*, whereas with an *in vitro* assay only one pH is used and the limited number of enzymes act simultaneously.
3. *The exact proportions of each enzyme involved in in vivo digestion*, and the ratio of enzyme to substrate, have been optimized for the *in vitro* system, but it is known that these may be significantly different

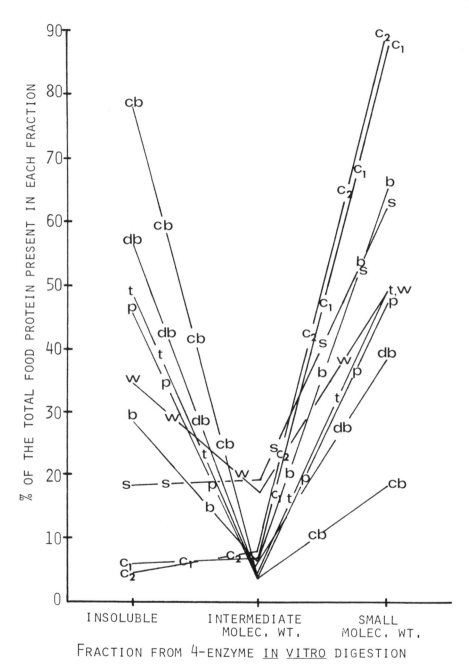

FIG. 18.3. THE PERCENTAGE OF THE ORIGINAL PROTEIN PRESENT IN A FOOD OR FOOD INGREDIENT THAT ENDS UP IN THE INSOLUBLE, INTERMEDIATE MOLECULAR WEIGHT (~ 13,700 DALTONS) AND SMALL MOLECULAR WEIGHT (~ 1,400 DALTONS) FOLLOWING 4 ENZYME *IN VITRO* DIGESTION

c_1 — ANRC casein, c_2 — Erie casein, s — soy flour, b — cooked beef, w — white wheat flour, p — pizza, t — textured soy protein, db — dry beans and cb — corn bread.

between humans and rats, and between *in vivo* and *in vitro* systems (Marshall *et al.* 1979).

4. *Secondary digestion by microorganisms* in large intestine is present in the *in vivo* system (Slump, 1975), but absent in the *in vitro* system.

When six of the food proteins (sodium caseinate, pizza, TSP, soy flour, dry beans and corn bread) were fed to humans in an *in vivo* protein digestibility study (Rich *et al.* 1980), fasting and two hour postprandial blood samples were also taken from the subjects in order to obtain serum free amino acid profiles. Earlier data (presented in Figure 18.3) showed an increase in the amount of small ml. wt. peptides produced with each increase in *in vitro* protein digestibility. The small mol. wt. ($\sim$ 1,400 daltons) peptides are those that would be readily hydrolyzed to amino acids by the intestinal peptidases, for subsequent transport across the intestinal wall into the bloodstream. These peptides should represent the surface regions of the food protein—those regions which are easily accessible to proteolytic enzymes, primarily hydrophylic and water soluble by nature. The insoluble core protein fraction, found after *in vitro* hydrolysis of a food protein, should represent the hydrophobic interior of the insoluble protein particles, and should be essentially apolar and insoluble by nature.

Figure 18.4 illustrates the hydrolphylic nature of:

1. the six food proteins tested
2. the various fractions obtained from the *in vitro* protein digestion assay
   a. insoluble protein peptide fraction
   b. soluble intermediate molecular weight peptide fraction
   c. soluble small molecular weight peptide fraction
3. the free amino acids present in the blood serum of 12 subjects that consumed a meal of the sample protein following a 12 hour fasting period.

The hydrophylic nature was determined by calculating the ratio of lysine to leucine for the protein, peptides and/or free amino acids in the fraction. If we consider lysine to be a typical hydrophylic amino acid and leucine as indicative of a hydrophobic amino acid, then a high lysine/leucine ratio should be indicative of a protein, peptide or amino acid system predominating in the hydrophylic amino acids, with the opposite being true for a system possessing a very low lysine/leucine ratio. Figure 18.4 utilizes the lysine/leucine ratio as a measure of hydrophylic character, but similar analyses using an arginine/phenylalanine ratio gave essentially the same results as discussed below.

Upon viewing Figure 18.4, one can readily see that the protein in the corn bread is very hydrophobic by nature, as is expected since we know that the major protein in corn endosperm is zein, an aqueous-alcohol soluble protein fraction. The five remaining food proteins are quite similar in lysine/leucine ratio, but what is definitely apparent with all six proteins is that:

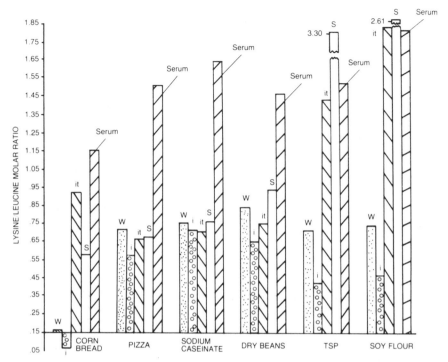

FIG. 18.4. MOLAR RATIOS OF LYSINE TO LEUCINE AS AN INDICATOR OF THE HYDROPHYLIC NATURE OF PROTEINS/PEPTIDES/OR AMINO ACIDS IN: whole food (w); insoluble (i); intermediate (it); and small (s) molecular weight fractions from 4 enzyme *in vitro* protein digestion; and the free serum amino acids (serum) from subjects consuming each respective protein.

1. The insoluble fraction left following *in vitro* protein digestion is *always* more hydrophobic than is the initial food protein.
2. The solubilized peptides (intermediate and small mol. wt. peptides), especially the small mol. wt. group from beans, TSP and soy flour, are very hydrophylic when compared to the insoluble fraction.
3. The character of the free amino acids found in the serum two hours after food consumption is always very hydrophylic in nature, when compared to the character of the food ingested by the subject two hours earlier.

The hypothesis made earlier has held true for these six food proteins, in that the nondigested protein core is definitely more hydrophobic, when compared to the moderately to highly hydrophylic small molecular weight peptide fraction from *in vitro* digestion, and serum free amino acids from *in vivo* digestion. It seems that, in general, *in vitro* digestion of food proteins is similar to that occurring *in vivo*, but very possibly not occurring with the speed or efficiency noted for the *in vivo* digestion system.

CONCLUSIONS

We previously discussed (p. 316) the need by the food industry for a single rapid assay for estimating protein nutritional quality that is capable of operating over a wide variety of foods. Currently, there are several rapid assays that have been described in the scientific literature, with two of those assays, the C-PER and DC-PER assays, undergoing collaborative testing for eventual consideration as an "Official Method" by the Association of Analytical Chemists (A.O.A.C.). These two rapid assays include data on protein digestibility in their final calculations for estimating protein quality. However, two questions need to be discussed. First, has it been determined that such data is needed? Secondly, is digestibility data obtained from such *in vitro* assays valid in its use in predicting how humans will respond nutritionally to the protein? These questions are addressed in the preceding sections.

We conclude that if a rapid assay for protein quality is to find use on a wide variety of foods, it must then include a protein digestibility component in order to be able to compensate for the wide range of apparent protein digestibilities (65–95%) seen in the normal range of foods and feed ingredients. It now appears that even though the data is limited, the abilities to digest a variety of food proteins by both the human and the rat are essentially equal. Therefore, rat bioassays for protein digestibility are able to yield comparable data to that achieved by the human bioassays. Thirdly, *in vitro* enzyme based or chemical based assays for protein digestibility can do a good to excellent job of estimating protein digestibility in the rat, which we feel can then be extrapolated to the human. But the *in vitro* assays, since they are lacking many of the characteristics and components found in *in vivo* digestion systems, are at their best when they use individual predictive equations for each class of foods (animal, plant, combination) under study. The selection of the appropriate predictive equation can be left up to the scientist or technician performing the digestibility assay, or perhaps better, can be performed by a mathematical model that will correctly calculate the digestibility for the sample based upon internal parameters rather than upon external inputs.

In conclusion, one could say that we are in many ways continually making the science of nutrition into a more exacting science. But this may in some ways be overdone, not with respect to the depth or quality of biological, chemical or physical knowledge acquired, but from the aspect of how precise we must determine and report the nutritive quality of a food or one of its components. For example, a PER of 2.0 ± 0.2 might give us as much information as we need, compared to the supposedly more "accurate" reported value of $2.07 \pm .05$. If that analogy can be used with respect to rapid assays for determining protein nutritional quality, then each component measured need not be so precise that it is accurate to the second decimal place. It need only have that accuracy that would allow us to use it to rank our food proteins into perhaps the four categories of poor—fair—good—excellent, having respective PER values of 0.2–0.6; 0.7–1.6; 1.7–2.3; 2.4–3.2. Nutrition on a day-to-day basis need not be any more precise than this.

336   PROTEIN QUALITY IN HUMANS

REFERENCES

ANDERSON, R.H. 1978. Protein quality testing: Industry needs. Food Technol. 32:65.

AKESON, W.R. and STAHMANN M.A. 1964. A pepsin pancreatin digest index of protein quality evaluation. J. Nutr. 83:257.

ALSMEYER, R.H., CUNNINGHAM, H.E. and HAPPICH, M.L. 1974. Equations predict PER from amino acid analysis. Food Technol. 28:34.

ANONYMOUS #1. 1977. Anonymous Article #1: Need for a rapid, accurate, widely applicable method for assessing the biological value of proteins in commercial processed foods. Food Technol. 31:69.

ANONYMOUS #2. 1977. Anonymous Article #3: Need for rapid assay for protein evaluation. Food Technol. 31:70.

BODWELL, C.E. 1977a. Problems in the development and application of rapid methods of assessing protein quality. Food Technol. 31:73.

BODWELL, C.E. 1977b. Application of animal data to human protein nutrition: A review. Cereal Chem. 54:958.

BODWELL, C.E. 1979a. Human versus animal assays. In Soy Protein and Human Nutrition, H.L. Wilcke, D.T. Hopkins and D.H. Waggle (Editors). Academic Press, New York, p. 331.

BODWELL, C.E. 1979b. The nutritive value of the same protein preparations as estimated by human, rat and chemical assays. J.A.O.C.S. 56:156.

BODWELL, C.E., SATTERLEE, L.D. and HACKLER, L.R. 1980. Protein digestibility of the same protein preparations by human and rat assays and by in vitro enzymic digestion methods. Am. J. Clin. Nutr. 33:677.

BOOTH, R.H. 1971. Problems in the determination of FDNB—available lysine. J. Sci. Food Agr. 22:658.

BUCHANAN, R.A. 1969. In vivo and in vitro methods of measuring nutritive value of leaf protein preparations. Br. J. Nutr. 23:533.

FLORIDI, A. and FIDANZA, F. 1975. The enzyme-ultrafiltrate-digest (EUD) assay for measuring protein quality. Riv. Sci. Tech. Alim. Nutr. Um. 5:13.

FORD, J.E. and SALTER, D.N. 1966. Analysis of enzymically digested food proteins by Sephadex-gel filtration. Br. J. Nutr. 20:843.

HAPPICH, M.L., SWIFT, C.E. and NAGHSKI, J. 1975. Equations for predicting PER from amino acid analysis—A review and current scope of application. In Protein Nutritional Quality of Foods and Feeds: Part I. Assay methods—biological, biochemical and chemical, M. Friedman (Editor). Marcel Dekker, Inc., New York, p. 125.

HSU, H.W., VAVAK, D.L., SATTERLEE, L.D. and MILLER, G.A. 1977. A multienzyme technique for estimating protein digestibility. J. Food Sci. 42:1269.

HSU, H.W., SUTTON, N.E., BANJO, M.O., SATTERLEE, L.D. and KENDRICK, J.G. 1978. The C-PER and T-PER assays for protein quality. Food Technol. 32:69.

JEWELL, D.K., KENDRICK, J.G. and SATTERLEE, L.D.   1980.   The DC-PER assay: A method for predicting protein quality solely from amino acid compositional data.   Nutr. Reports Int. 21:25.

LEE, Y.B., ELLIOTT, J.G., RICKANSRUD, D.A. and HAGBERG, E.C.   1978. Predicting protein efficiency ratio by the chemical determination of connective tissue content of meat.   J. Food Sci. 43:1359.

MAGA, J.H., LORENZ, K. and ONAYEMI, O.   1973.   Digestive acceptability of proteins as measured by the initial rate of in vitro proteolysis.   J. Food Sci. 38:173.

MARIANI, A. and SPADONI, M.A.   1979.   Rat models in protein quality evaluation.   J.A.O.C.S. 56:154.

MARSHALL, H.F., WALLACE, G.W. and SATTERLEE, L.D.   1979.   Prediction of protein digestibility by an in vitro procedure using human, porcine and rat pancreatin preparations.   Nutr. Reports Int. 19:901.

MOTTU, F. and MAURON, J.   1967.   The differential determination of lysine in heated milk. II. Comparison of the in vitro methods with biological evaluation.   J. Sci. Food Agr. 18:57.

RICH, N.   1978.   A comparison of human and rat in vivo and in vitro digestibility with purified enzyme in vitro digestibility of selected foods.   M.Sc. Thesis. University of Nebraska, Lincoln, NE.

RICH, N., SATTERLEE, L.D. and SMITH, J.L.   1980.   A comparison of in vivo apparent protein digestibility in man or rat to in vitro protein digestibility as determined using human and rat pancreatins and commercially available proteins.   Nutr. Reports Int. 21:285.

RIESEN, W.H., CLANDININ, D.R., ELVEHJEM, C.A. and CRAVENS, W.W. 1947.   Liberation of essential amino acids from raw, properly heated and overheated soybean oil meal.   J. Biol. Chem. 167:143.

SATTERLEE, L.D., KENDRICK, J.G. and MILLER, G.A.   1977.   Rapid in vitro assays for estimating protein quality.   Food Technol. 31:78.

SATTERLEE, L.D., MARSHALL, H.F. and TENNYSON, J.M.   1979.   Measuring protein quality.   J.A.O.C.S. 56:103.

SAUNDERS, R.M., CONNOR, M.A., BOOTH, A.N., BICKOFF, E.M. and KOHLER, G.O.   1973.   Measurement of digestibility of alfalfa protein concentrates by in vivo and in vitro methods.   J. Nutr. 103:530.

SCHUSTER, E.M. and BODWELL, C.E.   1979.   Relationship between protein nutritive value as estimated in young men and rats, essential amino acid composition and protein digestibility.   Federation Proc. 38:313.

SLUMP, P. and van BEEK, L.   1975.   Amino acids in feces related to digestibility of food proteins.   In Protein Nutritional Quality of Foods and Feeds: Part I. Assay methods—biological, biochemical and chemical, M. Friedman (Editor).   Marcel Dekker, Inc., New York, p. 67.

STAHMANN, M.A. and WOLDEGIORGIS, G.   1975.   Enzymatic methods for protein quality determination.   In Protein Nutritional Quality of Foods and Feeds: Part I. Assay methods—biological, biochemical and chemical, M. Friedman (Editor).   Marcel Dekker, Inc., New York, p. 211.

STEINKE, F.H. 1979. Measuring protein quality of foods. *In* Soy Protein and Human Nutrition, H.L. Wilcke, D.T. Hopkins and D.H. Waggle (Editors). Academic Press, New York, p. 307.

DISCUSSION

DR. HUTNER: There are certain hints from the comparative biochemistry of phagocytosis or the more general term, *endocytosis*. The spotlight on bowel chemistry has now moved (as I intimated earlier) to the cathepsins, the characteristic proteolytic enzymes of the brush border. The implications are that a certain unreality attends any enzymatic procedure that neglects God's gift to us: accessible cathepsins. The first such gift, papain, we had found in tetrahymena work to be rather poor but then emerged the extraordinary efficiency of bromelain. I have no stock in any bromelain company nor am I especially fond of pineapples. The hint we acted on is that in Europe, bromelain is popular as a means of digesting away dead tissue. In this country, we use streptokinase or Dornase or whatever. In Europe, they use a very cheap bromelain. Bromelain is so powerful that even the semi-purified bromelain sold by Sigma or Calbiochem gives a very satisfactorily low blank. Our data then, are in quite pretty agreement with Dr. Satterlee's. I've mentioned that in presenting foodstuffs to *Tetrahymena* after bromelain treatment of the sample, we discard the residue and just assay the supernatant; that correlates with the rat results. So the implication, I would suggest, is that one might get more realistic results by including an enzyme like bromelain, inactivated in a nontoxic way with a bulky reducing agent that, like mercaptosuccinic acid (or its other name, thiomalic acid) that would match those from *Tetrahymena* directly. We don't really need the amino acid score with *Tetrahymena*. We go right to the point; we throw the foodstuff at it.

DR. SATTERLEE: You are using the *Tetrahymena* instead of the analyzer to estimate essential amino acid quantity and availability?

DR. HUTNER: Exactly.

DR. SATTERLEE: There is one last slide that shows the hydrophylic nature (lysine to leucine ratio) of the peptide's being released from *in vitro* protein digestion. You see, in the case of each of the foods, the insoluble fraction is very hydrophobic compared to the initial food itself. When the insoluble fraction is compared to the small peptides or the intermediate sized solubilized peptides, in almost every case the solubilized peptides are more hydrophylic. We compared the levels of the two amino acids (lysine and leucine) found in the serum of human subjects having consumed six food protein sources, and we were amazed to also find that the serum aminograms show the ratio of hydrophylic to hydrophobic amino acids (as indicated by lysine and leucine) to be similar to those of the soluble peptides from the *in vitro* assay. We believe that the proteolytic enzymes are stripping off the outer hydrophylic shell of amino acids very rapidly, leaving behind a fairly insoluble hydrophobic amino acid core.

DR. HUTNER: Thereby, I see a very valuable suggestion! It would be valuable to do assays with *Tetrahymena* or however you like, on the makeup, in detail, of that insoluble fraction and so ascertain why *Tetrahymena* is behaving like your assistant.

DR. PELLETT: I would like to congratulate Dr. Satterlee on the work you are doing at present which seems to be producing impressive results. I have one question, though. You are predicting PER, that is you are predicting a value for a single protein. What account do you take of the fact that many of the sources have an excess of essential amino acids? It seems that you're missing valuable data from the amino acid analyses by predicting a value of a protein by itself.

DR. SATTERLEE: When one takes two proteins and mixes them together, the so-called synergistic effect of cereal based and legume based proteins on nutritional quality is rapidly detected by the C-PER methodology. This synergism is just pure amino acid chemistry. For example, a protein which is lysine deficient (cereal) is completely balanced by combining it with a protein rich in lysine (legume). But problems associated with imbalances of amino acids, have not been addressed by the C-PER methodology. That's a problem that can arise *in vivo* that we can not detect with *in vitro* assays.

DR. PELLETT: The point that I'm getting at is that when we consider protein quality as an entity, we are looking at the value of a *single* foodstuff by itself. To my mind, this is of very limited value. We eat diets containing mixed food proteins. In fact, while you are predicting the individual values and getting nice agreement with a PER, you've still got the data for the amino acids which almost of necessity must be neglected because you don't know which foods are going to be consumed with which real dietaries.

DR. SATTERLEE: The way you could do this is to have an amino acid profile on a 24 hour diet composite, but I do not know how valuable that data would be. That would be a much better indication of how we are doing on a day-to-day protein nutritional basis and this would allow for synergistic effects between various proteins to be accounted for.

19

Use of Amino Acid Data to Predict Protein Nutritive Value for Adults

C.E. Bodwell[1]

For the evaluation of the protein nutritional quality of foods for regulatory purposes, the desirability of developing or implementing procedures that are based, at least in part, on amino acid composition data has been indicated in Chapters 4, 5 and 6. The general purpose of my discussion is to assess the potential of such procedures for predicting the nutritional value of the protein in foods for adults. More specifically, the following key questions will be addressed: (1) Can protein nutritive value be predicted from amino acid composition data? If so, how? (2) Are scores based on reference patterns or amino acid requirements more useful than composition data *per se*? If so, which scores? (3) How do corrections for differences in nitrogen digestibility affect predictions of protein nutritive value? How can allowances be made for significant differences in digestibility? (4) If procedures based on amino acid composition data are used for prediction of protein nutritional quality, what provisions should be made for the possible presence of physiologically toxic factors (e.g., trypsin inhibitors, gossypol, glucosinylates, etc.) and for possibly deleterious amino acid imbalances?

DATA FROM HUMAN STUDIES

Sources and Selection of Data

In previous comparisons (e.g., FAO/WHO, 1973) of estimates of nutritional value derived from amino acid scores and estimates obtained in humans, Net Protein Utilization (NPU) values derived from human studies have been used. The usefulness or validity of NPU values obtained at a single level of protein intake can be questioned (e.g., see Young and Scrim-

---

[1]Opinions expressed are those of the author and do not reflect endorsement or approval by the U.S. Dept. of Agriculture or of any other government agency.

shaw, 1977; Rand *et al.* 1977). For this discussion, only data from those studies with adult humans who were fed multiple levels of protein intake have been used. The data were grouped (Data Sets 1, 2 and 3). Data for five protein sources from studies at MIT and for seven sources from studies at Beltsville comprise Data Set 1 (Table 19.1). Amino acid compositions of samples of the twelve protein preparations fed to the human subjects were determined in Beltsville. Amino acid scores were calculated before and after corrections for digestibility by use of values for each protein for human true digestibility, human apparent digestibility, and *in vitro* digestibility. *In vitro* digestibility was estimated by use of the 4-enzyme method of Satterlee *et al.* (1979). The digestibility data used in correcting the amino acid compositions are described in Table 19.2.

Data from the evaluation of 16 protein sources in various laboratories comprise Data Set 2 (Table 19.3). For calculating amino acid scores for these sources, amino acid composition values from the original studies and other publications (FAO, 1973) or from analyses at Beltsville of similar protein sources were used.

In the long-term studies from which the data in Data Sets 1 and 2 were obtained, N balance periods of 6–11 days (or longer) were used for each of several protein intake levels. Data from studies in which short-term nitrogen balance periods (see Chapter 8) were used comprise Data Set 3 (Table 19.4). Published values or values from analyses of similar protein sources at Beltsville were used for calculating amino acid scores.

Selection of Data.—In view of our current knowledge, some of the values obtained in the studies listed (Data Sets 1–3) cannot be considered to be reasonable estimates of protein nutritive value. In our own studies, the values for wheat gluten (Data Set 1; protein #7; Table 19.1) are in this category (the regression coefficient for N balance vs N intake is much too high). Proteins #1–6 were fed to subjects in random order in six periods of 11 days each while the wheat gluten bread was fed in a single period of 11 days to all of the subjects who consumed this source. This difference alone should not have resulted in invalid data. For whatever reason, however, the data for wheat gluten are suspect and all calculations were done with or without these data. In Data Sets 1 and 2 (Tables 19.3 and 19.4), similarly invalid values (for the purposes of these comparisons) also occur. The high calorie levels fed probably explain the high regression coefficients and low estimates of the N requirement for zero balance for several of the proteins (e.g., protein sources #5, 7, 9 and 12; Data Set 2; Table 19.3). The high upper protein intake level (1.6 gm/kg body wt/day) fed in the Peru studies might explain the unusually high estimate of N requirement for zero balance in comparison to the value of the regression coefficient of N balance vs N intake. Other values, in both Data Sets 2 and 3, that might be questioned for the purposes of the present comparisons are indicated (Tables 19.3 and 19.4). Amino acid scores were calculated for these sources and are given (see below) but were not included in the comparisons of amino acid scores and estimated nutritive value.

TABLE 19.1.   DATA SET 1: RESULTS FROM STUDIES (USDA; MIT) ON TWELVE PROTEIN SOURCES WITH ADULT HUMAN SUBJECTS FED MULTIPLE-LEVELS OF NITROGEN INTAKES[1]

Protein Source	Regression (Y = N balance [mg N/kg body wt/day]; X = mg N intake/kg body wt/day)	Requirement for "zero N balance" (mg N/kg body wt/day)	"Efficiency Score"[2]	"Requirement Score"[2]
(1) Egg white bread	$Y = 0.46 \times X - 45.24$	97.9	(100)[2]	(100)[2]
(2) Lactalbumin bread	$Y = 0.37 \times X - 40.13$	109.2	80	90
(3) Casein bread	$Y = 0.30 \times X - 31.86$	107.4	65	91
(4) Textured soy (Supro 50) bread	$Y = 0.40 \times X - 42.96$	108.4	87	90
(5) Soy Isolate (Promine F) bread	$Y = 0.39 \times X - 50.17$	127.3	85	77
(6) Peanut flour bread	$Y = 0.39 \times X - 47.84$	123.9	85	79
(7) Wheat gluten bread	$Y = 0.41 \times X - 55.43$	135.2	89	72
(8) Beef bologna	$Y = 0.39 \times X - 39.60$	102.6	93	95
(9) Beef + soy isolate (Supro 620) bologna	$Y = 0.29 \times X - 28.50$	97.3	69	100
(10) Soy isolate (Supro 620)	$Y = 0.40 \times X - 47.41$	121.0	95	81
(11) Soy isolate (Supro 710)	$Y = 0.43 \times X - 45.23$	105.2	102	93
(12) Non-fat dried skim milk	$Y = 0.42 \times X - 41.36$	97.6	(100)	(100)

[1]Values for protein sources 1–7 from USDA (Bodwell, 1979; Schuster and Bodwell, 1979; Bodwell et al., 1979); for protein sources 8–12 derived from MIT data (Scrimshaw and Young, 1979; Young et al., 1979; and V.R. Young, personal communication).

[2]Values for proteins 1–7 calculated by assigning a value of 100 for egg white bread as the reference protein; values for proteins 8–12 calculated by assigning a value of 100 for non-fat dried skim milk as the reference protein; "efficiency score" = 100 × (regression co-efficient for each test protein divided by co-efficient for reference protein); "requirement score" = 100 × (requirement for zero N balance for reference protein divided by requirement for zero N balance of each test protein).

TABLE 19.2.   DESCRIPTION OF DIGESTIBILITY DATA USED IN CORRECTING AMINO ACID COMPOSITIONS (DATA SET 1)[1]

Protein Source	N Intake Levels (mg/kg body wt/day)	Apparent Digestibility (%)		True Digestibility (%)		In Vitro "Apparent" Digestibility[2]
		Value (or Range of Values)	Value Used for Correcting Amino Acid Composition Data	Value (or Range of Values)	Value Used for Correcting Amino Acid Composition Data	
(1) Egg white bread	51–131	81.5±3.8[3](74–88)	81.5	93.0±1.5[3](90–95)	93.0	88.3±1.7[3]
(2) Lactalbumin Bread	51–131	81.6±5.1(72–89)	81.6	93.2±4.0(86–102)	93.2	85.2±0.1
(3) Casein bread	51–125	85.8±3.8(81–90)	85.8	96.7±2.4(94–102)	96.7	84.0±1.0
(4) Textured soy (Supro 50) bread	54–132	75.8±6.4(57–84)	75.8	87.7±4.7(74–95)	87.7	81.5±0.1
(5) Soy Isolate (Promine F) bread		82.0±5.5(74–89)	82.0	93.4±3.7(84–97)	93.4	84.3±0.3
(6) Peanut flour bread	49–146	81.6±4.9(67–87)	81.6	90.9±2.9(83–93)	90.9	81.0±0.5
(7) Wheat gluten bread	58–168	87.3±5.2(77–93)	87.3	96.2±2.4(91–99)	96.2	86.4±0.2
(8) Beef bologna	53–196	(81.1±4.4–90.4±0.9)	87.0	(96.8±5.2–98.8±1.7)	98.0	85.2±0.4
(9) Beef + soy isolate (Supro 620) bologna	56, 72, 88, 104	(84.9±1.9–92.1±2.9)	86.9	(100.6±2.3–96.7±3.1)	97.9	85.0±0.3
(10) Soy isolate (Supro 620)	56, 72, 88, 104	(77.5±3.1–86.1±2.1)	83.7	(93.2±3.7–94.6±2.3)	94.7	92.2±0.1
(11) Soy isolate (Supro 710)	56, 72, 88, 104	(75.5±2.4–85.4±3.0)	82.0	(91.2±2.9–93.5±3.3)	93.0	87.3±0.2
(12) Non-fat dried skim milk	56, 72, 88, 104	(82.1±2.9–88.0±2.6)	85.8	(98.0±3.8–96.7±2.6)	96.8	82.9±0.1

[1] Based on USDA and MIT data; see footnotes of Table 1.
[2] Determined according to 4-enzyme method of Satterlee et al., 1979; mean values based on 2 or more determinations.
[3] Means ± S.D.

TABLE 19.3. DATA SET 2: ESTIMATES (CONVENTIONAL N BALANCE METHOD) OF PROTEIN NUTRITIVE VALUE INCLUDED OR EXCLUDED[1,2]

Protein Source	Regression (Y = N balance) [mg N/kg body wt/day]; X = mg N intake/kg body wt/day)	Requirement for "zero N balance" (mg N/kg body wt/day)	"Efficiency Score"	"Requirement Score"	"Reasonable" Values (In View of Current Knowledge)
INCAP					
(1) Egg	$Y = 0.70 \times -57.58$	82.3	91	111	Yes[3]
(2) Milk	$Y = 0.82 \times -72.34$	88.2	106	104	Yes[3]
(3) Milk	$Y = 0.77 \times -70.38$	91.4	(100)[4]	(100)	Yes[3]
Japan					
(4) Egg (Maintenance Calories)	$Y = 0.41 \times -37.03$	90.3	(100)	(100)	Yes
(5) Egg (Excess Calories)	$Y = 0.37 \times -25.62$	69.2	90	131	No
(6) Rice (Maintenance Calories)	$Y = 0.27 \times -31.98$	118.4	66	76	Yes
(7) Rice (Excess Calories)	$Y = 0.47 \times -38.29$	81.5	114	111	No
(8) Wheat Gluten	$Y = 0.13 \times -26.16$	201.2	32	45	Yes
MIT					
(9) Egg	$Y = 0.65 \times -46.3$	71.2	(100)	(100)	No
(10) Whole Wheat	$Y = 0.27 \times -33.6$	124.4	42	57	Yes[5]
(11) Canned Beef	$Y = 0.51 \times -41.9$	82.2	79	87	Yes[5]
Berkeley					
(12) Egg Protein	$Y = 0.66 \times -44.9$	68		76	No
(13) Egg White Protein		89		(100)	Yes
Purdue					
(14) Rice, Milk, White Flour (50:25:25; N Basis)	$Y = 0.25 \times -15.89$	64.1	86	67	No
(15) Rice, White Flour, Chicken, Milk (25:15:30:30; N Basis)	$Y = 0.29 \times -27.49$	95.0	(100)	(100)	No
Peru					
(16) Bulgar Wheat, Soy Flour, Non-Fat Dried Skim Milk (57:16:28; Wt. Basis) (Young Adults)	$Y = 0.39 \times -54.53$	139.8	(−)	(−)	No
(Elderly)	$Y = 0.39 \times -47.78$	122.5	(−)	(−)	No

[1] Values for protein sources 1–3, from Bressani et al., 1979a,b; for 4–7, from Inoue et al., 1973; for 8, from Inoue et al., 1974; for 9–11, from Young et al., 1975; for 12 from Calloway and Margen, 1971; for 13, from Calloway, 1975; for 14, from Clark et al., 1972; for 15, from Clark et al., 1974; and for 16, from Cheng et al., 1978.
[2] See Table 1 for definition of "Requirement Score" and "Efficiency Score".
[3] Values for regression coefficients high.
[4] Values in parantheses used as the reference protein within each group.
[5] Although regression coefficients for reference egg protein is high and the value for zero N balance low, *relative* values for wheat and beef protein were obtained under similar conditions and are considered "reasonable".

TABLE 19.4. DATA SET 3: ESTIMATES (SHORT-TERM N BALANCE METHOD) OF PROTEIN NUTRITIVE VALUE INCLUDED OR EXCLUDED[1,2]

Protein Source	Regression (Y = N balance [mg N/kg body wt/day]; X = mg N intake/kg body wt/day)	Requirement for "zero N balance" (mg N/kg body wt/day)	"Efficiency" Score	"Requirement" Score	"Reasonable" Values (In View of Current Knowledge)
(1) Whole Egg	$Y = 0.86 \times -55.21$	64.2	95	135	No
(2) Whole egg	$Y = 0.57 \times -63.64$	111.7	63	78	No
(3) Casein	$Y = 0.64 \times -60.28$	94.2	70	92	Yes[3]
(4) Milk	$Y = 0.98 \times -73.61$	75.1	108	115	No
(5) Milk	$Y = 1.03 \times -81.30$	78.9	113	110	No
(6) Milk	$Y = 1.00 \times -70.60$	70.6	110	122	No
(7) Milk	$Y = 0.91 \times -78.80$	86.6	(100)	(100)	Yes[3]
(8) Beans	$Y = 0.54 \times -62.69$	114	60	75	Yes[3]
(9) Beans:Maize (40:60; protein basis)	$Y = 0.89 \times -86.81$	97.6	98	88	Yes[3]
(10) Rice:Bean (60:40; protein basis)	$Y = 0.75 \times -71.6$	95.5	82	91	Yes[3]
(11) Rice:Bean:Milk (55:35:10; protein basis)	$Y = 0.95 \times -74.6$	78.5	104	110	Yes[3]
(12) Textured Veg. Protein (TVP)	$Y = 0.68 \times -65.73$	95.2	76	91	Yes[3]
(13) Beef	$Y = 0.87 \times -74.40$	84.6	97	103	Yes[3]
(14) TVP + Beef (50:50, protein basis [?])	$Y = 0.87 \times -79.90$	90.8	97	95	Yes[3]

[1]Values for protein sources 1–9 and 12–14 from Bressani et al., 1979a,b; for 10 and 11, from Bressani et al. (Chapter 8).
[2]See Table 1 for definition of "Requirement Score" and "Efficiency Score".
[3]High regression coefficient values were obtained for all sources.

Estimates of Protein Nutritive Value

Both "efficiency score" (regression coefficient for the test protein divided by the coefficient for the reference protein [× 100]) and "requirement score" (N intake level required for "zero" N balance obtained with the reference protein divided by the level required for the test protein [× 100]) derived from the regression equation relating N balance and N intake, as discussed in Chapters 7, 8 and 10, were used as estimates of protein nutritive value.

AMINO ACID COMPOSITION DATA

Relationships to Nutritive Value and N Digestibility

We consume dietary protein to meet our requirements for essential amino acids and indispensable N and it is thus obvious that amino acid compositions *per se* of different proteins would be related to estimated nutritive value. This is so even though some proteins contain large excesses of some amino acids that are metabolized and are either used as energy sources or excreted in the urine.

Much of the variability in protein nutritive value can be associated with groups of amino acids (Table 19.5). It is also possible to demonstrate a high

TABLE 19.5.   EXAMPLES OF VALUES FROM R^2 ANALYSES SHOWING THE RELATION-SHIP BETWEEN "N REQUIREMENT SCORE" AND VARIOUS AMINO ACIDS (DATA SET 1)

Amino Acids	12 Proteins	11 Proteins[1]
Lys, Met + Cys, Try	0.729[2]	0.626
Lys, Met + Cys, Thr, Ile, Phe + Tyr, Try	0.746	—
Lys, Met, Cys, Ile, Leu, Val	—	0.832
Lys, Met, Thr, Ile, Leu, Try	—	0.857
Lys, Cys, Thr, Ile, Leu, Try	—	0.855

[1]With data for wheat gluten omitted.
[2]R^2 values determined by SAS procedure (Barr et al., 1976).

degree of association between groups of amino acids and estimated N digestibility. For the sources included in Data Set 1, this degree of association is high (Table 19.6). Although the number of protein sources is small, these data offer indirect support for the model proposed by Satterlee and his co-workers (Chapter 18).

The significance of the high R^2 values listed in Tables 19.5 and 19.6, however, must not be overemphasized. The number of sources was small. Even if the values were based on data from a much larger number of samples, because of the empirical nature of the observed relationships, I would not be completely convinced that such data provide a satisfactory basis for predicting either protein nutritional value or N digestibility (see below).

TABLE 19.6. EXAMPLES OF VALUES FROM R^2 ANALYSES SHOWING THE RELATIONSHIP BETWEEN NITROGEN DIGESTIBILITY AND VARIOUS AMINO ACIDS FOR 12 PROTEINS (DATA SET 1)

Amino Acids	Digestibility		
	Human True	Human Apparent	In Vitro "Apparent"[1]
Try, Thr, Ile, Pro, Asp, Gly	—	0.977	0.855[2]
Met, Thr, Val, Phe + Tyr, Asp, Arg	0.958	—	—
Met, Ile, Val, Asp, Ser, Pro	0.993		
Met, Thr, Val, Phe + Tyr, Asp, Gly	0.807	—	—
Met, Val, Asp, Ile, Ser, Pro		0.787	0.522

[1]Determined by 4-enzyme method of Satterlee et al. (1979).
[2]R^2.

AMINO ACID SCORES

Reference Patterns

Various amino acid patterns used for calculating amino acid scores are shown in Table 19.7 and compared to the FAO/WHO (1973) patterns suggested for meeting the amino acid requirements of infants, children, and adults. It should be noted that, for the most part, the level of total essential amino acids and the levels of individual amino acids (or of sulfur or aromatic amino acids) in both the FAO/WHO (1973) and NRC (1974) scoring patterns are equivalent to the highest levels of estimated requirements among the three population groups (an exception is the NRC value for methionine + cystine). Thus, when these patterns are used, a protein with a high amino acid score should certainly meet or exceed the requirements of adults if the protein were reasonably digestible and were consumed at an adequate level. Similar observations are true for proteins that have high scores when compared with amino acid patterns of egg or milk protein as the reference pattern.

TABLE 19.7. COMPARISON OF SCORING PATTERNS AND FAO/WHO (1973) SUGGESTED PATTERNS OF REQUIREMENTS[1]

Amino Acid	Suggested Patterns of Requirements			FAO/WHO (1973) Scoring Pattern	NRC (1974) Scoring Pattern	Egg	Human Milk
	Infant	Child	Adult				
His	1.4	—	—	—	1.7	2.3	2.2
Ile	3.5	3.7	1.8	4.0	4.2	4.9	5.5
Leu	8.0	5.6	2.5	7.0	7.0	8.8	9.1
Lys	5.2	7.5	2.2	5.5	5.1	6.7	6.6
Met + Cys	2.9	3.4	2.4	3.5	2.6	5.5	4.1
Phe + Tyr	6.3	3.4	2.5	6.0	7.3	10.3	9.5
Thr	4.4	4.4	1.3	4.0	3.5	5.4	4.5
Try	0.9	0.5	0.7	1.0	1.1	1.5	1.6
Val	4.7	4.1	1.8	5.0	4.8	6.1	6.2
Total							
(+ His)	37.3	—	—	—	37.3	51.5	49.3
(− His)	—	32.6	15.2	36.0	35.6	49.2	47.1

[1]Values expressed as gm amino acid/16 gm N.

Scoring Procedures Used

In addition to the usual amino acid scores (chemical scores based on the limiting amino acid, total essential amino acid scores, modified essential amino acid index scores; reviewed by Sheffner, 1967, and Hackler, 1977), three other scores have been calculated (only two for the data in Data Sets 2 and 3). Using the FAO/WHO (1973) and NRC (1974) reference patterns, two scores have been calculated by use of values for only lysine, tryptophan and methionine + cystine. In one of these ("lysine-tryptophan-methionine + cystine" score), the total amount of cystine present was used in deriving the methionine + cystine value. For the other ("lysine-tryptophan-adjusted cystine" score), levels of cystine in excess of the level of methionine were not included.

A further score was calculated by the procedure of Satterlee *et al.* (1977) in which all of the essential amino acids (including total sulfur and total aromatic amino acids) are incorporated into a single score. However, values for those essential amino acids below the "100% level" of the FAO/WHO (1973) amino acid scoring pattern are multiplied by a negative weighting factor that markedly decreases the contribution to the total score of a given amino acid as the level of that amino acid decreases (e.g., the weighting factor for an amino acid present at 25% of the level in the FAO/WHO pattern is about six times as great as the factor used when the amino acid is present at a level equivalent to 75% of the FAO/WHO pattern).

RELATIONSHIPS TO ESTIMATED PROTEIN NUTRITIVE VALUE

The N requirement or efficiency scores and values for the various amino acid scores calculated for the protein sources in Data Set 1 are compared in Tables 19.8–19.13. Data are listed for the individual sources as well as mean values (with or without the data for wheat gluten). Scores calculated from compositions without and with corrections for digestibility are included. Similar data, based on compositions uncorrected for digestibility are given in Tables 19.14 and 19.15 for the protein sources in Data Sets 2 and 3, respectively.

The mean differences (± S.D.) between protein nutritive value as estimated by the amino acid scores (calculated without corrections for digestibility) and as estimated by the two scores derived from the data from human studies for Data Set 1 are given in Table 19.16. The mean differences (−3.2 to +8.7) are not large. However, some of the standard deviations (S.D.) are quite high, ranging up to ± 10.5 for the mean difference between the N requirement scores and two of the amino acid scores and up to 22 for the mean difference between efficiency score and one of the amino acid scores. The magnitude of the difference between each amino acid score and the estimates of nutritive value from the N requirement (or efficiency) scores can also be assessed by inspection of the data for individual protein sources (Tables 19.8–19.13).

Comparisons of values for differences between three of the scores (all with the FAO/WHO pattern as the reference) for the sources in each data set and for the three sets combined, are given in Tables 19.17 and 19.18. With some exceptions, the difference values for the protein sources in the three data sets are in general agreement.

From these results, the amino acid scores based on the levels of lysine, tryptophan and methionine + cystine (without adjustment) appear to be as useful as any of the other scores. Scores, based on only these four amino acids, and nitrogen requirement or efficiency scores for individual protein sources are compared in Table 19.19. For some proteins, differences are

TABLE 19.8. COMPARISONS OF "N REQUIREMENT SCORES", "EFFICIENCY SCORES", AND CHEMICAL SCORES BASED ON AMINO ACID COMPOSITIONS UNCORRECTED AND CORRECTED FOR DIGESTIBILITY (DATA SET 1)[1]

Protein Source	N Requirement Scores	Efficiency Scores	Chemical Scores								
			No Correction For Digestibility		Corrected for Digestibility						
					Human True		Human Apparent		In Vitro "Apparent"		
			FAO/WHO	NRC	FAO/WHO	NRC	FAO/WHO	NRC	FAO/WHO	NRC	
(1) Egg white bread	100	100	100	100	93	93	82	82	88	88	
(2) Lactalbumin bread	90	80	100	95	93	88	82	78	85	81	
(3) Casein bread	91	65	100	100	97	97	86	86	84	84	
(4) Textured soy (Supro 50) bread	90	87	88	91	77	80	67	69	72	74	
(5) Soy Isolate (Promine F) bread	77	85	79	91	93	85	65	75	67	77	
(6) Peanut flour bread	79	85	60	65	54	59	49	53	49	53	
(7) Wheat gluten bread	72	89	31	33	30	32	27	29	27	28	
(8) Beef bologna	95	93	85	89	83	87	74	77	72	76	
(9) Beef + soy isolate (Supro 620) bologna	100	69	85	94	83	92	74	82	72	80	
(10) Soy isolate (Supro 620)	81	95	76	92	72	87	64	77	70	85	
(11) Soy isolate (Supro 710)	93	102	88	91	82	85	72	75	77	79	
(12) Non-fat dried skim milk	100	100	100	100	97	97	86	86	83	83	
Mean (n = 12)	89	88	83	87	80	82	69	72	70	74	
S.D.	10	12	20	19	20	19	17	16	17	17	
Mean (n = 11)	91	87	87	92	84	86	73	76	74	78	
S.D.	8	12	13	10	13	10	11	9	11	9	

[1]FAO/WHO and NRC denote reference patterns; values for 11 proteins calculated with data for wheat gluten omitted.

TABLE 19.9. COMPARISONS OF "N REQUIREMENT SCORES", "EFFICIENCY SCORES" AND "LYSINE-TRYPTOPHAN-METHIONINE-CYSTINE" SCORES BASED ON AMINO ACID COMPOSITIONS UNCORRECTED AND CORRECTED FOR DIGESTIBILITY (DATA SET 1)[1]

Protein Source	N Requirement Scores	Efficiency Scores	"Lysine-Tryptophan-Methionine-Adjusted Cystine"							
			No Correction For Digestibility		Corrected for Digestibility					
					Human True		Human Apparent		In Vitro "Apparent"	
			FAO/WHO	NRC	FAO/WHO	NRC	FAO/WHO	NRC	FAO/WHO	NRC
(1) Egg white bread	100	100	100	100	93	93	82	82	88	88
(2) Lactalbumin bread	90	80	100	100	93	93	82	82	85	85
(3) Casein bread	91	65	100	100	97	97	86	86	84	84
(4) Textured soy (Supro 50) bread	90	87	94	100	82	88	71	76	77	82
(5) Soy Isolate (Promine F) bread	77	85	79	100	74	93	65	82	67	84
(6) Peanut flour bread	79	85	60	65	54	59	49	53	49	53
(7) Wheat gluten bread	72	89	31	33	30	32	27	29	27	28
(8) Beef bologna	95	93	100	91	98	89	87	79	85	78
(9) Beef + soy isolate (Supro 620) bologna	100	69	85	100	83	98	74	87	72	85
(10) Soy isolate (Supro 620)	81	95	76	100	95	95	64	84	70	92
(11) Soy isolate (Supro 710)	93	102	94	97	87	90	77	80	82	85
(12) Non-fat dried skim milk	100	100	100	100	97	96	86	86	83	83
Mean (n = 12)	89	88	85	91	82	85	71	75	72	77
S.D.	10	12	21	21	21	20	18	17	18	18
Mean (n = 11)	91	87	90	96	87	90	75	80	77	82
S.D.	8	12	13	11	13	11	12	9	12	10

[1]FAO/WHO and NRC denote reference patterns; values for 11 proteins calculated with data for wheat gluten omitted.

TABLE 19.10. COMPARISONS OF "N REQUIREMENT SCORES", "EFFICIENCY SCORES", AND "LYSINE-TRYPTOPHAN-METHIONINE-ADJUSTED CYSTINE" SCORES BASED ON AMINO ACID COMPOSITIONS UNCORRECTED AND CORRECTED FOR DIGESTIBILITY (DATA SET 1)[1]

			"Lysine-Tryptophan-Methionine-Adjusted Cystine"							
			No Correction For Digestibility		Corrected for Digestibility					
					Human True		Human Apparent		In Vitro "Apparent"	
Protein Source	N Requirement Scores	Efficiency Scores	FAO/WHO	NRC	FAO/WHO	NRC	FAO/WHO	NRC	FAO/WHO	NRC
(1) Egg white bread	100	100	100	100	93	93	82	82	88	88
(2) Lactalbumin bread	90	80	100	100	93	93	82	82	85	85
(3) Casein bread	91	65	100	100	97	97	86	86	84	84
(4) Textured soy (Supro 50) bread	90	87	89	100	78	88	67	76	72	82
(5) Soy Isolate (Promine F) bread	77	85	79	100	74	93	65	82	67	84
(6) Peanut flour bread	79	85	60	65	54	59	49	53	49	53
(7) Wheat gluten bread	72	89	31	33	30	32	27	29	27	28
(8) Beef bologna	95	93	100	91	98	89	87	79	85	78
(9) Beef + soy isolate (Supro 620) bologna	100	69	85	100	83	98	74	87	72	85
(10) Soy isolate (Supro 620)	81	95	74	100	70	95	62	84	68	92
(11) Soy isolate (Supro 710)	93	102	75	97	70	90	62	80	66	85
(12) Non-fat dried skim milk	100	100	100	100	97	97	86	86	83	83
Mean (n = 12)	89	88	83	90	78	85	69	75	70	77
S.D.	10	12	21	21	20	20	18	17	18	18
Mean (n = 11)	91	87	87	96	82	90	72	80	74	82
S.D.	8	12	14	11	14	11	13	9	12	10

[1]FAO/WHO and NRC denote reference patterns; values for 11 proteins calculated with data for wheat gluten omitted.

TABLE 19.11. COMPARISONS OF "N REQUIREMENT SCORES", "EFFICIENCY SCORES", AND TOTAL ESSENTIAL AMINO ACID (TEAA) SCORES BASED ON AMINO ACID COMPOSITIONS UNCORRECTED AND CORRECTED FOR DIGESTIBILITY (DATA SET 1)[1]

Protein Source	N Requirement Scores	Efficiency Scores	No Correction For Digestibility				Corrected for Digestibility											
							Human True				Human Apparent				In Vitro "Apparent"			
			FAO/WHO	NRC	Egg	Human Milk	FAO/WHO	NRC	Egg	Human Milk	FAO/WHO	NRC	Egg	Human Milk	FAO/WHO	NRC	Egg	Human Milk
(1) Egg white bread	100	100	100	100	100	100	93	93	93	93	82	82	82	82	88	88	88	88
(2) Lactalbumin bread	90	80	100	100	100	100	93	93	93	83	82	82	82	82	85	85	85	85
(3) Casein bread	91	65	100	100	100	100	97	97	97	97	86	86	86	86	84	84	84	84
(4) Textured soy (Supro 50) bread	90	87	100	95	82	85	88	83	72	75	76	72	62	65	82	77	66	70
(5) Soy Isolate (Promine F) bread	77	85	100	100	80	84	93	93	74	78	82	82	65	69	84	84	67	70
(6) Peanut flour bread	79	85	86	87	62	65	79	79	57	59	70	71	51	54	69	70	51	53
(7) Wheat gluten bread	72	89	93	94	69	72	90	91	66	69	82	82	60	63	81	82	59	62
(8) Beef bologna	95	93	100	100	84	87	98	98	82	85	87	87	73	76	85	85	72	74
(9) Beef + soy isolate (Supro 620) bologna	100	69	100	100	82	86	98	98	80	84	87	87	71	75	85	85	68	73
(10) Soy isolate (Supro 620)	81	95	100	100	81	84	95	95	77	80	84	84	68	70	92	92	75	77
(11) Soy isolate (Supro 710)	93	102	100	100	82	85	93	93	76	79	82	82	67	70	87	87	72	74
(12) Non-fat dried skim milk	100	100	100	100	99	100	97	97	96	97	86	86	85	86	83	83	82	83
Mean (n = 12)	89	88	98	98	85	87	93	92	80	82	82	82	71	73	84	84	73	75
S.D.	10	12	4	4	13	11	6	6	13	12	5	5	11	10	6	6	11	10
Mean	91	87	99	98	87	89	93	93	82	84	82	82	72	74	84	84	74	76
S.D.	8	12	4	4	12	11	6	6	12	11	5	6	11	10	6	6	11	10

[1]FAO/WHO, NRC, Egg and Human Milk denote reference patterns; values for 11 proteins calculated with data for wheat gluten omitted.

TABLE 19.12. COMPARISONS OF "N REQUIREMENT SCORES", "EFFICIENCY SCORES", AND MODIFIED ESSENTIAL AMINO ACID INDEX (MEAA) SCORES BASED ON AMINO ACID COMPOSITIONS UNCORRECTED AND CORRECTED FOR DIGESTIBILITY (DATA SET 1)[1]

Protein Source	N Requirement Scores	Efficiency Scores	Modified Essential Amino Acid Index (MEAA) Scores															
			No Correction For Digestibility				Corrected for Digestibility											
							Human True				Human Apparent				In Vitro "Apparent"			
			FAO/WHO	NRC	Egg	Human Milk	FAO/WHO	NRC	Egg	Human Milk	FAO/WHO	NRC	Egg	Human Milk	FAO/WHO	NRC	Egg	Human Milk
(1) Egg white bread	100	100	100	100	98	98	93	93	92	93	82	82	80	82	88	88	87	88
(2) Lactalbumin bread	90	80	100	95	93	93	93	88	87	87	82	78	76	76	85	81	79	79
(3) Casein bread	91	65	100	100	94	98	97	97	91	97	86	86	80	86	84	84	79	83
(4) Textured soy (Supro 50) bread	90	87	98	99	82	85	86	87	72	74	74	75	62	64	80	81	67	69
(5) Soy Isolate (Promine F) bread	77	85	95	99	79	82	89	93	73	77	76	82	64	68	81	84	66	70
(6) Peanut flour bread	79	85	78	81	59	62	70	73	53	57	63	66	34	51	62	65	48	51
(7) Wheat gluten bread	72	89	81	82	64	66	78	80	62	63	71	72	56	57	70	72	55	57
(8) Beef bologna	95	93	98	98	79	82	95	96	78	81	84	85	69	72	83	84	68	70
(9) Beef + soy isolate (Supro 620) bologna	100	69	97	99	80	83	95	97	78	81	84	86	69	72	82	84	68	70
(10) Soy isolate (Supro 620)	81	95	94	99	78	82	89	94	74	78	79	83	65	69	86	91	72	76
(11) Soy isolate (Supro 710)	93	102	97	99	79	82	90	92	73	76	80	81	65	67	85	86	69	71
(12) Non-fat dried skim milk	100	100	100	100	93	96	96	97	90	94	86	86	79	83	83	81	77	80
Mean (n = 12)	89	88	95	96	82	84	89	91	77	80	79	80	67	70	81	82	70	72
S.D.	10	12	8	7	12	12	8	7	12	12	7	6	13	10	7	7	11	11
Mean (n = 11)	91	87	96	97	83	86	90	92	78	81	79	81	67	72	82	83	71	73
S.D.	8	12	6	6	11	10	8	7	11	11	7	6	13	10	7	7	10	10

[1] FAO/WHO, NRC, Egg and Human Milk denote reference patterns; values for 11 proteins calculated with data for wheat gluten omitted.

TABLE 19.13. COMPARISONS OF "N REQUIREMENT SCORES", "EFFICIENCY SCORES", AND SCORES CALCULATED BY USE OF THE NEBRASKA PROCEDURE (SATTERLEE ET AL., 1979; FAO/WHO PATTERN USED AS REFERENCE) BASED ON AMINO ACID COMPOSITIONS UNCORRECTED AND CORRECTED FOR DIGESTIBILITY (DATA SET 1)[1]

Protein Source	N Requirement Scores	Efficiency Scores	Scores		Corrected for Digestibility	
			No Correction For Digestibility	Human True	Human Apparent	In Vitro "Apparent"
(1) Egg white bread	100	100	100	100	100	100
(2) Lactalbumin bread	90	80	100	99	94	96
(3) Casein bread	91	65	100	100	100	99
(4) Textured soy (Supro 50) bread	90	87	95	88	74	83
(5) Soy Isolate (Promine F) bread	77	85	90	86	78	80
(6) Peanut flour bread	79	85	83	65	58	58
(7) Wheat gluten bread	72	89	44	40	39	40
(8) Beef bologna	95	93	95	94	88	87
(9) Beef + soy isolate (Supro 620) bologna	100	69	94	92	85	84
10) Soy isolate (Supro 620)	81	95	89	86	78	82
11) Soy isolate (Supro 710)	93	102	94	90	82	86
12) Non-fat dried skim milk	100	100	100	100	98	97
Mean (n = 12)	89	88	90	87	81	83
S.D.	10	12	16	18	18	18
Mean (n = 11)	91	87	95	91	85	87
S.D.	8	12	5	10	13	12

Values for 11 proteins calculated with data for wheat gluten omitted.

TABLE 19.14. COMPARISONS OF "N REQUIREMENT SCORES", "EFFICIENCY SCORES", AND AMINO ACID SCORES (DATA SET 2)[1]

Protein Source	Efficiency Score	N Requirement Score	Chemical		Lys-Try-Met-Cys		MEAA Index		Nebraska
			NRC	FAO/WHO	NRC	FAO/WHO	NRC	FAO/WHO	FAO/WHO
INCAP									
(1) Egg	91	111	100	100	100	100	100	100	100
(2) Milk	106	104	100	100	100	100	100	100	100
(3) Milk	100	100	100	100	100	100	100	100	100
Japan									
(4) Egg (Maintenance Calories)	100	100	100	100	100	100	100	100	100
(5) Egg (Excess Calories)	(90)	(131)	100	100	100	100	100	100	100
(6) Rice (Maintenance Calories)	66	76	79	73	79	73	97	95	83
(7) Rice (Excess Calories)	(114)	(111)	79	73	79	73	97	95	83
(8) Wheat Gluten	32	45	33	31	33	31	82	81	44
MIT									
(9) Egg	(100)	(100)	100	100	100	100	100	100	100
(10) Whole Wheat	42	57	54	50	54	50	90	87	64
(11) Canned Beef	79	87	100	100	100	100	100	100	100
Berkeley									
(12) Egg Protein	—	(76)	100	100	100	100	100	100	100
(13) Egg White Protein	—	100	100	100	100	100	100	100	100
Purdue									
(14) Rice, Milk, White Flour (50:25:25; N basis)	(86)	(67)	91	92	91	100	99	99	98
(15) Rice, White Flour, Chicken, Milk, (25:15:30:30; N basis)	(100)	(100)	100	95	100	100	100	99	99
Peru									
(16) Bulgar Wheat, Soy Flour, Non-Fat Dried Skim Milk (51:14:25; Wt. Basis)	(—)	(—)	100	98	100	100	100	100	98

[1]Scores calculated from amino acid compositions *not* corrected for digestibility; composition values used in calculations for proteins 6, 10, 11, 14, 15 and 16 from references given in Table 3 or from FAO (1970); all other compositions used from USDA (Beltsville) analyses. See Table 3 for sources of data. Values for efficiency and N requirement scores within parentheses may not be useful estimates (see Table 3). NRC and FAO/WHO denote reference patterns used in calculating scores.

TABLE 19.15.   COMPARISONS OF "N REQUIREMENT SCORES", "EFFICIENCY SCORES", AND AMINO ACID SCORES (DATA SET 3)[1]

Protein Source	Efficiency Score	N Requirement Score	Chemical		Lys-Try-Met-Cys		MEAA Index		Nebraska
			NRC	FAO/WHO	NRC	FAO/WHO	NRC	FAO/WHO	FAO/WHO
(1) Whole Egg	(95)	(135)	100	100	100	100	100	100	100
(2) Whole Egg	(63)	(78)	100	100	100	100	100	100	100
(3) Casein	70	92	100	100	100	100	100	100	100
(4) Milk	(108)	(115)	100	100	100	100	100	100	100
(5) Milk	(113)	(110)	100	100	100	100	100	100	100
(6) Milk	(110)	(122)	100	100	100	100	100	100	100
(7) Milk	100	100	100	100	100	100	100	100	100
(8) Beans	60	75	77	57	77	57	95	92	72
(9) Beans:Maize (40:60; protein basis)	98	88	67	74	64	70	94	91	84
(10) Rice:Bean (60:40; protein basis)	82	91	89	75	89	77	99	97	90
(11) Rice:Bean:Milk (55:35:10; protein basis)	104	110	71	78	71	78	96	94	87
(12) Textured Veg. Protein (TVP)	76	91	91	88	100	94	99	98	95
(13) Beef	97	103	100	100	100	100	100	100	100
(14) TVP + Beef (50:50, protein basis)	97	95	100	99	100	100	100	100	100

[1]Scores calculated from amino acid compositions *not* corrected for digestibility; literature composition values used in calculations for proteins 8, 9, 10, 11, and 13; other compositions used from USDA (Beltsville) analyses; see Table 4 for sources of data. Values for N requirement and Efficiency scores within parenthesis may not be useful estimates for scoring purposes (see Table 4); NRC and FAO/WHO denote reference patterns used in calculating scores.

TABLE 19.16. AMINO ACID SCORES MINUS "N REQUIREMENT SCORES" OR "EFFICIENCY SCORES"[1]

	Difference (N Requirement)		Difference (Efficiency)	
	Mean	+ S.D.	Mean	+ S.D.
Chemical Score				
FAO/WHO	−3.2	9.0	0.0	17.8
NRC	+1.09	8.3	+4.3	15.7
Lys-Try-Met-Adj. Cys				
FAO/WHO	−3.1	10.3	+0.1	19.4
NRC	+5.2	10.5	+8.4	16.3
Lys-Try-Met-Cys				
FAO/WHO	−0.7	9.1	+2.5	17.3
NRC	+5.2	10.5	+8.4	16.3
MEAA Index				
FAO/WHO	+5.5	6.6	+8.7	13.8
Egg	−7.1	8.6	−4.3	16.5
Nebraska Scoring Procedure				
FAO/WHO	+4.0	5.6	+7.2	13.7

[1]Data Set 1; data for wheat gluten omitted. FAO/WHO and NRC denote reference patterns used in calculating amino acid scores.

TABLE 19.17.   AMINO ACID SCORES MINUS "N" REQUIREMENT SCORES"[1]

	Chemical Score	Lys-Try-Met-Cys	Nebraska
Data Set No. 1			
Mean Difference (± S.D.)	−3.2 (9.0)	−0.7 (9.1)	+4.0 (5.6)
Data Set No. 2			
Mean Difference (± S.D.)	−2.9 (7.8)	−2.9 (7.8)	+1.2 (7.0)
Data Set No. 3			
Mean Difference (± S.D.)	−8.2 (12.7)	−7.7 (13.4)	−1.9 (8.9)
Data Sets No.'s 1+2+3			
Mean Difference (± S.D.)	−4.7 (9.9)	−3.6 (10.4)	−1.4 (8.3)

[1]FAO/WHO (1973) reference pattern used; data for 11 proteins used in Data Set 1 (data for wheat gluten omitted), nine in Set 2, nine in Set 3, and 29 for Sets 1+2+3.

TABLE 19.18.   AMINO ACID SCORES MINUS "EFFICIENCY SCORES"[1]

	Chemical Score	Lys-Try-Met-Cys	Nebraska
Data Set No. 1			
Mean Difference (± S.D.)	0.0 (17.8)	+2.5 (17.3)	+7.2 (13.7)
Data Set No. 2			
Mean Difference (± S.D.)	−4.8 (8.4)	+4.8 (8.4)	+9.4 (10.5)
Data Set No. 3			
Mean Difference (± S.D.)	−1.4 (17.1)	−0.9 (18.5)	+4.9 (14.8)
Data Sets No.'s 1+2+3			
Mean Difference (± S.D.)	+0.9 (15.1)	+2.0 (15.3)	+7.1 (12.9)

[1]FAO/WHO (1973) reference pattern used; data for 11 proteins used in Data Set 1 (data for wheat gluten omitted), eight in Set 2, nine in Set 3, and 29 for Sets 1+2+3.

TABLE 19.19. DIFFERENCES BETWEEN PROTEIN NUTRITIVE VALUES AS ESTIMATED BY "LYSINE + TRYPTOPHAN + METHIONINE + CYSTINE" SCORE, AND BY NITROGEN REQUIREMENT OR EFFICIENCY SCORES

	Lysine-Tryptophan-Methionine-Cystine (LTMC) Score	Nitrogen Requirement (NR) Score	Efficiency (EFF) Score	LTMC Score - NR Score	LTMC Score - EFF Score
Data Set No. 1 [1]					
USDA					
(1) Egg White	(100)	(100)	(100)	0	0
(2) Lactalbumin	90	100	80	-10	+10
(3) Casein	100	91	65	+9	+35
(4) Textured Soy Protein	94	90	87	+4	+7
(5) Soy Isolate	79	77	85	+2	-6
(6) Peanut Flour	60	79	85	-19	-25
MIT					
(8) Beef Bologna	100	95	93	+5	+7
(9) Beef + Soy Bologna	85	100	69	-15	+16
(10) Soy Isolate (S-620)	76	81	95	-5	-19
(11) Soy Isolate (S-710)	94	93	102	+1	-8
(12) NFDSM	(100)	(100)	(100)	0	0
Means ± S.D.	88.9 ± 12.9	91.5 ± 8.9	87.4 ± 12.4	-2.5 ± 8.8	1.5 ± 16.5
Data Set No. 2					
INCAP					
(1) Egg	100	111	91	-11	+9
(2) Milk	100	104	106	-4	-6
(3) Milk	100	100	100	0	0
JAPAN					
(4) Egg	100	100	100	0	0
(6) Rice	73	76	66	-3	+7
(8) Wheat Gluten	31	45	32	-14	-1
MIT					
(10) Whole Wheat	50	57	42	-7	+8
(11) Canned Beef	100	87	79	+13	+21
BERKELEY					
(13) Egg White Protein	100	100	—	0	—
Means ± S.D. n = 8[2]	81.8 ± 27.6	85.0 ± 23.8	77.0 ± 28.0	-2.0 ± 8.1	4.8 ± 8.4
n = 9	83.8 ± 26.5	86.8 ± 22.8	—	-2.9 ± 7.8	—

TABLE 19.19. (Cont'd.)

	Lysine-Tryptophan-Methionine-Cystine (LTMC) Score	Nitrogen Requirement (NR) Score	Efficiency (EFF) Score	LTMC Score - NR Score	LTMC Score - EFF Score
Data Set No. 3					
INCAP					
(3) Casein	100	92	70	+8	+30
(7) Milk	100	100	100	0	0
(8) Beans	57	75	60	-18	-3
(9) Beans + Maize	70	88	98	-18	-28
(10) Rice + Beans	77	91	82	-14	-5
(11) Rice + Beans + Milk	78	110	104	-32	-26
(12) T V Protein	94	91	76	+3	+18
(13) Beef	100	103	97	-3	+3
(14) TVP + Beef	100	95	97	+5	+3
Mean ± S.D.	86.2 ± 16.2	93.9 ± 10.0	87.1 ± 15.6	-7.7 ± 13.4	-0.9 ± 18.5
Data Sets 1+2					
Mean ± S.D. n = 19[2]	86.4 ± 20.3	88.2 ± 16.3	83.0 ± 20.4	-1.8 ± 8.6	3.4 ± 14.0
n = 20	87.1 ± 20.0	88.8 ± 16.1	—	-1.7 ± 8.4	—
Data Sets 1+2+3					
Mean ± S.D. n = 28[2]	86.4 ± 18.7	90.0 ± 14.6	84.3 ± 18.8	-3.7 ± 10.5	2.0 ± 15.3
n = 29	86.8 ± 18.6	90.4 ± 14.5	—	-3.6 ± 10.4	—

[1] Data for wheat gluten omitted.
[2] Data for egg white protein (Data Set No. 2, protein No. 13) omitted.

large between the amino acid scores and either of the two scores based on data from the human studies, but differences also are large between the two scores based on the data from the human studies. For the most part, however, the LTMC ("lysine + tryptophan + methionine + cystine") scores agree reasonably well with one or the other of the two scores based on the human data. Exceptions occur. The LTMC score is much lower for peanut flour (Data Set 1, Table 19.19), beans + maize (Data Set 3) or TVP + beef (Data Set 3) than either the nitrogen requirement or efficiency scores. The LTMC score is considerably higher than one of the scores and lower than the other score (based on the human data) for beef + soy bologna (Data Set 1). For canned beef (Data Set 1), the LTMC is much higher than either of the two estimates based on the data for humans. For the proteins for which the LTMC score and the scores based on the human data are in marked disagreement, it cannot be stated with certainty which values are in error.

Potential for Use of Amino Acid Scores for Nutritional Labeling

Some differences were large between protein nutritive value as estimated by amino acid scores and as estimated by the data from the human studies. Inspection of the data for individual protein sources (Tables 19.8 – 19.15 and Table 19.19) suggests that, in several instances (e.g., as noted above with respect to the LTMC scores), the values derived from the human studies are suspect. Given a coefficient of variation of 15% which is usually accepted as being applicable to estimates of protein nutritive value in human studies (FAO/WHO, 1973), and a coefficient of at least 5 to 10% for estimates of amino acid levels from chemical analyses, the agreement is not altogether unsatisfactory. A reasonable estimate of protein nutritional value can be obtained by use of amino acid scores.

Harper (Chapter 2) distinguished between evaluations for determining protein nutriture (protein or amino acid requirements, protein nutritive status, etc.) and evaluations for estimating protein nutritional quality for nutritional labeling or other regulatory purposes. For the latter, particularly in populations whose daily diets contain adequate levels of protein derived from a broad mixture of sources, amino acid scores would appear to be potentially useful and practical.

DIGESTIBILITY AND OTHER FACTORS

Effects of Correcting Compositions for Digestibility Prior to the Calculation of Amino Acid Scores

The effects of correcting the amino acid composition data for digestibility (human true, human apparent, *in vitro* "apparent") prior to calculating scores for the individual protein sources in Data Set 1 are given in Tables 19.8 – 19.13. Mean values, calculated with or without the data for wheat gluten are compared to the various scores calculated without corrections for digestibility (Table 19.20). In general, correcting for digestibility decreased the relationship between nutritive value as estimated by the scores and as estimated by the two scores derived from the data for humans.

For use of amino acid scores for estimating protein nutritive value for nutritional labeling, on the basis of these data, corrections for digestibility probably would not be necessary for *most* proteins. A threshold level of amino acid availability or of N digestibility (e.g., equivalent to a true digestibility of 80–85% or so) could be required for a protein source to quality for scoring by a routine procedure. Values for most proteins would exceed the defined threshold level and estimates of bioavailability or digestibility would not be needed. For sources known to have digestibilities or amino acid bioavailabilities near or below the threshold, quantitative evidence of the level extant could be required. This evidence could be derived from any of several approaches (rat or *in vitro* enzymic digestibility; microbiological or chemical [i.e., for lysine] amino acid bioavailability). A minimal level could also be specified below which a protein source would not qualify as a source of dietary protein. At levels above the minimal level but below the threshold level, prior to calculating the amino acid score, the amino acid composition of the source would be adjusted according to its analytically determined level of digestibility or amino acid bioavailability.

The use of amino acid composition data for predicting digestibility is an initially appealing concept. As developed by the Nebraska workers (Satterlee *et al.*, Chapter 18), it would also appear to have considerable usefulness in their procedures which utilize amino acid composition data for rapidly predicting rat PER values. The procedure apparently gives useful information about the *inherent* digestibility (possibly associated with protein peptide structure?) of different proteins. I question, however, whether or not a digestibility value predicted from amino acid composition data gives much information about possible decreases in digestibility or in amino acid bioavailability that might be caused by heat damage or overprocessing. For detecting processing effects, methods based on *in vitro* enzymic analyses, rat digestibility assays, or especially estimates (microbiological and/or chemical) of amino acid bioavailability, would seem to have some advantages.

Digestibility vs Amino Acid Bioavailability

Digestibility, whether estimated by rat, human or *in vitro* assays, is at best only a general indicator of possible effects of heat damage, overprocessing, etc. on amino acid bioavailability or of a possibly inherent lack of availability. In particular, *in vitro* enzymic assays may greatly overestimate the availability of specific amino acids (e.g., methionine, lysine) compared to values from *in vivo* assays based on rat growth (Ford and Salter, 1966; also, see Chapters 15 and 16). For purposes such as nutritional labeling, if only a general estimate of amino acid bioavailability is required, then an *in vitro* enzymic method such as that of Satterlee *et al.* (1979) would appear to be adequate (also, see Chapter 18). If the assessment

TABLE 19.20. EFFECTS OF CORRECTING AMINO ACID COMPOSITIONS FOR DIGESTIBILITY ON MEAN AMINO ACID SCORES OF 12 PROTEINS; MEAN "N REQUIREMENT SCORE" = 89 (91); MEAN "EFFICIENCY SCORE"[2] = 88 (87)[1]

Reference Pattern:	FAO/WHO (1973)	NRC (1974)	Whole Egg	Human Milk
Chemical Scores				
No Correction	83 (87)	87 (92)	—	—
Corrected for Digestibility				
Human, True	80 (84)	82 (86)	—	—
Human, Apparent	69 (73)	72 (76)	—	—
In Vitro, "Apparent"	71 (74)	74 (78)	—	—
Lysine-Tryptophan-Methionine-Adjusted Cystine				
No Correction	83 (87)	91 (96)	—	—
Corrected for Digestibility				
Human, True	78 (83)	85 (90)	—	—
Human, Apparent	70 (73)	75 (80)	—	—
In Vitro, "Apparent"	71 (75)	77 (82)	—	—
Lysine-Tryptophan-Methionine-Cystine				
No Corrections	85 (90)	91 (96)	—	—
Corrected for Digestibility				
Human, True	82 (87)	85 (90)	—	—
Human, Apparent	71 (75)	75 (80)	—	—
In Vitro, "Apparent"	72 (76)	77 (82)	—	—
Total Essential Amino Acids				
No Correction	98 (99)	98 (98)	85 (87)	87 (89)
Corrected for Digestibility				
Human, True	93 (93)	92 (93)	80 (82)	82 (84)
Human, Apparent	82 (82)	82 (82)	71 (72)	(74)
In Vitro, "Apparent"	84 (84)	84 (84)	73 (74)	75 (76)
Modified Essential Amino Acid Index				
No Correction	95 (96)	96 (97)	82 (83)	84 (86)
Corrected for Digestibility				
Human, True	89 (90)	91 (92)	77 (78)	80 (81)
Human, Apparent	79 (80)	80 (81)	67 (68)	70 (72)
In Vitro, "Apparent"	81 (82)	82 (83)	70 (71)	72 (73)
Nebraska Procedure for Scoring[2]				
No Correction	90 (95)	—	—	—
Corrected for Digestibility				
Human, True	87 (91)	—	—	—
Human, Apparent	81 (85)	—	—	—
In Vitro, "Apparent"	83 (87)	—	—	—

[1]Values in parentheses calculated with data for wheat gluten omitted; see Table 1 for protein sources.
[2]Satterlee et al., 1979.

of nutritive value is raised to a more critical level, however, assays for bioavailability of one or more "indicator" amino acids would be essential. Chemical assays for available lysine (Chapter 14) or microbiological assays for one or more other amino acids (Chapter 16) would appear to be quite adequate for this purpose.

Evaluation by Amino Acid Score Plus Rat Assays

If use of *in vitro* enzymic, chemical or microbiological methods were unacceptable to one or more regulatory agencies for estimating amino acid bioavailability or nitrogen digestibility, use of short-term rat assays would be indicated. Due to the labor required and difficulties (technical) associated with determining digestibility *per se* in the rat, a simple growth assay could be used to screen new foods or protein sources to detect and quantitate effects of low digestibility or poor amino acid bioavailability. For this purpose, the 10-day NPR (Net Protein Ratio) assay probably would be the most suitable. The NPR assay is of shorter duration than the PER or RPV assays and yet retains the advantage of linearity between relative values of 0 to 100.

Amino Acid Imbalances and Other Possibly Deleterious Factors

In any approach for protein nutritional labeling in which amino acid composition data were used, appropriate limits should be established to ensure the absence of possibly deleterious amino acid imbalances or of significant levels of factors such as trypsin inhibitors, gossypol, or glucosinylates.

A POSSIBLE APPROACH FOR PROTEIN NUTRITIONAL LABELING

An approach for protein nutritional labeling based on amino acid scores would indeed meet many of the requirements of what I consider to be an ideal assay (Bodwell, 1976, 1977). In a general way, some quantitative information would be provided about the nutritive value of a protein for humans either as a single source or when consumed with other sources. Those with knowledge of the amino acid compositions on which the scores were based (or in the case of scores based on the limiting amino acid, knowledge of the level of the amino acid which was limiting in each source) would be able to predict the estimated protein nutritive value of a mixture of two or more proteins that had been evaluated (scored) separately. This advantage, however, would not be realized by the less knowledgeable (i.e., the consumer). However, an approach based on amino acid scores would be simple in application and would reflect, as differences in protein quality, only differences that might be of significance under practical conditions.

One of the most serious disadvantages of the current procedure is the use of "breakpoints." This artificially creates large differences in apparent protein nutritive value when such differences do not exist. The use of amino acid scores to classify proteins into groups, or of any other procedure in which protein sources would be classified into groups, according to estimated protein nutritive value (e.g., high, intermediate and low quality groups) would probably involve the use of arbitrary breakpoints. As I have previously noted (Bodwell, 1977), any procedure used for nutritional labeling purposes in which breakpoints are incorporated in the application of the assay would be inadequate.

A further deficiency in the current procedure that would likewise not be eliminated by the use of an approach involving amino acid scores *per se* (nor by any of the other alternative methods proposed) is that of providing a valid means for allowing for *both quality and quantity* in labeling for protein nutritional value.

I would like to suggest an approach that would essentially preclude the problems discussed above. It was noted previously that amino acid scores based only on levels of lysine, tryptophan and methionine plus cystine appeared to be as useful as scores based on levels of limiting amino acids or of all essential amino acids. Accordingly, it seems reasonable to suggest that protein nutritional labeling could be based on the levels of these amino acids (and possibly, also total essential amino acids). As shown in Table 19.21, the levels of lysine, tryptophan, and methionine plus cystine (± total

TABLE 19.21. AMOUNTS AND % OF A HYPOTHETICAL U.S. RDA FOR SPECIFIC AMINO ACIDS PROVIDED BY 20 GMS OF PROTEIN FROM 3 DIFFERENT SOURCES[1]

	In 20 gms of Protein					
	Canned Meat		Beans		Whole Wheat	
	gms	% "U.S. RDA"	gms	% "U.S. RDA"	gms	% "U.S. RDA"
Lysine	1.66	46	1.48	41	0.55	15
Tryptophan	0.22	34	0.19	29	0.25	39
Methionine + Cystine	0.72	32	0.40	18	0.87	38
Total Essential Amino Acids	8.22	34	8.88	36	7.10	29

[1]Hypothetical "U.S. RDA" for lysine, tryptophan, methionine + cystine, and total essential amino acids of 3.58, 0.65, 2.28 and 24.5 gms, respectively. Levels of total essential amino acids could be omitted from the label. However, a minimal level of each essential amino acid would be required regardless of whether or not levels of total essential amino acids were listed.

essential amino acids) contained in a protein source would be expressed in grams and as a percent of a "U.S. RDA" for these amino acids. Whether or not levels of total essential amino acids were listed on the label, the presence of a minimum level of each essential amino acid could be required. Total protein would also be listed on the label but not expressed as a percent of a U.S. RDA. For new or "problematic" protein sources (see above), an allowance for the effects of poor amino acid availability or low nitrogen digestibility could be made by one or more of the approaches discussed above. As also discussed above, provisions to preclude amino acid imbalances or the presence of significant levels of other possibly deleterious factors also should be included.

This proposed approach (or some similar approach based on levels of specific essential amino acids) would allow for a possibly greater public awareness of the value of protein sources when consumed alone and when consumed with other protein sources, and would encourage consumer selection of protein sources on the basis of amino acid complementation. The problem of the inadequacies associated with the use of "breakpoints" in the current U.S. nutritional labeling procedure would be eliminated. Such an approach would provide useful information to the consumer which would not be provided by any approach based on the concept of "utilizable protein."

SUMMARY

As indicated by R^2 analyses, groups of amino acids from amino acid composition data of several proteins are highly correlated with estimates of protein nutritional value and nitrogen digestibility derived from data obtained with young adults in USDA and MIT studies. Any practical usefulness or application of the observed relationships is limited both by the small number of protein sources tested and by the empirical nature of the relationships. It is difficult to demonstrate a high degree of accuracy in predicting protein nutritional value from amino acid scores because of the variability in estimates of protein nutritional value in humans. However, different amino acid scoring procedures do give reasonable estimates of protein nutritional value. On the basis of the data discussed, the use of different estimates of digestibility (human, true; human, apparent; *in vitro*, apparent) to correct composition data prior to the calculation of scores does not appear to improve the relationship between amino acid scores *per se* and protein nutritive value as estimated in human adults. Thus, if an amino acid scoring procedure were used for nutritional labeling, inclusion of estimates of digestibility for *most* protein sources would probably be unnecessary. Provisions should be included, however, for making adjustments for sources with marginal amino acid bioavailability or digestibility and for disregarding foods with poor amino acid bioavailability or very low digestibility as sources of protein. In any approach which uses amino acid composition data, appropriate limits should be established to ensure the absence of possibly deleterious amino acid imbalances and of significant levels of factors such as trypsin inhibitors, gossypol, or glucosinylates. Amino acid scores based only on levels of lysine, tryptophan and methionine plus cystine appear to be as useful as scores based on levels of the limiting amino acid or of all essential amino acids. A possible procedure for nutritional labeling is proposed in which lysine, tryptophan, and methionine plus cystine (and possibly total essential amino acids as well) would be expressed in grams and as a percent of a "U.S. RDA" for these amino acids.

Whether or not a value for total essential amino acids were listed on the label, the presence of a minimum level of each essential amino acid would be required. Total protein would also be listed on the label but would not be expressed as a percent of a U.S. RDA. This proposed approach, or a similar approach, would allow for a possibly greater public awareness of the value of protein sources when consumed either alone or with other protein sources, and would encourage consumer selection of protein sources on the basis of amino acid complementation. The inadequacies associated with the use of "breakpoints" in the current U.S. nutritional labeling procedure (and which also would be present in any system in which protein sources were classified into groups; e.g., high, intermediate, or low quality groups) would be eliminated. Useful information would be provided to the consumer which would not be provided by any approach which was based on the concept of "utilizable protein."

ACKNOWLEDGEMENTS

The assistance of Dr. James Koch (Agriculture Research, USDA, Beltsville) and of Mr. Ernest Schuster, Mr. Robert Staples and Mr. James Shultz of the Protein Nutrition Laboratory in statistical analyses and computations and of Mrs. E.C. Cole and Mrs. M. Kincius in manuscript preparation is gratefully acknowledged.

REFERENCES

BARR, A.J., GOODNIGHT, J.H., SOLI, J.P. and HELWIG, J.T.   1976.   "A User's Guide to SAS".   SAS Institute, Raleigh, N. Carolina.

BODWELL, C.E.   1976.   Application of animal data to human protein nutrition: A Review.   Cereal Chem. *54*, 958.

BODWELL, C.E.   1977.   Problems associated with the development and application of rapid methods of assessing protein quality.   Nutr. Reports Int. *16*, 163.

BODWELL, C.E.   1979.   Human versus animal assays.   In "Soy Protein & Human Nutrition".   H.L. Wilcke, D.T. Hopkins, and D.H. Waggle (Eds). Academic Press, N.Y., p. 331.

BODWELL, C.E.   1979.   The nutritive value of the same protein preparations as estimated by human, rat, and chemical assays.   J. A. Oil Chem. Soc. *56*, 156.

BODWELL, C.E., SCHUSTER, E.M., BROOKS, B. and WOMACK, M.   1979. Relative protein nutritive value of seven protein breads estimated in young men and in rats.   Fed. Proc. *38*, 772.

BRESSANI, R., NAVARRETE, D.A., DeDAQUI, A.L., ELIAS, L.G., OLIVARES, J. and LACHANCE, P.A.   1979a.   Protein Quality of spraydried whole milk and of casein in young adult humans using a short term nitrogen balance index assay.   J. Food Sci. *44*, 1136.

BRESSANI, R., NAVARRETE, D.A., ELIAS, L.G. and BRAHAM, J.E.   1979b. A critical summary of a short-term nitrogen balance index to measure protein quality in adult human subjects.   In "Soy Protein & Human Nutrition", H.L. Wilcke, D.T. Hopkins and D.H. Waggle (Eds).   Academic Press, N.Y., p. 313.

CALLOWAY, D.H.   1975.   Nitrogen balance of men with marginal intakes of protein and energy.   J. Nutr. *105*, 914.

CALLOWAY, D.H., and MARGEN, S.   1971.   Variation in endogenous nitrogen excretion and dietary nitrogen utilization as determinants of human protein requirements.   J. Nutr. *101*, 205.

CHENG, A.H.R., GOMEZ, A., BERGAN, J.G., TUNG-CHING, L., MONCKEBERG, F. and CHICHESTER, C.O.   1978.   Comparative nitrogen balance study between young and aged adults using three levels of protein intake from a combination wheat-soy-milk mixture.   Am. J. Clin. Nutr. *31*, 12.

CLARK, H.E., HOWE, J.M., MAGEE, J.L. and MALZER, J.L.   1972.   Nitrogen balances of adult human subjects who consumed four levels of nitrogen from a combination of rice, milk and wheat.   J. Nutr. *102*, 1647.

CLARK, H.E., MOON, W.H., MALZER, J.L. and PANG, R.L.   1974.   Nitrogen retention of young women who consumed between six and eight grams of nitrogen from a combination of rice, wheat, chicken and milk.   Am. J. Clin. Nutr. *27*, 1059.

FAO/WHO (Food and Agriculture Organization/World Health Organization).   1973.   "Energy and Protein Requirements. Report of a Joint FAO/WHO ad hoc expert committee."   Wld. Hlth. Org. Techn. Rep. Ser., No. 522. World Health Organization, Geneva.

FAO (Food and Agriculture Organization).   1970.   Amino acid content of foods and biological data on proteins.   Food and Agriculture Organization, Rome, Italy.

FORD, J.E. and SALTER, D.N.   1966.   Analysis of enzymically digested food proteins by Sephadex gel filtration.   Br. J. Nutr. *20*, 843.

HACKLER, L.R.   1977.   *In vitro* indices: Relationships to estimating protein value for the human.   In "Evaluation of Proteins for Humans," C.E. Bodwell (Ed.).   AVI Publishing Co., Inc., Westport, Conn., p. 55.

INOUE, G., FUJITA, Y., NIIYAMA, Y.   1973.   Studies on protein requirements of young men fed egg protein and rice proteins with excess and maintenance energy intakes.   J. Nutr. *103*, 1673.

INOUE, G., FUJITA, Y., KISHI, K., YAMAMOTO, S. and NIIYAMA, Y.   1974.   Nutritive values of egg protein and wheat gluten in young men.   Nutr. Reports Int. *10*, 201.

NRC (Natl. Acad. Sci./Natl. Res. Council).   1974.   "Improvement of Protein Nutriture".   Natl. Acad. Sciences, Washington, D.C.

RAND, W.M., SCRIMSHAW, N.S. and YOUNG, V.R.   1977.   Determination of protein allowances in human adults from nitrogen balance data.   Am. J. Clin. Nutr. *30*, 1129.

SATTERLEE, L.D., MARSHALL, H.F. and TENNYSON, J.M.   1979.   Measuring Protein Quality.   J. Am. Oil Chem. Soc. *56*, 103.

SCHUSTER, E.M. and BODWELL, C.E.   1979.   Relationship between protein nutritional value as estimated in young men, rats, essential amino acid composition and protein digestibility.   Fed. Proc. *38*, 284.

SCRIMSHAW, N.S. and YOUNG, V.R.   1979.   Soy protein in adult human nutrition: A review with new data.   In "Soy Protein & Human Nutrition," H.L. Wilcke, D.T. Hopkins and D.H. Waggle (Eds.).   Academic Press, N.Y., p. 121.

SHEFFNER, A.L.   1967.   *In vitro* protein evaluation.   In "Newer Methods of Nutritional Biochemistry," A.A. Albanese (Ed.), Vol. III.   Academic Press, N.Y., p. 125.

YOUNG, V.R., TAYLOR, Y.S.M., RAND, W.M. and SCRIMSHAW, N.S. 1973.   Protein requirements of man: Efficiency of egg protein utilization at maintenance and submaintenance levels in young men.   J. Nutr. *103*, 1164.

YOUNG, V.R., FAJARDO, L., MURRAY, E., RAND, W.M. and SCRIMSHAW, N.S.   1975.   Protein requirements of man. Comparative nitrogen balance response within the submaintenance to maintenance range of intakes of wheat and beef proteins.   J. Nutr. *105*, 534.

YOUNG, V.R., RAND, W.M. and SCRIMSHAW, N.S.   1977.   Measuring protein quality in humans. A review and proposed method.   Cereal Chem. *54*, 929.

YOUNG, V.R., SCRIMSHAW, N.S., TORUN, B. and VITERI, F.   1979.   Soybean protein in human nutrition: An overview.   J. Am. Oil Chem. Soc. *56*, 110.

DISCUSSION

DR. SATTERLEE: I become very concerned when a Nebraska weighting process which was developed to predict PER for the rat is used in the prediction of protein quality for the human. I know Dr. Kendrick and I, in thinking about the eventual estimation of protein quality for the human, would never consider using the CPER method unless we could get Dr. Scrimshaw to agree that the rat is predicting human protein quality.

DR. BODWELL: The point, in using the procedure you use for your initial scoring, is that I like the idea of trying to attempt to use a weighted factor. When low levels of a specific amino acid are present, such as only 10% of the FAO/WHO pattern, your weighting factor is quite large. The reverse is true, that is the weighting factor is small, for those amino acids which are present at levels approaching or exceeding the FAO/WHO level. I have used the procedure simply to see what effects such a weighting procedure might have. The resulting scores are probably the second or third best.

DR. SATTERLEE: One other thing, because just about an hour ago I mentioned it and I really think I need to clarify it. Predicting digestibility from the amino acid profile alone is being used primarily to provide a model to see if, from two different chemical approaches, a similar phenomenon is happening. I would be very afraid to move to that as my sole method for determining digestibilities. There are many factors which will interfere with that.

DR. SAMONDS: Could you explain again your fifth suggestion? I understood the first four.

DR. BODWELL: This is a suggestion that, with any procedure based on amino acid data, certain limits should be specified for factors other than amino acid composition *per se* which may influence nutritive value. For example, a maximum level of trypsin inhibitor permitted could be specified, large imbalances of amino acids defined and prohibited, and/or minimum levels of bioavailability of key amino acids or of nitrogen digestibility could be defined and specified.

DR. HARPER: I'm quite impressed with how well the predictions came out from the limiting amino acid scores. I think it's a step in the right direction. I'll leave it to John Vanderveen as to whether it's the solution for the nutritional labeling or not, but I do think we should be talking about some recommended or at least tentative allowances for the few amino acids that may be limiting in human diets. This might stimulate a harder look at how significant the problem of digestibility is and how well the predictions would apply with and without a correction for bioavailability. It looked to me as if the digestibility factor wasn't too critical in the products, or in the combinations of foodstuffs that you selected. Did I interpret that correctly?

DR. BODWELL: I think you interpreted it correctly. However, with some protein sources with low levels of digestibility, there could be problems. It could be specified that with marginal sources that may have low digestibility or with sources which are certain to have low digestibility levels, these sources must be assayed in some manner. Appropriate corrections could then be made as needed.

DR. HARPER: I suppose it would be out of place for me to ask how you would define an imbalance which you wanted to preclude from these products?

DR. BODWELL: I think the "Bible" on this is talking through the other microphone.

DR. HARPER: Seriously, how would you go about defining such a situation, an imbalance in an individual food? That's really what you're doing, isn't it?

DR. BENTON: The FDA faced us with the sulfur amino acid problem and, as I recall, they accepted the protein with the highest value as their limit for methionine and so they used a protein. One might use corn as a protein that has, at least as far as I can recall, the greatest degree of imbalance and say it shouldn't exceed that. I'm not sure whether that's a solution or not.

DR. HOPKINS: As a practical problem, are imbalances a problem when the protein requirement is met?

DR. BODWELL: I haven't been convinced that they are.

DR. SCRIMSHAW: As to comparisons with the rat, the problem does not lie in differences between the metabolism of the rat and that of the human, but in taking data for a young, rapidly growing animal with hair, studied in a state of deficiency and extrapolating them to slowly growing children or adults. Adult rats would be more appropriate for evaluation of protein utilization in humans, but then growth methods could not be used.

There is a possible drawback to our using the requirement level for the evaluation of protein quality in humans. It is appropriate for maintenance and preadolescent children because their average requirement for growth is no more than about 10% above maintenance. However, considering proteins for catch-up growth at rates of 4 to 8 times higher, or optimal recovery from deficiency states, protein values measured at suboptimal levels, either in the rat or in the human, may be useful. There have been successive minor adjustments in the FAO/WHO amino acid reference pattern in the 1957, 1965, and 1973 reports, and I believe that we've probably reached the point of diminishing returns in tinkering with the pattern. Any further improvement is unlikely to be of practical significance from the standpoint of human nutrition. The FAO/WHO scoring procedure is satisfactory as long as there is not a problem of digestibility, because a narrower range of proteins has been examined than is used in diets worldwide. This is not readily apparent from the data presented. The 1975 consultation (Food and Nutr. *1* (2):11, FAO, Rome, 1975) to review experiences with the 1973 report, concluded that when they were available, digestibility should be taken into account in utilizing the amino acid scores and that seems a matter of common sense.

Finally, do amino acid imbalances occur at requirement levels of amino acid intake? In the feeding of maize protein to children, the INCAP group has demonstrated that the addition of methionine at the level called for in the 1957 FAO/WHO reference pattern resulted in a decrease in nitrogen retention (J. Nutr. *66*: 501; 1958). This is a warning not to fortify with methionine unless there's a clear indication for it. Methionine improves the value of soy protein in rat assays. In humans, soy protein is used at requirements levels or above and usually in a mixed diet. Under these circumstances, there is no evidence that methionine confers any additional value.

DR. BODWELL: Dr. Bressani, what was the digestibility of your bean diet? I didn't get that from your paper.

DR. BRESSANI: A bean diet as those used in our studies, showed an apparent digestibility of 64%. This increases in corn and beans, or rice and bean diets to around 72–75%.

DR. BODWELL: On digestibility, certainly there are products which are a problem. With something like millet or sorghum, if products such as these enter the food supply, they certainly should be screened with a rat bioassay or something which is a direct or indirect estimate of "digestibility".

DR. SOLBERG: I like your approach in terms of the three amino acids for setting a label requirement but I fail to see the rationale of a value for your percent of total essential amino acids. In no way could that be helpful from what I can see.

DR. BODWELL: As I said, there should be a minimum defined for each of the essential amino acids. This is to get around problems such as that of protein powders based on hydrolyzed collagen. The idea of including total essential amino acids is to give an overview of what you're getting other than those amino acids which would be listed (lysine, methionine plus cystine, and tryptophan). It could be "hidden" but I don't know why we hide that sort of thing.

DR. SOLBERG: Do you feel that you could in any way, add up these totals to come to some decision as to whether you had enough essential amino acids to fulfill a requirement?

DR. BODWELL: Not exactly. The idea is that the consumer could look at the label, for instance, and see how much lysine from wheat they were getting and how much they were getting from cottage cheese. They could make a determination as to whether or not they were getting an adequate level of lysine. I am also trying to get around the "quantity" versus "quality" issue.

DR. EDWARDS: If I take off my hat as a scientist and put on my hat as a consumer, I was wondering how I would react in a store in which I picked up a container with a label giving the information that we're talking about. I also was mulling over how a person who perhaps has not finished high school would react to such a label. I think that the simpler the presentation on the label, the better for me as a housewife, and I would be inclined to look at only three amino acids. I wonder whether or not the relative value of including tryptophan would be as great as simply listing lysine and the methionine plus cystine.

I have trouble only with tryptophan. I'm sure that one could communicate, on the basis of volumes of data, the concern for lysine and the concern for the sulfur containing amino acids. I'm not sure that one could convey with ease the concern regarding tryptophan. I agree with Dr. Harper that now there seems to be a need to establish requirements for the amino acids, perhaps I should say Recommended Dietary Allowances. I believe probably that the work of Rose and Leverton could at this point help us go further. So in summary, as a consumer, I would initially be sort of depressed by seeing so much unfamiliar material and I would recommend that the inclusion of tryptophan be reconsidered.

As a consumer I would welcome such information, particularly on foods to be used by vegetarians. One of my horrors is that people, who don't know what they're doing, don't have a scientific base from which they can make selections. I am aware of an additional group who now move from health stores and the organic food stores into the supermarkets with their calculators right along with them. This approach offers, I think, a tremendous possibility for undercutting what is, as I see it, trends that will undermine the nutrition of the population. The approach would provide a stimulus for additional information. One of the beauties of labeling has been that the housewife now wants to know more about nutrition. We as nutrition educators have failed all these years, until we got nutritional labeling, to stimulate that kind of interest. As a consumer I vote yes with the possible suggestion that tryptophan be reconsidered. I'm speaking only as a consumer today. I'll mull over tonight how I would respond as a nutritionist.

DR. VANDERVEEN: I think that the comments made relative to putting amino acid information on the label are worth discussing further. There are two things we ought to think about: The first is that labeling is something that is not very quickly changed. I have a feeling that we should think ten years downstream. We have to start thinking about the population in terms of where they will be at ten years down the stream. To say that the public won't understand it at this point in time is short-sighted. What you have to think about is whether sixth graders, who are presently being taught nutrition about amino acids, are going to understand it ten years hence and will they teach their parents, et cetera. The second point is that half the value of nutritional labeling is not that the public understands it, but that the public benefits from it as a quality assurance factor. I must qualify my remarks to indicate that I am not speaking for the agency now because I find some people in the agency don't agree with me on this point. But I feel half the value of nutritional labeling is the fact that the manufacturer puts on the label, a guarantee to the public as to what exactly is in his product and he states that he is willing to stand behind it. Putting nutrition information on the label, even if the public doesn't understand it, benefits the consumer.

20

Use of Amino Acid Composition Data to Predict Protein Nutritive Value for Children With Specific Reference to New Estimates of Their Essential Amino Acid Requirements

B. Torun, O. Pineda, F.E. Viteri, and G. Arroyave

Studies conducted recently at the Institute of Nutrition of Central America and Panama (INCAP), have provided experimental information on the needs of 2-year old children for six essential amino acids. The details of these studies are described in Chapter 3 (Pineda *et al.* 1981) in which we also noted that the requirements that have been accepted since 1973 were derived from assumptions based on the scarce information which existed at that time on amino acid requirements for other age groups (FAO/WHO, 1973). Table 20.1 shows the essential amino acid intake levels recommended by the FAO/WHO Expert Committee on Energy and Protein Re-

TABLE 20.1. AMINO ACID INTAKES (MG/KG/DAY) TO SATISFY THE REQUIREMENTS OF 2-YEAR OLD CHILDREN

Amino Acid	Studies at INCAP[1]	Suggested by FAO/WHO[2]
Isoleucine	32	48
Leucine	n.i.[3]	84
Lysine	66	66
Methionine + Cystine	28	42
Phenylalanine + Tyrosine	n.i.	72
Threonine	>37, <53	48
Tryptophan	13	12
Valine	39	60

[1]Using semi-synthetic diets as described by Pineda et al. (1981).
[2]Based on the FAO/WHO (1973) amino acid scoring pattern (see Table 6) and a safe level of protein intake of 1.2 g/kg/day.
[3]Not investigated.

quirements (1973) for 2-year old children. They could be termed "safe levels of amino acid intakes" since they were calculated from the amino acid scoring pattern and the "safe level" of milk protein intake (1.2 g/kg/day) suggested by the FAO/WHO Committee. Table 20.1 also shows the amounts of 6 essential amino acids necessary to satisfy preschool children's needs, as established in our studies. These amounts may be higher than the average requirements since the criteria to select them were based on the lowest intakes that would not alter the free amino acid plasma concentration, the nitrogen balance and the urinary urea/creatinine ratio in *all* children under study (Pineda *et al.*, 1981). The results of other studies conducted to determine total protein requirements of preschool aged children (Torún *et al.*, 1981a, 1981b; Cabrera-Santiago and Torún, 1980) suggested that the values shown in Table 20.1 may, in fact, be "safe levels of intake", at least for lysine, methionine + cystine, threonine and tryptophan, and also probably for isoleucine and valine. These investigations are summarized in the following section.

TOTAL PROTEIN REQUIREMENTS

Nitrogen Balance Studies

The protein requirements of 10 children between 17 and 31 months of age (mean: 23 months) were investigated using either cow's milk or a soybean protein isolate (Purina Protein 220, Ralston-Purina Co.), as the only dietary protein source. Table 20.2 shows the essential amino acid composition of the two proteins based on 88-hour hydrolyses for isoleucine and valine and on 24-hour hydrolyses for the other amino acids. A multiple level N balance technique was used (Viteri and Bressani, 1972; Young *et al.*, 1977) based on

TABLE 20.2.   ESSENTIAL AMINO ACIDS (MG OF AMINO ACID PER GRAM OF PROTEIN) IN COW'S MILK AND SOYBEAN PROTEIN ISOLATE USED TO STUDY PROTEIN REQUIREMENTS[1]

Amino Acid		Milk		Soy[2]
Histidine		38.8		32.1
Isoleucine		58.1		51.7
Leucine		98.5		79.0
Lysine		83.9		62.4
Total sulfur a.a.		38.2		26.6
Methionine	27.4		13.3	
Cystine	10.8		13.3	
Total aromatic a.a.		91.8		89.4
Phenylalanine	52.1		53.6	
Tyrosine	39.7		35.8	
Threonine		44.6		36.8
Tryptophan		17.5		16.6
Valine		65.6		53.2

[1]From Torún et al. (1981a, 1981b) and Cabrera-Santiago and Torún (1980). Amino acid analyses performed at the Ralston-Purina Company's Research Laboratories, based on 24- and 88-hour hydrolyses.
[2]Purina-Protein 220, Ralston-Purina Co., St. Louis, Mo.

four levels of protein intake (0.5, 0.75, 1.0 and 1.25 g/kg/day). Some children followed a descending experimental design, beginning with the 1.25 level, and others followed an ascending design, beginning with 0.5 g/kg/day. Each protein level was fed for 9 days and excreta were collected for nitrogen analyses during the last 4 days. The protein intake required by each child for retention of 24 mg N/kg/day was calculated. This level of N retention provides allowances both for integumental N losses and for growth (FAO/WHO, 1973). The mean requirements thus calculated were 0.61 and 0.75 g protein/kg/day for milk and soy protein, respectively. The value for milk is about 40% lower than the mean N requirement estimated by the FAO/WHO Expert Committee. "True" protein digestibilities were 94 and 93%, respectively.

Factorial Calculations

In a separate study, the obligatory nitrogen losses through urine and feces were determined in five of the children who took part in the nitrogen balance studies. These were, on the average, 34 and 20 mg N/kg/day for urine and feces, respectively. The addition of nitrogen allowances for growth and integumental losses and the application of the correction factor of 1.3 suggested by FAO/WHO (1973) gave an estimate of 0.61 g/kg/day as the mean requirement for milk protein, coinciding with the results of the nitrogen balance studies.

Table 20.3 shows the coefficients of variation of the various indicators used in the studies described above. The large variability of the mean

TABLE 20.3.   COEFFICIENTS OF VARIATION IN STUDIES TO DETERMINE PROTEIN REQUIREMENTS OF COW'S MILK AND A SOYBEAN ISOLATE FOR CHILDREN OF PRE-SCHOOL AGE[1]

Test Protein	Method	Parameter	Coefficient of Variation
	Factorial	Obligatory N losses	15.1
Cow's milk	N balance	Slope[2]	16.4
Cow's milk	N balance	Mean Requirement[3]	27.8 (16.3)[4]
Soy Isolate	N balance	Slope[2]	14.8
Soy Isolate	N balance	Mean Requirement[3]	14.2

[1]Adapted from Torún et al. (1981a, 1981b) and from Cabrera-Santiago and Torún (1980).
[2]Regression coefficient of N balance (Y) on N intake (X).
[3]N intake required to retain 24 mg N/kg/day, therefore allowing for growth needs and miscellaneous insensible N losses.
[4]Variability is reduced (value in parenthesis) if 2 children with high requirements are excluded.

requirements for milk protein were mainly due to the 2 children with the highest requirement estimates, which were even greater than their requirements for soy protein. If these two children were excluded from the calculations, the coefficient of variation would be 16.3%. Therefore, it was decided to use a correction of 30% to allow for inter-individual variation, as suggested in the FAO/WHO report of 1973 and reiterated more recently (FAO/WHO, 1979). The safe levels of protein intake calculated in this manner correspond to 0.79 g and 0.98 g/kg/day for milk and soybean isolate, respectively.

PROTEIN DIGESTIBILITY

Not all proteins are digested, absorbed and utilized to the same extent. The amounts of amino acids provided by a diet are not reliable indicators of amino acid adequacy unless the protein is absorbed and utilizable at a level equal to about 100%. This is seldom the case when a mixed diet is eaten and this is specially true when vegetable foods are the main or sole protein sources. For example, only 77–83% of the protein nitrogen in mixed vegetable diets studied by Viteri *et al.* (1971) was absorbed by children. In recognition of this fact a recommendation was made by a joint FAO/WHO informal gathering of experts (1975) to adjust for "true" digestibility in calculating amino acid scores to derive a safe level of dietary protein. This "true" digestibility is calculated subtracting the obligatory (or endogenous) fecal nitrogen from total fecal nitrogen:

$$\text{"true" protein digestibility} = \frac{\text{N ingested} - (\text{total fecal N} - \text{obligatory fecal N})}{\text{N ingested}} \times 100$$

The correction for digestibility indicated by the FAO/WHO experts (1975) was in the context of deriving safe levels of dietary protein intake and since the safe level was based upon the ingestion of milk or egg protein, they recommended that the correction for digestibility should be made relative to that of egg and milk. However, corrections for digestibility must also be made to compare the amino acids biologically contributed (i.e., those that are absorbed by the human gut) by the ingestion of different specific proteins. In other words, if a comparison is made with a pattern of amino acid requirements (which does not necessarily correspond to the amino acid composition of milk and which has already been corrected for absorption), the capacity of the protein under evaluation to satisfy amino acid requirements should include a correction for its "true" digestibility and not its digestibility relative to that of milk. For example, if the digestibility of milk is 94% and that of a corn + beans diet is 77%, the amino acids provided by the proteins ingested should be multiplied by 0.94 for milk and by 0.77 for corn + beans, rather than by 0.82 for the latter (i.e., digestibility of corn + beans relative to milk).

It should be recognized that these corrections for digestibility imply two assumptions which may not always be true: a) that all dietary amino acids are absorbed in the same proportions, and b) that all subjects have the same endogenous nitrogen losses since an average figure of fecal obligatory losses is used to calculate "true" digestibility. Such assumptions, however, allow a more accurate estimation of the amino acids contributed by food proteins than if no correction for digestibility were made.

"True" digestibilities were calculated using 20 mg/kg/day as the obligatory losses, based on investigations of protein requirements by the factorial method (Torún *et al.*, 1981a). Table 20.4 shows the amounts of essential amino acids provided by the safe levels of protein intake for milk and soybean isolate suggested by our nitrogen balance studies and the amounts absorbed based on "true" digestibilities of 94 and 93% respectively, for the two protein sources. It also shows the values derived from our studies on specific amino acid requirements corrected for the digestibility (97%) of the semi-synthetic diets used.

AMINO ACID REQUIREMENTS

Table 20.4 indicates that the amounts of lysine and sulfur amino acids absorbed from the safe level of intake of the two protein sources were similar (milk) or 11% lower (soy isolate) than the amounts determined in the studies with semi-synthetic diets. As discussed by Pineda *et al.* (1981), the investigations on specific amino acids resulted in a range of 36 to 52 mg of threonine absorbed/kg/day. The values derived from the protein requirement studies (33 mg/kg/day) suggest that the safe level for this amino acid is closer to 36 than 52 and a level of 37 mg/kg/day agrees with the results of our various investigations.

The data in Table 20.4 also indicate that the amounts of leucine and the aromatic amino acids absorbed from the safe intake levels of milk or soy protein were somewhat lower than the "safe levels of amino acid intakes" suggested by FAO/WHO (1973).

Based on all of these considerations, we propose in Table 20.5 a provisional pattern of essential amino acid intakes for 2-year old children. These intakes provide a safety margin above the mean requirements and could be termed "safe levels of amino acid intakes", following the FAO/WHO recommendation for protein intakes (FAO/WHO, 1973). These conclusions may be further refined by studies on the specific requirements for leucine and the aromatic amino acids and by additional experiments with threonine.

Additional analyses of the data in Tables 20.1 and 20.5 indicate the following:

a) The amino acids supplied by both the milk and soybean proteins support the conclusions from our studies with semi-synthetic diets of lower requirements than those based on the 1973 FAO/WHO report for the sulfur amino acids, valine and possibly threonine for children of this age group.

b) The requirement for isoleucine also seems to be lower than suggested by the FAO/WHO report.

c) Threonine appears as the first limiting amino acid in the cow's milk *used in this study* with either tryptophan, lysine or the sulfur amino acids in a close second as the next limiting amino acid.

d) Threonine and lysine may be as important as methionine in soybean protein.

TABLE 20.4. ESSENTIAL AMINO ACIDS INGESTED AND ABSORBED (MG/KG/DAY) WITH THE SUGGESTED SAFE LEVELS OF MILK AND SOYBEAN ISOLATE PROTEIN INTAKES[1]

Amino Acid	mg Amino Acid provided by				Needs suggested by specific amino acid studies mg/kg/day[2]
	0.79 g milk protein/kg/day		0.98 g soybean protein/kg/day		
	Intake	Absorbed[2]	Intake	Absorbed[2]	
Isoleucine	46	43	51	47	31
Leucine	78	73	77	72	—
Lysine	66	62	61	57	64
Methionine + Cystine	30	28	26	24	27
Phenylalanine + Tyrosine	73	69	88	82	—
Threonine	35	33	36	33	36–52
Tryptophan	13.8	13.0	16.3	15.1	12.5
Valine	52	49	52	48	38

[1]From Torún et al., (1981a, 1981b) and Cabrera-Santiago and Torún (1980).
[2]Corrected for "true" digestibilities of 94, 93 and 97% for milk protein, soybean protein and amino acid mixtures, respectively.

TABLE 20.5.   AMINO ACID INTAKES RECOMMENDED FOR 2-YEAR OLD CHILDREN

Amino Acid	mg/kg/day[1]	Criteria to suggest amino acid intake
Isoleucine	31	Specific amino acid study[2]
Leucine	73	Safe level of cow's milk protein intake[2]
Lysine	64	Specific amino acid study
Methionine + Cystine	27	Specific amino acid study
Phenylalanine + Tyrosine	69	Safe level of cow's milk protein
Threonine	37	Within range of specific amino acid study (36−52) and similar to safe level of intake of cow's milk (33)
Tryptophan	12.5	Specific amino acid study
Valine	38	Specific amino acid study

[1]Figures assume 100% absorption and must be corrected for the dietary protein's "true" digestibility.
[2] Corrected for "true" digestibility of the experimental amino acid and milk diets (97% and 94%, respectively).

AMINO ACID SCORING PATTERN

Based on considerations of the requirements for essential amino acids and for total protein nitrogen, amino acid scoring patterns have been devised for proteins in foods. The Joint FAO/WHO Expert Committee on Energy and Protein Requirements (1973) proposed a reference scoring pattern based on estimates of amino acid requirements for school-children (10−12 years old) and non-pregnant adults and on the amounts of amino acids contributed by milk and various cereal proteins when fed at levels adequate for the normal growth of young children. The pattern was then "adjusted in the light of experience gained in the application of the 1957 FAO reference pattern of amino acids", so that amino acid requirements would be met when protein was ingested at the safe level of intake. A standard reference pattern could then be used for application to persons of either sex and all age groups, based on their total protein requirements. Although the Committee concluded that the proportions of many essential amino acids required by adults fell gradually after early childhood, they decided that the standard pattern of reference should satisfy the needs of children in the pre-school age group, even though its application might overestimate adult protein requirements by underestimating protein quality for adults. The reference scoring pattern proposed as "provisional" in 1973, but still used without modification, is shown in Table 20.6. It is quite similar to that proposed for infants based on a safe level of intake of 2 g protein/kg/day and a composite of the amino acid requirements estimated by Holt and Snyderman (1967) and by Fomon and Filer (1967) (although the Committee agreed that "for infants breast milk was the appropriate food and the amino acid requirements of infants should be excluded from the application of any tentative guide to protein scoring that might be developed for older children and adults"; FAO/WHO, 1973).

TABLE 20.6.   AMINO ACID SCORING PATTERNS (MG/G OF PROTEIN) SUGGESTED BY FAO/WHO (1973)

Amino Acid	"Provisional" Pattern	Pattern for Infants
Histidine	—	14
Isoleucine	40	35
Leucine	70	80
Lysine	55	52
Methionine + Cystine	35	29
Phenylalanine + Tyrosine	60	63
Threonine	40	44
Tryptophan	10	8.5
Valine	50	47

In accordance with the FAO/WHO recommendations, a protein with amino acid concentrations similar to or higher than those in the provisional pattern would have a nutritive value of 100%, implying that it would satisfy the amino acid and protein needs of a 2-year old child when consumed at a level of 1.2 g/kg/day, after making corrections for its digestibility relative to milk or egg digestibility (FAO/WHO, 1975; 1979). A protein with a score below 100 would, therefore, be required in larger amounts to satisfy the child's needs. The soybean protein isolate in the studies mentioned earlier can be used as an example of the application of the amino acid pattern to estimate the safe level of intake of a protein. The protein's amino acid score is based on the most limiting amino acid relative to the FAO/WHO reference pattern, which according to the data shown in Tables 20.2 and 20.6 would be methionine + cystine. The score would then be 26.6/35 = 76%. Since the "true" digestibility of this soybean isolate was 99% in relation to that of milk (Cabrera-Santiago and Torún, 1980), the safe level of intake for a 2-year old child would be estimated as:

$$1.2 \times \frac{100}{76} \times \frac{100}{99} = 1.59 \text{ g protein/kg/day (FAO/WHO, 1979)}.$$

Since all essential amino acids in milk (Table 20.2) are in higher concentrations than those of the FAO/WHO reference amino acid pattern, it has a score of 100 and since milk is the reference for protein digestibility the safe level of intake for this protein would be:

$$1.2 \times \frac{100}{100} \times \frac{100}{100} = 1.2 \text{ g kg/day}$$

Our studies, however, indicated that the needs of a 2-year old child for both total protein and essential amino acids can be satisfied with lower dietary intakes. The safe levels of protein intake for the milk and the soybean protein isolate *used in this study* were set at 0.79 and 0.98 g/kg/day, respectively. The safe level of intake of a batch of cow's milk with the average amino acid composition determined by FAO (1970) from the analysis of various lots of milk might be higher, as shown below in terms of its relative nutritive value, but still less than the 1.2 g protein/kg/day recommended by FAO/WHO.

These discrepancies are partly due to the results of our investigations on total protein requirements and partly due to the differences between our measurements of amino acid requirements and those estimated by FAO/WHO for 2-year old children. The amino acid intakes established at INCAP suggest a different amino acid scoring pattern which may provide a better assessment of protein nutritive value since it is based on experimental evidence. The use of the scoring pattern, however, should be mainly to indicate whether a protein has a better quality than another in terms of its amino acid composition and not necessarily to indicate safe levels of intake. The latter may be true if the level of intake necessary to satisfy all essential amino acid requirements will also satisfy total nitrogen (or protein) needs. This concept is clear in the definition proposed by Arroyave (1975a) for *nutritive value of a protein* as "the extent to which, when ingested in enough quantity to satisfy nitrogen requirements, it will also meet a person's requirements for each essential amino acid". The FAO/WHO Expert Committees recognized this concept in their approach to estimating the dietary needs for the protein being assessed (a protein's relative amino acid score is multiplied by the safe level of intake of the protein used as a reference; see the formula used above).

The amino acid scoring patterns suggested by the FAO/WHO Expert Committees have been based on specific protein intakes. For example, those suggested in 1973 were based on milk or egg protein intakes of 2.0, 1.2, 0.8 and 0.55 g/kg/day, respectively for infants, pre-school children, school-children and non-pregnant adults. We suggest the use of an amino acid scoring pattern based exclusively on the amounts of essential amino acids that must be absorbed from the diet by a 2-year old child to satisfy his requirements for each amino acid. The pattern corresponds to the intake shown in Table 20.5 but the amino acids are expressed as mg of amino acid per g of protein. We propose that this pattern be used to calculate a protein's *amino acid quality index* (AAQI), which can be used to compare its nutritive quality with that of another protein (i.e., its *relative nutritive quality* or *RNQ*). If the safe level of intake of one protein is known, the safe level of intake of the other protein may be estimated from the RNQ values of the two proteins.

In the following description of the sequence to calculate the AAQI and RNQ, reference is made to the examples shown in Table 20.7:

a) A protein's essential amino acid concentrations (Table 20.7, Column B) are divided by the amino acids in the scoring pattern (Column A).

b) The lowest ratio thus obtained (Column C) indicates the protein's first (or most) limiting amino acid. As shown in Table 20.7, this corresponds to threonine in the case of milk and to lysine in the case of the soybean isolate.

c) The ratio of the first limiting amino acid is multiplied by the "true" digestibility of the protein to obtain the AAQI.

TABLE 20.7. CALCULATIONS OF AMINO ACID QUALITY INDICES (AAQI) AND RE-
LATIVE NUTRITIVE QUALITY (RNQ) OF TWO PROTEINS BASED ON THE ESSENTIAL
AMINO ACID REQUIREMENTS OF 2-YEAR OLD CHILDREN

Amino Acid	Scoring Pattern[1]	Milk from Table 20.2		Soy isolate from Table 20.2	
		AA comp.[1]	Ratio B/A	AA comp.[1]	Ratio B/A
	A	B	C	B	C
Isoleucine	31	58	1.87	52	1.68
Leucine	73	98	1.34	79	1.08
Lysine	64	84	1.31	62	0.97
Sulfur aa	27	38	1.41	27	1.00
Aromatic aa	69	92	1.33	89	1.29
Threonine	37	44	1.19	37	1.00
Tryptophan	12.5	17.5	1.40	16.6	1.33
Valine	38	66	1.74	53	1.39
True digestibility		0.94[2]		0.93[3]	
AAQI		$1.19 \times 0.94 = 1.12$		$0.97 \times 0.93 = 0.90$	

RNQ_{milk} for soy isolate = $(0.90/1.12) \times 100 = 80\%$
Safe level of intake of soy isolate protein based on 0.79 g/kg/day for milk[2] = 0.79/(80/100)
= 0.99 g protein/kg/day

[1]Scoring pattern and amino acid compositions (AA comp.) expressed as mg amino acid
per g protein.
[2]From Torún et al. (1981a).
[3]From Cabrera-Santiago and Torún (1980).

d) The RNQ of a protein ("test" protein) compared to another ("reference"
protein) is calculated as:

$$RNQ = (AAQI \text{ of test protein}/AAQI \text{ of reference protein}) \times 100$$

We suggest writing the name of the reference protein as a subscript of
RNQ. Table 20.7 shows that the RNQ_{milk} of the soybean protein isolate is
80%.
  e) The safe level of intake of the "test" protein can be estimated from the
safe intake level of the "reference"/ (RNQ/100). The safe intake level of the
milk used in the example was determined as 0.79 g/kg/day. The safe level of
intake of the soybean isolate can then be estimated as 0.79/(80/100) = 0.99 g
protein/kg/day.
  The calculations shown in Table 20.7 coincide very well with the soybean
protein isolate's nutritive quality relative to milk and its safe level of intake
determined by Cabrera-Santiago and Torún (1980) using nitrogen balance
techniques. These were 82% and 0.98 g protein/kg/day, respectively. Table
20.8 shows other examples of the AAQI of several protein sources. Table
20.9 summarizes their RNQ, which may vary depending on the protein used
as reference, and compares them with estimates derived from nitrogen
balance studies performed with children of the same age group. These
estimates were calculated as the average of the relative nitrogen requirement of each protein (i.e., mean nitrogen intake to retain 24 mg N/kg/day to

allow for growth needs and insensible N losses) and relative protein value (i.e., regression coefficient of N balance on N intake), using cow's milk as the reference protein (Viteri and Bressani, 1972; Young et al., 1977). With the exception of the values for the bean and corn mixture, there is good agreement between the protein quality assessment based on N balance and the RNQ values. Several investigations on the growth and N balance of children fed the bean and corn mixture suggest that its protein quality and nutritive value are closer to 67 than 42% (compared with milk; Viteri et al., 1980). The low RNQ based on amino acid composition and protein digestibility could be due to an overestimation of tryptophan requirements, since a small decrease in such requirements would increase significantly the AAQI of the mixture (Table 20.8), or it could be due to an underestimation of the proportion in which the most limiting amino acids in the mixture were absorbed. It was mentioned before that the correction for digestibility assumes that all amino acids in a protein are absorbed (and utilized) in the same proportional amounts. However, if this were not true and most of the tryptophan and lysine of the corn + bean mixtures were absorbed (and utilized), the AAQI would increase. This non-proportional absorption (or utilization) of the constituent amino acids of a protein is purely speculative, but if the phenomenon does exist it would be more important in proteins with low digestibilities.

ACCURACY OF PREDICTION OF PROTEIN NUTRITIVE VALUE

These results suggest that protein nutritive value for pre-school children (or other age groups) can be predicted from their total nitrogen (or protein) requirements, their essential amino acid requirements, the amino acid composition of the dietary protein in question and its digestibility. The determination of all these factors influences the accuracy of the prediction. Inter-individual variability, methodological errors and precision of analysis, test conditions and physiological situation of the human subjects all play a role.

A working group of the International Union of Nutritional Sciences and the United Nations University recently reviewed assays and procedures used in studying protein quality including various factors related to amino acid data, such as the biological availability of the amino acids which are chemically determined in a dietary protein and the importance of the proportions in which amino acids are present in a protein (Pellett and Young, 1981). The accuracy with which such factors can be measured or established and their variability are not known. Furthermore, various features of body protein metabolism, including the diversity in the physiological responses or "adaptation" to changes in the dietary levels of different essential amino acids fed to humans (Young et al., 1972; Özalp et al., 1973) or rats (Yamashita and Ashida, 1969; Said et al., 1974; Chu and Hegsted, 1976), suggest that, in addition, the accuracy obtained in determining requirements may vary from one amino acid to another. Therefore,

TABLE 20.8.   AMINO ACID QUALITY INDEX (AAQI) OF SEVERAL PROTEINS BASED ON THE ESSENTIAL AMINO ACID REQUIREMENTS OF 2-YEAR OLD CHILDREN

Amino Acid	Scoring Pattern[5]	Average Cow's Milk[1]		Supro 710[2]		Vegetable Mixture[3]		Corn: Bean 76:24[4]	
		AA comp.[5]	Ratio B/A	AA comp.[5]	Ratio B/A	AA comp.[5]	Ratio B/A	AA comp.[5]	Ratio B/A
	A	B	C	B	C	B	C	B	C
Isoleucine	31	47	1.52	49	1.58	41	1.32	38	1.23
Leucine	73	95	1.30	82	1.12	76	1.04	114	1.56
Lysine	64	78	1.22	65	1.02[6]	57	0.89[6]	38	0.59
Sulfur aa	27	33	1.22	28	1.04	33	1.22	31	1.15
Aromatic aa	69	102	1.48	94	1.36	75	1.09	85	1.23
Threonine	37	44	1.19	38	1.03	33	0.89	37	1.00
Tryptophan	12.5	14	1.12[6]	14	1.12[6]	12	0.96	7	0.56[6]
Valine	38	64	1.68	49	1.29	41	1.08	48	1.26
True digestibility		0.94		0.94		0.77		0.78	
AAQI		$1.12 \times 0.94 = 1.05$		$1.02 \times 0.94 = 0.96$		$0.89 \times 0.77 = 0.69$		$0.56 \times 0.78 = 0.44$	

[1] Composition reported by FAO (1970); value for tryptophan by microbiological assay.
[2] Soybean protein isolate, Ralston-Purina Co. (Torun, 1979).
[3] Corn 58%, cottonseed flour 38%, torula yeast 3%, $CaCO_3$ 1% (Bressani et al., 1961; Scrimshaw et al., 1961).
[4] From Arroyave (1975b).
[5] Scoring pattern and amino acid compositions (AA comp.) expressed as mg amino acid per g protein.
[6] Ratio corresponds to the most limiting amino acid.

TABLE 20.9.   ESTIMATES OF PROTEIN QUALITY BASED ON AMINO ACID SCORING
PATTERN (FROM TABLES 20.7 AND 20.8) AND ON NITROGEN BALANCE TECHNIQUES

	Milk A[1]	Milk B	Soy A	Soy B	Vegetable Mix	Corn: Bean
AAQI	1.12	1.05	0.90	0.96	0.69	0.44
$RNQ_{milk\ A}$, %	—	94	80	86	62	39
$RNQ_{milk\ B}$, %	107	—	86	91	66	42
Based on N balance, %	—	—	82	90	63	67

[1]Milk A: composition shown in Tables 20.2 and 20.7; Milk B: composition shown in Table
20.8; Soy A: soybean protein isolate; composition shown in Tables 20.2 and 20.7; Soy B:
soybean protein isolate (SUPRO 710, Ralston-Purina Co.), calculated from Torún (1979);
Vegetable Mix: INCAP formula 9 (corn, cottonseed flour and torula yeast), calculated
from Viteri and Bressani (1972); Corn + Bean mixture (76:24 in terms of protein),
calculated from Arroyave (1975b).

it is difficult to define the degree of accuracy or the precision with which
protein nutritive value for children (or any other age or sex group) can be
predicted from amino acid composition data.

The coincidence of the amino acid requirements assessed from our studies
with semi-synthetic diets and with milk or soybean protein isolate, and the
variability for the latter shown in Table 20.3 suggest that 15% may also be a
reasonable approximation of the coefficient of variation in the assessment
of amino acid requirements. However, the degree of accuracy to predict
protein nutritive value from amino acid data may vary more, especially
when proteins with low digestibility are involved as previously discussed
for the corn + bean mixture. This indicates the need for direct studies in
humans to support the conclusions from indirect predictions of some proteins' nutritive qualities.

The use of a standard amino acid pattern to assess protein quality for
humans of different age groups is attractive, mainly to simplify the estimations, but it can lead to erroneous predictions of protein nutritive value.
There are disproportionate age-related differences in the needs for both
total protein and essential amino acids which explain why a protein that
may be nutritionally inadequate for individuals of one age group may not be
so for others. This has been shown, for example, to be the case with corn
protein which has a low nutritive value for infants but is adequate for adults
(Arroyave, 1975b). If more sound scientific information on the amino acid
requirements of various age groups (e.g., young school-children, pubertal
and older adolescents) becomes available it will be better to use age-specific
amino acid scoring patterns to assess protein quality, unless it is demonstrated that essential amino acid requirements per unit of body weight is
constant at all ages beyond infancy.

In the meantime, the recommendation of the FAO/WHO Expert Committee (1973) to use a scoring system that will favor small children is adequate
in theory since it is based on the nutritional safety of children. However,
this is not a reasonable approach from the point of view of food policy
planning since the use of the FAO/WHO standard scoring pattern underestimates protein quality for adults by overestimating their needs for es-

sential amino acids. Therefore, its application for the population at large may lead to estimated needs of protein production or food supplementation programs that may be grossly exaggerated and expensive. In this regard, we believe that protein nutritive value has little importance for adults, except in the elderly and pregnant or nursing women, or when the major protein source is grossly deficient in one or more essential amino acids (e.g., cassava or wheat gluten). Therefore, in most instances the major emphasis in food policy planning under the context of protein needs *for adults* should be on the fulfillment of their total dietary nitrogen requirements.

SUMMARY

Studies on the obligatory nitrogen losses and on the total protein requirements of children 17–31 months-old confirmed the results of investigations on amino acid requirements of preschool children using semi-synthetic diets (Pineda *et al.*, Chapter 3). Daily amino acid requirements were determined with a safety margin, similar to the "safe levels" of protein intake. Such levels were achieved by the absorption of 31 mg isoleucine, 73 mg leucine, 64 mg lysine, 27 mg methionine + cystine, 69 mg phenylalanine + tyrosine, 37 mg threonine, 12.5 mg tryptophan and 38 mg valine/kg body weight, although the values given for leucine and the aromatic amino acids may be an overestimation. An amino acid scoring pattern to evaluate protein quality is proposed based on these values and taking into account the protein's "true" digestibility. This will lead to the calculation of an amino acid quality index (AAQI) and to the assessment of a protein's nutritive quality relative to another protein (relative nutritive quality, RNQ). A protein's nutritive *value*, however, must also include its ability to satisfy total nitrogen needs. It can be calculated from its RNQ and the safe level of intake of the reference protein. Proteins with high digestibilities show good agreement between RNQ predicted from amino acid composition data and protein quality assessed by multi-level nitrogen balance studies. Some proteins with low digestibilities may yield erroneous predictions from the amino acid composition values and their RNQ must be assessed by other means, probably by direct studies in humans. It may be better to use age-specific amino acid scoring patterns to assess AAQI and RNQ, unless it is shown that amino acid requirements per unit of body weight are constant at all ages beyond infancy.

ACKNOWLEDGEMENTS

We want to express our appreciation to Dr. José Héctor Aguilar, Dr. Patricio Aycinena, Dr. Salvador García and Ms. María Isabel Cabrera-Santiago, who assisted us in various portions of the studies discussed in this presentation. These studies were partly conducted through the financial support of the United States' National Institutes of Health (NIH Grant R22 AM 17086) and the Ralston-Purina Company's contribution of some of the materials used.

REFERENCES

ARROYAVE, G.   1975a.   Nutritive value of dietary proteins: For whom? Proc. 9 Int. Cong. Nutr., Mexico 1972, Vol. 1, p. 43, Karger, Basel, Switzerland.

ARROYAVE, G.   1975b.   Amino Acid requirements and age.   In: Olson, R., ed., "Protein-Calorie malnutrition", Academic Press, New York, pp. 1 .

BRESSANI, R., ELÍAS, L.G., AGUIRRE, A., and SCRIMSHAW, N.S.   1961. All-vegetable protein mixtures for human feeding. III Development of INCAP Vegetable Mixture 9.   J. Nutr. 74: 201.

CABRERA-SANTIAGO, M.I., and TORÚN, B.   1980.   Informe Anual 1979. Inst. Nutr. Central Amer. and Panamá (INCAP), Guatemala, Guatemala.

CHU, S.H.W., and HEGSTED, D.M.   1976.   Dietary regulation of lysine-ketoglutarate reductase activity in the rat.   Fed. Proc. 35: 257.

FAO.   1970.   Amino acid content of foods and biological data on proteins. FAO Nutritional Studies No. 24. Rome, Italy.

FAO/WHO.   1973.   Energy and protein requirements: Report of a Joint FAO/WHO ad hoc Expert Committee.   WHO Techn. Rep. Series No. 522. Geneva, Switzerland.

FAO/WHO.   1975.   Energy and protein requirements: Recommendations by a Joint FAO/WHO informal gathering of experts.   Food and Nutrition 1 (2):11.

FAO/WHO.   1979.   Protein and energy requirements: a Joint FAO/WHO Memorandum.   Bull. Wld. Hlth. Org., 57:65.

FOMON, S.J., and FILER, L.J.   1967.   Amino acid requirements of normal growth.   In: Nyan, W.L. ed.   "Amino Acid Metabolism and Genetic Variation", McGraw Hill, New York, p. 391.

HOLT, L.E., and SNYDERMAN, S.E.   1967.   The amino acid requirements of children.   In: Nyan, W.L., ed., "Amino Acid Metabolism and Genetic Variation", McGraw Hill, New York, p. 381.

ÖZALP, J., YOUNG, V.R., NAGCHAUDHURI, J., TONTISIRIN, K., and SCRIMSHAW, N.S.   1973.   Plasma amino acid response in young men given diets devoid of single essential amino acids.   J. Nutr., 102: 1147.

PELLETT, P.L., and YOUNG, V.R. (Editors).   1981.   Nutritional Evaluation of Protein Foods. Report of a working group representing the International Union of Nutritional Sciences and the United Nations University. United Nations University Press, Tokyo, Japan.

PINEDA, O., TORÚN, B., VITERI, F.E., and ARROYAVE, G.   1981.   Protein quality in relation to estimates of essential amino acid requirements.   In: "Protein Quality in Humans: Assessment and In Vitro Estimation", C.E. Bodwell, J.S. Adkins, and D.T. Hopkins, Editors, AVI Publishing Co., Westport, Conn.

SAID, A.K., HEGSTED, D.M., and HAYES, K.C.   1974.   Response of adult rats to deficiencies of different essential amino acids.   Brit. J. Nutr. 31: 47.

SCRIMSHAW, N.S., BÉHAR, M., WILSON, D., VITERI, F., ARROYAVE, G., and BRESSANI, R. 1961. All-vegetable protein mixtures for human feeding. V. Clinical trials with INCAP mixtures 8 and 9 and with corn and beans. Am. J. Clin. Nutr. 9: 196.

TORÚN, B. 1979. Nutritional quality of soybean protein isolates: Studies in children of preschool age. In: Wilcke, H.L., Hopkins, D.T., and Waggle, D.H., eds. "Soy Protein and Human Nutrition", p. 101, Academic Press, New York.

TORÚN, B., CABRERA-SANTIAGO, M.I., and VITERI, F.E. 1981a. Protein requirements of preschool children: obligatory N losses and N balance measurements using cow's milk. Arch. Latinoam. Nutr. (in press).

TORÚN, B., CABRERA-SANTIAGO, M.I., and VITERI, F.E. 1981b. Protein requirements of preschool children: milk and soybean protein isolate. Food Nutr. Bull. (in press).

VITERI, F.E., ARROYAVE, G., and TORÚN, B. 1980. Unpublished data. Inst. Nutr. Central Amer. and Panamá (INCAP), Guatemala, Guatemala.

VITERI, F.E., BRESSANI, R. and ARROYAVE, G. 1971. Fecal and urinary nitrogen losses. Presented at the Joint FAO/WHO Expert Committtee Meeting on Energy and Protein Requirements, March 22–April 2, Rome, Italy.

VITERI, F.E., and BRESSANI, R. 1972. The quality of new sources of protein and their suitability for weanlings and young children. Bull. Wld. Hlth. Org. 46: 827.

YAMASHITA, K., and ASHIDA, K. 1969. Lysine metabolism in rats fed a lysine-free diet. J. Nutr. 99: 267.

YOUNG, V.R., SCRIMSHAW, N.S. and RAND, W. 1977. Measuring protein quality in humans: A review and proposed method. Cereal Chem. 54: 929.

YOUNG, V.R., TONTISIRIN, K., ÖZALP, J., LAKSHMANAN, F., and SCRIMSHAW, N.S. 1972. Plasma amino acid response curve and amino acid requirements: valine and lysine. J. Nutr. 102: 1159.

DISCUSSION

DR. YOUNG: This is not a question but initially a comment. I'm really intrigued how so many of your observations, generated from a study in two-year-old children, compare so closely with data that have been generated in adults and about your closing statement, that protein quality, if it is of significance, has equal significance in relation to adult nutrition as it does in relation to the nutritional concerns for the younger age groups. I'm beginning to think when we compare our data, those of Dr. Margen's together with yours, and of yours and Dr. Bressani's also generated, in part, in young adults, that the quality issue is of significance throughout the growth and development and maintenance phase of life.

Now my question relates to a matter of clarification basically. You showed us on one of your slides, the essential amino acid intakes that would be provided by your determined safe level of milk and of soy isolate. You compared these essential amino acid intakes with what I think was your estimates of the *mean requirements* for the essential amino acids based on your own studies. If this is the case, I can't understand how the safe level of milk and soy isolate was determined to be safe since the level of intake, for example, of methionine divided by the safe level of milk intake approximated your *mean* requirement for methionine.

DR. TORUN: Thank you, Dr. Young. I was expecting such a question. I inadvertently skipped the repetition of one of the slides when I was going to refer to that point. We decided to call "requirement" that level of amino acid intake where there were still no problems in terms of nitrogen balance, plasma amino acid concentration or urinary urea/creatinine excretion. For example, in the case of lysine, we began to see an increase in the urinary urea/creatinine ratio with an intake of 53 milligrams of lysine per kilogram. We then decided to suggest the value of 66 mg/kg as adequate. I referred to that level as "requirement", but you are right in the sense that this is higher than the mean requirement and could be better called "a recommended level of intake."

DR. SCRIMSHAW: If your scores for milk are 100 and for two samples of soy isolate 94 and 97, that would imply that at the requirement level, milk will be providing more of the essential amino acids than are needed. In this age group, INCAP studies show that you can dilute milk with glycine-diammonium citrate by about 10%, which matches pretty well. In metabolic studies at the requirement level, it is impossible to distinguish between the protein value of milk and soy. Only when you went to lower levels would you begin to distinguish between the two. Do you agree with that interpretation?

The second question deals with this 1.3 figure that we've been using, or the 30% difference between NPU and efficiency of utilization at requirement level of intake. You implied that you took this from the 1973 report, and my question is: Did you have experimental verification in your own data for that 1.3? As you know from Inoue's data, it would be 1.4, and I've heard Doris Calloway say a number of times that she felt 1.3 was too low a figure. When they were obtained, her data were not given as much weight as some of the older published studies in which there may have been methodological difficulties.

DR. TORUN: In terms of the 30% correction, we do not have any experimental evidence to back us up. We decided to use that figure based precisely on the analysis done by Doris Calloway, George Beaton and John Waterlow (Bull. Wld. Hlth. Org. 57: 65, 1979). They included in their report the studies and analysis done at Berkeley and the ones done at MIT, as well as others. Their conclusion was that the discrepancy between some of the data was such that up to that moment, there was no sound reason to alter the value of 1.3. We don't have anything new to add to that and we decided to go along these lines.

In terms of the sensitivity or the capacity to differentiate between milk and soybean isolate at the level of requirement, I think that we can say that there is a difference. The requirement level was about 20% higher for soy than milk. If we accept a coefficient of variation of 15% for the two diets, we will find a difference between the two proteins at the requirement level.

Finally, the reason why we included the final statement in our presentation was that if adults have similar essential amino acid requirements as younger individuals and they are consuming mixed diets, by fulfilling their total protein requirements they will satisfy their amino acid requirements. The only exceptions would be populations who eat staples of very poor protein quality, such as cassava.

DR. MARGEN: First, Dr. Torun, I think everyone here has failed to congratulate you and your group on this work. Dr. Young certainly suggested it, but I think it is important to point out that this is not the kind of work that takes place in one day. As far as I'm aware, this has taken you at least four or almost five years.

I would like to comment on the questions that were brought up by Dr. Young and particularly by Dr. Scrimshaw. As I understood you, you stated that when you carried out your balance study with milk protein, you came up with a value of 0.61 grams as the requirement. And then you multiplied by the factor 1.3. Now I did a very rough calculation here, and it seems to me that the total amount of amino acids, essential amino acids, for your two-year-old children, utilizing your pattern, comes out to about 360 milligrams. So that means that you are suggesting that well over 50%, pretty close to 60%, of the amino acids must be essential amino acids to meet requirements for a two-year-old child.

Now the question becomes, what happens with growth? You have two possibilities. One is that the ratio of amino acids may remain the same but the quantity needed might change, and the other is that the ratio or actual pattern might change. This really has to be worked out very carefully. We should, at this point, say we do not know what is going to happen with growth and then use this question as a starting point for further work.

Now the other thing that I am a little disturbed about is something which is a trap that we as nutritionists fall into. You talked about your plus or minus a C.V. of 15% or more in terms of variability. Yet we keep on seeking to come up with some absolute value. We worry about small differences yet the variability of our data is great. In fact the individual variability is of at least the magnitude of that between individuals. The alterations in measurement variability, at least in the same individual, from time to time also varies considerably. The plea I am making is that we don't try to act as if we have the type of precision that we do not have. If we had such precision it's perfectly clear that something was wrong with your experiment because you could not have achieved balance at 0.61 grams, in spite of multiplying by 1.3. Now as far as the 1.3 is concerned, I think that Dr. Scrimshaw raised a very interesting point. Whether it's 1.3, or whether its 1.4, or 1.2 or 1.5, we really do not know because again we are dealing with a range. We are dealing with variability and we have to consistently remind ourselves of this fact. I hope that ultimately nutritionists will start talking and thinking more in terms of ranges and variability than we have up to this time.

DR. TORUN: As I mentioned before, the figures that we have given for the essential amino acids tested corresponded to those intakes which did not produce any alterations of the various indicators explored in any of the six children who received each amino acid mixture. These figures are closer to "safe levels of intake" than to "mean requirements". The value of 0.61 g protein/kg/day obtained in our studies with milk is the mean requirement for that protein. If we want to compare the total amount of essential amino acids with the mean requirement of total dietary nitrogen, we have to think in terms of mean requirements of each amino acid and that will lower the 360 mg in our pattern. On the other hand, we did not test the needs of leucine and the aromatic amino acids and we included in the pattern the amounts of those amino acids present in the "safe levels" of milk protein intake. Their sum accounts for 40% of the 360 mg. Therefore, if the figures for these amino acids are overestimated, their real requirement values will further lower the 360 mg.

I do not know whether the pattern of amino acid requirements changes with age. I agree with you that this is a very important issue that has to be investigated.

DR. HARPER: This has been an excellent contribution. All I want to discuss is the interpretation. It's important to emphasize that the FAO/WHO and the NRC patterns were based on the requirements of infants four to six months of age. Snyderman's work with phenylketonuric patients provided the possibility of following closely the change in phenylalanine requirement with age. Over 24 months phenylalanine intakes were adjusted to maintain normal blood phenylalanine values in children unable to metabolize phenylalanine. The fall in phenylalanine with age was similar to the fall in protein requirement with age. We do have some evidence, although it's very limited, that the amino acid requirements fall a little bit more rapidly than the protein requirement.

What the NRC and the FAO/WHO did was to express the infant requirement as milligrams per gram of protein, assuming that the requirements could be met by one and a half grams of protein. They further assumed that protein intake should be at least two grams per kg body weight for young infants. This would exceed the requirements for older age groups, but as the protein requirement falls, amino acid requirements will fall at least as rapidly, so the appropriate protein intake with this amino acid pattern would still meet amino acid requirements. If you look at the values for the NRC and FAO/WHO scoring patterns they are not very far off from the values that you have. For methionine, one of the values we had was 28, the other was 35, yours is 27. For phenylalanine, you have 69; the others had 73 and 60. For threonine you have 37; the others had 35 and 40. I don't think there's that much discrepancy between the values. Your lysine value is a little high; it's 64 versus 51 and 55. I have some reservations about that. In interpolating requirements directly from milk composition, the milk amino acid values were based on the average of something over 100 samples by the USDA and FAO. I have little reservation about the precision of the value for lysine.

Arroyave calculated that sulfur containing amino acids were first limiting in milk protein and that lysine and tryptophan were the first limiting amino acids in a corn-bean mixture. Then he used the amino acid scoring standard of 10 milligrams of tryptophan per gram of protein for two- to three-year old infants to calculate the amount of protein needed to meet the growth requirement. The prediction, using 10 milligrams of tryptophan per gram of protein, corresponded closely with the intake of 1.33 grams/kg of body weight of that mixture needed to support satisfactory growth rates.

I shall leave the question about a discrepancy in the tryptophan value open and shall come back to the point that Dr. Margen and Dr. Scrimshaw made. There's a tendency, when we set requirement patterns, to assume a degree of refinement and a degree of precision that we don't really have. I would prefer to stay with the somewhat higher values of the FAO/WHO or NRC patterns until we have more information to suggest lowering the values, and assume that, as the protein requirement falls, the amino acid requirements will fall proportionately. It is unrealistic to think in terms of the adult amino acid requirements for a standard pattern. The values are so low that we would have to remove amino acids from protein if the objective was to conserve amino acids.

21

Statistical Considerations in Estimating Protein Nutritional Value for Humans from *In Vitro* Assays

K.W. Samonds

The quest for an accurate assay . . . . one which predicts, or is highly correlated with, human protein quality estimates by the nitrogen balance procedure . . . . remains unresolved primarily because of the lack of appropriate comparative data. *In vitro* assays of protein quality, however, have several definite advantages over *in vivo* procedures in humans, rodents, or other animal models. Their shorter duration and lower cost make them attractive alternatives *if* these assays can be shown (1) to provide the required degree of precision, (2) to be reproducible in the hands of different investigators, and (3) to result in quality estimates which may be extrapolated to human protein needs.

The selection of an assay procedure involves the trade-off of the relative advantages and disadvantages of a particular method depending upon the use to which the analysis will be put. This is an important point which has frustrated many who have tried to establish "the method" for protein quality evaluation. Whether a method is intended to detect small changes in a protein's quality due to a processing procedure, for example, or rather is meant to detect gross differences between widely differing proteins will have a great influence upon the imprecision which can be tolerated and the number of replicate analyses which will be required. A technique which demonstrates excellent repeatability in independent determinations by the same analyst or the same laboratory, but may not give identical results in another laboratory, could be used effectively as an in-house screening tool but would not be appropriate for interlaboratory comparisons or regulatory purposes.

Firstly, I shall examine the precision, reproducibility, and accuracy required for a variety of applications, and then evaluate the existing methods for these qualities.

PRECISION

The precision of a method is a composite quantity depending upon (1) variation in the results within a laboratory . . . . the variance of an estimate and the repeatability of the procedure when carried out by the same analyst, and (2) variation specific to individual laboratories or to particular samples analyzed in those laboratories. The equation for calculating the overall precision of a method, $s_{\bar{x}}$, is:

$$s_{\bar{x}} = \sqrt{s_L^2 + s_{LS}^2 + \frac{s_0^2}{r}} \qquad \text{(Equation 1)}$$

where s_L^2 is the variance between laboratories

s_{LS}^2 is the variance due to laboratory-sample interaction

s_0^2 is the within-laboratory variance

and   r   is the number of replicates performed by a laboratory.

The estimation of these three components of variability requires carefully controlled collaborative studies analyzed by the Analysis of Variance procedure to partition the total variance into its three (or more) components. A thorough description of this procedure is given by Steiner (1975) and the American Society for Testing and Materials (1976). Statements about the precision of an estimate, the 95% confidence interval about an estimate, or the testing of differences between the qualities of two proteins should be based upon this composite estimate of precision, *not* the within-laboratory variability as is commonly done. The 95% confidence interval about a quality estimate may be calculated approximately as $\pm 2s_{\bar{x}}$. For a more accurate estimate of the confidence interval, the formula of Welch (1974) should be employed to determine the appropriate number of degrees of freedom.

How precise must our estimates of protein quality be? The answer to this question depends upon the magnitude of the difference in quality we wish to be able to detect. Relatively small changes in quality may be of scientific interest but may have little practical effect. On the other hand, it does not take much methodological precision to distinguish grossly different protein qualities . . . . gelatin versus egg albumin, for example. Rather than test individual differences in quality, some investigators have wished simply to rank proteins according to their quality without placing much importance in the actual quality of an individual protein. On the surface it might appear that this process would require less precision, but in fact the ability to rank *consistently* a group of proteins with qualities within a narrow range would require considerable precision; any method which will rank consistently will also discriminate between unlike proteins. It does not require much precision to rank widely differing proteins or to discriminate between their qualities. The two processes are related.

Some investigators have proposed that proteins be divided into three categories . . . . high, medium, and low quality, ... arguing that methods do not need to be very precise to accomplish this categorization. Difficulties arise, however, with proteins whose qualities fall near the breakpoints. I have demonstrated this problem in Figure 21.1 where the probabilities of categorizing a protein as low quality (less than 40 on a zero to 100 scale),

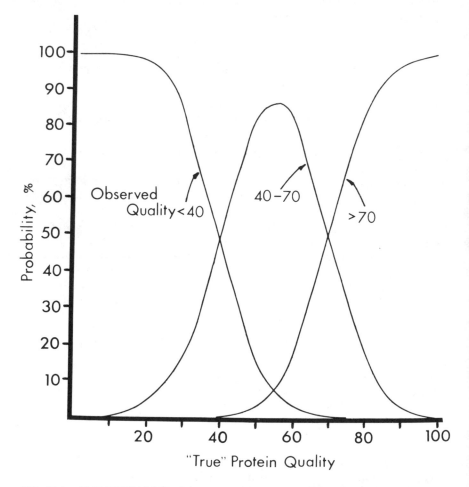

FIG. 21.1.   THE PROBABILITY OF CATEGORIZING A PROTEIN AS "GOOD" (QUALITY SCORE GREATER THAN 70), "FAIR" (QUALITY SCORE BETWEEN 40 AND 70), OR "POOR" (QUALITY SCORE LESS THAN 40) AS A FUNCTION OF THE "TRUE" PROTEIN QUALITY, ASSUMING A MEASUREMENT ERROR OF ± 10 SCORE UNITS

medium quality (40 to 70), or high quality (greater than 70) are plotted versus a hypothetical "true" relative quality score. I have assumed a standard error of ± 10 score units in the estimate of quality (not an unreasonable assumption as will be demonstrated below). The probability of miscategorizing a protein is quite large, particularly around the breakpoints. For example, if the true quality of a protein is 30, there is a 16% chance that it will be incorrectly categorized in the medium quality group. For proteins with qualities of 39 or 41 the chance of making an error is near 50%. If we could increase the precision of the assay the probability of a correct categorization would increase above or below the breakpoints, but the validity of the categorization close to the breakpoints would still be similar to the flip of a coin! I think the difficulties of a categorizing system and the unavoidable errors in categorization outweigh its simplicity. We should avoid this approach.

It is interesting to carry our evaluation of precision one step farther to see the effect of variability in a protein quality estimator upon the precision of an estimate of human protein requirements. Quality and quantity are related according to Equation 2:

$$\text{Requirement for Protein A} = \frac{\begin{array}{c}\text{Requirement for a}\\ \text{High Quality Protein (such as egg)}\end{array}}{\begin{array}{c}\text{Relative Quality of Protein A}\\ \text{(expressed as a decimal fraction)}\end{array}} \qquad (2)$$

This relationship is plotted in Figure 21.2 for a 70 kilogram man. A ten unit change in quality from 90 (or 0.9) to 80 would increase the recommended intake from 47 to 52 grams per day, an increase of 5 grams which would be attained easily. However, a ten unit change from 40 to 30 would increase the requirement from 105 to 140 grams per day, an increase of over 33%. Certainly it would be of importance to detect the latter difference, but not the former. The problem is compounded when one looks at the variability in an estimate of a protein requirement resulting from variability in our estimate of quality, which resides in the denominator of Equation 2. A rough approximation of this variability can be derived from Equation 3.

$$s_{\frac{1}{\overline{X}}} = \frac{1}{\overline{X}} \sqrt{\frac{s_x^2}{\overline{X}^2}} \qquad \text{(Equation 3)}$$

In Figure 21.2 it is apparent that, at low qualities, the variability of the requirement estimate increases substantially making it particularly difficult to detect differences in the requirements of two proteins within the range where protein quality becomes an important factor.

Before examining further the precision of available methods, I shall examine the individual variance components . . . repeatability, reproducibility, and interaction . . . for their relative influence on the precision of various methods and for the means by which they may be improved.

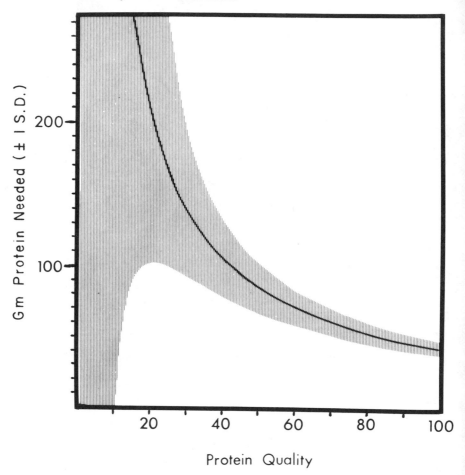

FIG. 21.2.   RELIABILITY OF AN ESTIMATE OF THE AMOUNT OF PROTEIN NEEDED TO MEET THE REQUIREMENTS OF A 70-KG MAN AS INFLUENCED BY A TEN-UNIT STANDARD DEVIATION FOR THE ESTIMATE OF RELATIVE PROTEIN QUALITY. THE SHADED AREA REPRESENTS ± 1 S.D. FOR THE ESTIMATE (SEE TEXT)

REPEATABILITY

Within-laboratory variability is a function of the random error encountered in all experimental work despite the most rigid controls of contributing variables. Examples of factors which may contribute to this variability are: weighing and pipetting errors, inaccuracies in the integration of peak areas in amino acid analyses, inherent variability in animal, microbial, or enzyme assays, non-homogeneity of the samples, etc. While inherent, random variability is unavoidable, the within-laboratory component of the overall precision can be reduced by increasing the number of replicates

(Equation 1) or by extending the range of the data in an assay involving a regression analysis.

How much variability can be tolerated or be overcome by increasing the number of replicates? The answer depends upon the magnitude of the difference in protein quality to be detected and the willingness of the investigator to make errors in this conclusion . . . either detecting a difference when none truly exists (Type I Error), or failing to detect a difference when one does exist (Type II Error). Ignoring for a moment the other components of variability, we can determine the number of replicates required to detect a specified difference when results from the same laboratory are compared:

$$r = \frac{2(t_0 + t_1)^2 s_0^2}{d^2} \qquad \text{(Equation 4)}$$

where r = the number of replications
s_0^2 = the within-laboratory variance of the method
t_0 = the Student-t value associated with Type I Error
t_1 = the Student-t value corresponding to the probability $2(1-P)$, where P is the probability of a Type II Error
and d = the difference to be detected

It should be clear that it is of little value to perform an experiment in which the number of replicates is insufficient to detect important differences on a regular basis. It is likewise pointless to use ten replicates to detect a difference that four replicates will find in most cases. Unfortunately, the within-laboratory variability is usually considerably smaller than the between-laboratory variability, so the benefits of increasing the number of replicates are frequently disappointing. If one wishes to establish a fixed number of replicates required for regulatory purposes, it is important to remember that laboratories may vary considerably in their repeatability; some analysts may be very precise while others may not. The variances of several laboratories may be pooled to estimate the "method variance," but it should be remembered that the results of some investigators may be considerably more variable than this estimate and appropriate precautions should be taken when specifying the required number of replicates.

How Repeatable are Currently Available Methods?

The most common *in vitro* procedure is the use of amino acid composition data to estimate Amino Acid Score. Seldom, if ever, does one find a standard error accompanying such data although there is certainly a degree of variability associated with the separation and quantitation of amino acids and with the determination of Kjeldahl nitrogen. It is my impression that few investigators realize the degree of uncertainty associated with Amino Acid Scores and they frequently have unwarranted confidence in their results.

For the purpose of a demonstration I have used the data of my colleague, Dr. Peter Pellett, who was one of the collaborators in the study reported by Dr. Sarwar (Chapter 13). (Dr. Pellett has assured me that his data would be representative even though the analyses were performed on an old analyzer and integration was done manually). For each amino acid, I have pooled the variances of the three replicates for the seven proteins, and the resulting standard deviations are reported in Table 21.1. Coefficients of variation were calculated by dividing these deviations by the grand mean of each amino acid averaged across all proteins. These coefficients of variation ranged from 3 to 8% ... within the range of within-laboratory coefficients of variation reported by Dr. Sarwar (Chapter 13). Expressing the variability as a percentage of the mean may, in fact, underestimate the variability in amino acid concentrations present in very small amounts in some proteins. Low, broad peaks are particularly difficult to integrate manually and are more seriously influenced by base-line shifts. The cystine data for the collaborative study discussed by Dr. Sarwar demonstrate this point; the coefficient of variation is greater in the samples with the lower cystine concentrations. Unfortunately, for amino acid scoring we are most interested in those amino acids which are present at low concentrations.

The variability in amino acid composition data results in a corresponding variation in the score estimate which can be determined by dividing the pooled standard deviation for each amino acid by the FAO/WHO standard value for that amino acid. For example, the standard deviation of lysine is ± 22.6 mg/gm N which is divided by the FAO/WHO value of 340 mg/gm N resulting in a standard deviation for the lysine score of 6.6 score units (see Table 21.1). Pooling these standard deviations over all amino acids results in an overall estimate of variability of ± 7.6 units. Tryptophan and total sulfur amino acids appear to be the least reliable analyses, reflected by high standard deviations of scores based upon these amino acids. This variability accounts for occasional difficulties in specifying the limiting amino acid in proteins consistently. Pooling the variability for the most frequently limiting amino acids . . . lysine, total sulfur, and tryptophan . . . results in a

TABLE 21.1.   VARIABILITY IN AMINO ACID ANALYSES[1] AND THE RESULTING VARIABILITY IN AMINO ACID SCORES[2]

Amino Acid	Standard deviation (mg/gm N)	Coefficient of variation (%)	Standard deviation of Amino Acid Score (score units)
Isoleucine	22.6	6.9	9.0
Leucine	16.4	3.1	3.7
Lysine	24.2	6.1	7.1
Total Sulfur	22.4	8.4	10.2
Total Aromatic	29.2	5.1	7.7
Threonine	16.6	6.2	6.6
Tryptophan	6.6	7.6	11.0
Valine	18.2	5.1	5.9

[1]Pellett, unpublished data.
[2]Scores calculated according to FAO/WHO (1973).

standard deviation of ± 9.4 score units. If we assume triplicate analyses, as in the collaborative studies, we can calculate that the 95% confidence interval for an amino acid score includes ± 23.4 score units and we could detect a difference of 15.3 units according to Equation 4. Or, we can determine the number of replicate analyses required to detect a specified difference in score for individual amino acids using Equation 4. I have calculated the number of replicates required to detect a 10-unit difference assuming a 5% chance of finding a difference when there is none (Type I Error) and a 10% chance of failing to find a difference when there is one (Type II Error), and the results are recorded in Table 21.2. If the limiting amino acid is leucine or valine . . . for which we have the most reliable score estimates . . . we may be reasonably sure of detecting a 10-unit difference with three replicates. Four replicates would be required if the limiting amino acid were lysine, and seven if the limiting factor were total sulfur amino acids.

TABLE 21.2.   NUMBER OF REPLICATES REQUIRED TO DETECT A 10-UNIT CHANGE IN AMINO ACID SCORE (A WITHIN-LABORATORY COMPARISON)[1]

Replicates	Maximum Allowable Standard Deviation of Score	Limiting Amino Acid
3	6.1	Leu, Val
4	7.1	Thr, Lys
5	8.6	Total Aromatic
6	9.5	Ile
7	10.4	Total Sulfur
8	11.1	Try

[1] 5% chance of finding a significant difference when there is none; 10% chance of finding no difference when a significant difference exists.

I should emphasize that the values in Table 21.2 apply to comparisons of proteins analyzed in the same laboratory because they are based upon s_0, the within-laboratory variability. If it were possible to reduce the coefficients of variation for all amino acids to 3% we could detect differences of 6 or 7 score units with three replicates. I think this is a reasonable goal for analytical technique and an adequate level of discrimination for most problems relating to protein quality. The collaborative assays, however, indicate that some laboratories are not attaining this degree of reliability even when one might expect them to use their best technique . . . when their results will be compared with those of others and their self-esteem, if not their reputation, is at stake! Tryptophan and total sulfur amino acids are a particular problem because of their variability and their importance in scoring, and more effort must be exerted to improve the precision of these assays.

Other assays for estimating total or available amino acids by microbiological methods have been reviewed by Ford (Chapter 16). If one is already certain of the limiting amino acid, these assays appear to provide a substantial degree of repeatability . . . coefficients of variation of 3 or 4%. They also have the advantage that the number of determinations can be increased to improve the within-laboratory component of the overall variability without as much time and expense as with chromatographic analyses.

Measurement of digestibility has been suggested as an adjunct to Amino Acid Score to correct for the availability of individual amino acids or of total protein. The simplest and most straight-forward of these methods is that of Hsu et al. (1977) relating the change in pH during enzymatic hydrolysis to the apparent digestibility of a protein in rats. I have re-examined this method to assess its variability. Satterlee et al. (1979) have subsequently modified the method and extended the sample base for the revised method to over 50 proteins, but insufficient data were included in the paper for my purposes. Hsu et al. (1977) report that the change in pH is correlated with digestibility with a coefficient of .90 and that the standard deviation between assays is ± .75% digestibility. These are *descriptive* statistics, however, describing the relationship tested in the assay. We are interested in the use of *predictive* statistics . . . given a single value for the pH change (or in the procedure of Hsu et al., the average of 2 or 3 replicates), how precisely can we predict the digestibility? For this we should use the equation for the variance of a prediction from a regression equation, Equation 5.

$$s_Y^2 = s_{y \cdot x}^2 \left(1 + \frac{1}{n} + \frac{(X - \bar{X})^2}{\Sigma x^2} \right) \qquad \text{(Equation 5)}$$

where s_Y^2 = the variance of our digestibility estimate, Y
$\quad\quad s_{y \cdot x}^2$ = the variance of Y for fixed X, the pH change
$\quad\quad$ n = the number of data points in the original regression analysis
$\quad\quad$ X = the pH for which we wish to make an estimate of digestibility
$\quad\quad \bar{X}$ = the average pH for samples included in the original analysis
$\quad\quad \Sigma x^2$ = the sum of the squared deviations from $\bar{X}$ of the pH values of all samples.

I have estimated the values from the data points from Figure 2 in the paper of Hsu et al. (1977) and have computed s_Y for a range of interpolated pH values. The results, listed in Table 21.3, are therefore only approximate. The standard deviations ranged from approximately 2.8% digestibility near the center of the regression line to 3.3% at the extremities, and the resulting coefficients of variation range from 3.4 to 4.5%. I have redrawn Figure 2 of Hsu et al. (1977) as my Figure 21.3 including the 95% confidence interval for an estimated digestibility. Using this method one can estimate apparent digestibility within 5 or 6%, an adequate degree of reliability for most purposes. Extrapolation of the regression line beyond the limits of the data seriously affect the reliability of the method and makes the unwarranted assumption that the response continues to be linear.

TABLE 21.3.   RELIABILITY OF A PREDICTED APPARENT DIGESTIBILITY

pH after 10 minutes[1]	Estimated Digestibility, $\hat{Y}$	$s_{\hat{Y}}$	C.V.[2]	95% Confidence Interval
	(%)	(%)	(%)	(%)
6.2	95.0	3.3	3.5	88-102
6.4	92.1	3.1	3.4	86-98
6.6	89.2	3.0	3.4	83-95
6.8	86.3	2.9	3.4	80-92
7.0	83.4	2.8	3.4	78-89
7.2	80.5	2.9	3.6	74-86
7.4	77.6	2.9	3.7	72-84
7.6	74.7	3.1	4.1	70-80
7.8	71.8	3.2	4.5	65-78

[1]Based on data of Hsu et al. (1977).
[2]Coefficient of variation.

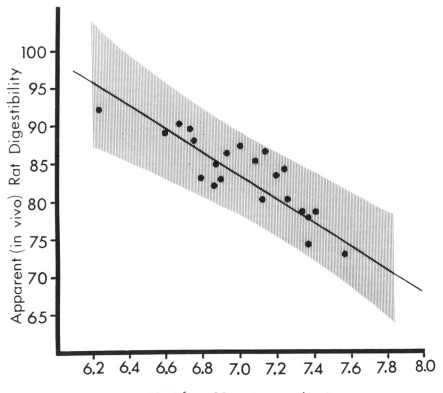

FIG. 21.3.   95% CONFIDENCE INTERVAL FOR AN ESTIMATE OF DIGESTIBILITY BY THE METHOD OF HSU *ET AL.* (REDRAWN FROM HSU *ET AL.*, 1977)

If one wishes to multiply an estimate of digestibility by an estimate of Amino Acid Score it is important to remember that there is variability in both estimates which must be reflected in the variability of the product according to Equation 6.

if $Y = X \cdot W$,

$$s_Y^2 = s_X^2 (\bar{W}^2) + s_W^2 (\bar{X}^2) \qquad \text{(Equation 6)}$$

where s_X^2 and s_W^2 are the variances of X and W, respectively and $\bar{X}$ and $\bar{W}$ are the respective means.

For example, if we have an Amino Acid Score of 80 with a coefficient of variation of 5% and a Digestibility of 80 with a coefficient of variation of 4%, the resulting product, an estimate of utilizable protein, would have a value of 64 ± 4.1, or a coefficient of variation of 6.4%. By performing the multiplication we derive a more useful estimate, but at the expense of a small degree of reliability.

BETWEEN LABORATORY REPRODUCIBILITY AND LAB-SAMPLE INTERACTION

"Do not forget," wrote W.J. Youden (1969), "that the between-laboratory error is the determining factor in evaluating a method!" While many investigators may be proud of their reliability and repeatability, it is important to note that, when results are compared in collaborative studies, laboratories may be arriving at precise, but *different* estimates of a protein's quality. If a laboratory's results are consistently higher or lower than the average results for other laboratories analyzing the same proteins, this bias must be considered when we compare results collected in different laboratories and is termed "between-laboratory" or "interlaboratory" variability, . . . s_L^2 in Equation 1. Factors which contribute to this type of systematic error include: variations in the concentrations of standard amino acid solutions used for calibration, the adjustment of an instrument, misinterpretation of a particular step in the analysis procedure, environmental factors, etc. If these factors cannot be identified and controlled, this component of error must be included as part of our estimate of the overall precision of a method. In fact, it is commonly the greatest contributor to the total variability (Youden, 1969).

In addition to systematic errors which affect all analyses, there may be greater discrepancies with some samples due to differences in handling, incorrect assumptions based upon prior treatment, etc., which may cause disagreements between laboratories for some samples but not for others. This component of variation is termed the "laboratory-sample interaction." For a method to be satisfactory for all types of samples in all laboratories there should be no inter-laboratory variability or interaction, but this is seldom if ever the case.

One way in which between-laboratory variability may be reduced is by the introduction of a reference sample which is analyzed along with the test materials and to which all laboratories compare their results by calculating the ratio of the test protein to the reference protein. However, if laboratory-sample interaction is significant, there is a possibility that the results for the reference material may be artificially high or low, and the correction of all values to this reference may actually introduce more laboratory bias than it removes. Also, calculating the ratio of two variables introduces imprecision in both the numerator and the denominator according to Equation 7:

$$\text{If } R = \frac{p}{q}$$

$$s_R = \frac{1}{q} \sqrt{s_p^2 + \frac{p^2}{q^2} s_q^2} \qquad \text{(Equation 7)}$$

where s_R is the standard deviation of the ratio of p and q and s_p and s_q are the corresponding standard deviations.

The usefulness of a reference protein should be evaluated carefully, with the ultimate goal of increasing overall precision.

How Repeatable are Currently Available Methods?

Inter-laboratory variabilities in amino acid determinations are substantial. Sarwar *et al.* (Chapter 13) report coefficients of variation for between-laboratory variability which range from 5 to 20%, depending upon the amino acid under consideration and the protein being analyzed. Lysine, methionine, and threonine, for example, have coefficients of variation of 5 to 10% while cystine and tryptophan are less repeatable, with coefficients of variation from 11 to 20%. For their "pre-test" samples, Happich *et al.* (Chapter 12) report many coefficients of variation greater than 10% . . . threonine, lysine, and methionine had coefficients of variation of 10 to 15%, and cystine of 77%. These interlaboratory variabilities were generally three to five times greater than the within-laboratory variability. The magnitude of the laboratory-sample interaction has not been reported by either investigator, but Happich *et al.* (Chapter 12) mention that, for their pre-test samples, it is significant for all amino acids except proline. It should be pointed out that both of the collaborative studies reported at this conference include close to the minimum number of collaborators for reliable inferences concerning the systematic laboratory variability. Therefore we should be cautious when making any extrapolation concerning how these procedures might perform if adopted by all laboratories. For both studies, a final evaluation should await the complete analyses of data from all of their collaborators.

It should be possible to improve between-laboratory reproducibility by identifying and controlling factors which contribute to this component of variability, but the study of Happich *et al.* shows that there may be considerable variation even when the methodology is standardized. Considerably more work is needed to measure the influence of each step in the methodology upon the ultimate reproducibility.

For the sake of making some estimates of overall precision of the amino acid scoring procedure, I will assume an average between-laboratory coefficient of variation of 12%, a within-laboratory coefficient of variation of 3%, and a coefficient of variation for interaction of 5% (an outright guess, but within reason). Inserting these values into Equation 1, we can calculate the estimated overall coefficient of variation as a function of the number of replicates (see Table 21.4). Increasing the number of replicates has a minor

TABLE 21.4.   PRECISION OF THE AMINO ACID SCORING PROCEDURE INCLUDING REPEATABILITY, REPRODUCIBILITY, AND LABORATORY-SAMPLE INTERACTION AS A FUNCTION OF THE NUMBER OF REPLICATES[1]

Replicates	Standard Deviation
1	13.34
2	13.17
3	13.11
4	13.09

[1]See Equation 1 (text).

effect on the overall precision because the other two components predominate and they are unaffected by the number of replicates. Transforming the variability in amino acid determination into variability in amino acid scores, as discussed earlier for within-laboratory repeatability, results in standard deviations of scores ranging from 14 to 19 score units, depending upon the limiting amino acid, with a pooled standard deviation for the four important amino acids of approximately 16 units and a 95% confidence interval encompassing ± 32 units. Using the procedure of Welch (1947) to estimate the appropriate degrees of freedom associated with overall precision, we can conclude that the method can detect differences of approximately 13 score units with a confidence of 95% when comparing results of triplicate analyses from different laboratories.

Other quality estimating procedures or digestibility measures which depend upon amino acid analysis, such as the rate of release of amino acids during enzymatic hydrolysis, will be equally affected by the variability in the analysis procedure. Other methods discussed earlier in this paper or elsewhere in this conference have not been subjected to rigorous, well-designed collaborative studies to determine their reproducibility or overall precision, but this should be a high priority if we propose to replace the current methodology.

ACCURACY

How well do *in vitro* methods estimate the "true" ability of a protein to meet human protein requirements? Methodological accuracy might depend for instance, upon the appropriateness of the amino acid scoring pattern (which in itself is a relatively imprecise estimate of amino acid requirements), the similarities between *in vitro* and *in vivo* amino acid availabilities, etc. The use of the word "accuracy" implies that we already know the true value in some cases and can compare the results of our assay to these values. This is not the case, however, and we usually use the long-term nitrogen balance response in humans as the target at which *in vitro* procedures should aim. But, as we have heard at this conference (Rand *et al.*, Chapter 7) the nitrogen balance method has a coefficient of variation of approximately 15% and is unable to distinguish a difference in quality of 20 score units using 8 subjects. Therefore it is not surprising that correlations between *in vitro* methods and nitrogen balance might be low; there is considerable variation in both measures. For the purposes of a demonstration I have fabricated the set of data shown in Table 21.5a, where X is the protein quality estimated by an *in vitro* procedure ($\bar{X}$ = 72.1, s_x = 18.76) and

TABLE 21.5a. HYPOTHETICAL DATA FOR *IN VIVO* AND *IN VITRO* MEASURES OF PROTEIN QUALITY FOR 12 PROTEINS.

in vivo Quality Y	*in vitro* Quality X	
40	45	$\bar{Y}$ = 72.1 s_y = 18.76
50	45	
55	62	$\bar{X}$ = 72.2 s_x = 18.67
60	70	
65	60	r = .922
70	72	
75	69	
80	85	
85	71	
90	92	
95	88	
100	108	

Y is the Relative Protein Value estimated by nitrogen balance ($\bar{Y}$ = 72.2, s_y = 18.67) and the two variables, measured *without* method variability, are correlated with a coefficient of 0.92. Now we may look at how the introduction of measurement imprecision influences the correlation according to Equation 8 from Brownlee (1965):

$$\rho_{x'y'} = \rho_{xy} \frac{1}{\sqrt{\left(1 + \dfrac{s_u^2}{s_x^2}\right)\left(1 + \dfrac{s_v^2}{s_y^2}\right)}} \qquad \text{(Equation 8)}$$

where $\rho_{x'y'}$ = the observed correlation of variables which include measurement error.

ρ_{xy} = the correlation of the variables if there were no errors of measurement

s_u^2 = the overall method variability for measurement of X

s_v^2 = the overall method variability for measurement of Y

In Table 21.5b I have included the estimated correlation of the two variables which would result from various degrees of measurement error. Standard deviations of ± 15 units in both variables would result in a relatively poor correlation of 0.56, for example. It is important that the relationship between *in vitro* and *in vivo* measures of protein quality be investigated more thoroughly in the future, but it is also important to make the assays as precise as possible so that any relationship between the two is not swamped in a sea of imprecision.

TABLE 21.5b. ESTIMATED CORRELATION BETWEEN *IN VIVO* AND *IN VITRO* PROTEIN QUALITY WHICH WOULD BE OBSERVED WITH VARYING DEGREES OF MEASUREMENT PRECISION FOR EACH VARIABLE

Measurement Error for Y, s_v	Measurement Error for X, s_u			
	±5	±10	±15	±20
±5	.861	.785	.695	.587
±10	.785	.717	.634	.555
±15	.695	.634	.561	.491
±20	.587	.555	.491	.430

CONCLUSIONS

In closing I would like to sound a note of caution and offer a few suggestions for the immediate future.

Replacing the current *in vivo* procedure, as bad as it is from a mathematical and statistical viewpoint, with an Amino Acid Score corrected for digestibility may be premature: while the theory is sound, the method, in practice, needs further investigation to "tighten-up" its precision, particularly with respect to between-laboratory variability. For regulatory purposes, the use of an assay which is not highly reproducible would be chaotic.

I think the following recommendations are in order:

(1) Investigators should report the precision of each *in vitro* analysis and the "predictability" of a method when it is to be applied to future analyses.

(2) Analysts should be vigilant for new ways to improve within-laboratory variability. Technicians should be trained carefully in the use of analytical procedures and their repeatability should be checked occasionally by unsuspected replicates.

(3) "Ruggedness tests" (Youden, 1975) should be undertaken by individual laboratories to determine the relative importance of specific steps and conditions in the analysis procedure in reducing variability.

(4) More collaborative studies with more participants should be undertaken to identify factors which cause laboratory biases and laboratory-sample interactions so that these may be reduced by the standardization of methodology. It is particularly important that collaborative studies be initiated for *in vitro* digestibility determinations or amino acid bioavailability assays since these methods have not been examined, to my knowledge, for between-laboratory variability.

(5) Regulatory agencies must decide upon the degree of discrimination an assay procedure should provide. While Dr. Vanderveen headed his list of essential elements for a new assay procedure with the requirement that it "must accurately estimate protein quality with respect to the population segments who consume the product" (Vanderveen and Mitchell, Chapter 4), he did *not* include the degree of precision with which an estimate should be made. As I have shown, a higher degree of precision is required for estimates of low quality proteins than for high quality proteins. The difficulty lies in incorporating this into a set of regulations.

REFERENCES

AMERICAN SOCIETY FOR TESTING AND MATERIALS. 1976. *Standard Recommended Practice for Developing Precision and Accuracy Data on ASTM Methods for the Analysis of Meat and Meat Products*. Prepared by ASTM Committee F-10 on Meat and Meat Products. Philadelphia

BROWNLEE, K.A. 1965. *Statistical Theory and Methodology in Science and Engineering*. Wiley and Sons, New York

FORD, J.E. Microbiological Methods for Protein Quality Assessment. (this conference)

HSU, H.W., VAVAK, D.L., SATTERLEE, L.D., and MILLER, G.A. 1977. A Multi-enzyme Technique for Estimating Protein Digestibility. J. Food Sci. 42:1269–1273

SATTERLEE, L.D., MARSHALL, H.F., and TENNYSON, J.M. 1979. Measuring Protein Quality. J.A.O.C.S. 56:103–109.

STEINER, E.H. 1975 Planning and Analysis of Results of Collaborative Tests. In: *Statistical Manual of the Association of Official Analytical Chemists*, p 66–88. The Association of Official Analytical Chemists, Washington, D.C.

WELCH, B.L. 1947. The generalization of "Student's" Problem When Several Different Population Variances are Involved. Biometrika, 34:28–35

YOUDEN, W.J. 1969 *Statistical Techniques for Collaborative Tests*. The Association of Official Analytical Chemists, Washington, D.C.

DISCUSSION

DR. SARWAR: I congratulate you. You did a very good job of analyzing my paper during the last couple of hours. I want to make two points. About your comment that these results are disappointing, I don't agree with you. There is a need to delay making conclusions about the results until after the final analyses are done but I don't expect there will be very many changes. When I compare our results with the other collaborative study (they are, hopefully, standardizing things), their coefficient of variation seems higher for lysine than in our lab. That certainly is an indication that standardization may not be working. The second point is that if you want to go ahead with some kind of recommendation of basing protein quality estimates on amino acids, we have to start worrying about the variability and we have to take it for granted that this is the variability we are going to get.

Part IV Task Force Reports and Conference Overview

22

Task Force I Report

Sheldon Margen

The participants of Task Force I were asked to address a series of specific questions that focused on the main problems raised during the first portion of the conference. These specific questions were to encompass certain key issues relating to the importance of protein quality, particularly in the U.S. diet, and the methods and factors involved in assessing protein nutritive value in humans.

The Task Force members were asked to address a series of questions. The questions and the responses of the conferees, as summarized, follow.

IS ASSESSMENT OF PROTEIN QUALITY IN THE U.S. DIET IMPORTANT FROM A NUTRITIONAL VIEWPOINT?

Yes, assessment is important! Although this question proved to be difficult because of its highly general nature and possible variable interpretations, it was recognized that the "average" U.S. diet as a whole, poses no problems of inadequacy in protein content and quality. (In fact, possible adverse effects of the high protein diet consumed by some Americans should be considered and investigated further.) However, there was general agreement that nutritional and food scientists should be concerned with the problem of protein quality, for at least the following reasons:

(a) Our food supply is undergoing rapid changes, which may significantly alter protein availability. As this occurs, the population needs to know. Scientists must be able to assess these alterations as a critical step, so policies and actions can be taken that will minimize or avoid deterioration in the quality of proteins and other nutrients in our food supply.

(b) There are subpopulations who either consume protein from a single food source, a small variety of foods, or foods with proteins of inferior quality. These people must be considered at risk and the quality of their dietary proteins must be known. Examples of such groups are (1) children consuming formulas, (2) people lowering their energy intake and therefore food consumption, generally for weight reduction, (3) people changing their eating patterns to foods of lower protein quantity and quality, e.g., vegetarians, and (4) individuals with various illnesses requiring therapeutic dietary deviations.

(c) There are subpopulations who, for economic, or other socio-cultural reasons, might become at risk because of their decreased ability to procure high-quality protein. Although there is insufficient data, these might include (1) the poor, most frequently ethnic minorities, (2) many recent immigrants who may not be acquainted with our food supplies, (3) the elderly, who are often economically strapped and relatively immobile, (4) teenagers who are undergoing rapid growth, and often eating unbalanced diets, and (5) pregnant and lactating women, especially the growing number of teenage mothers.

IS ASSESSMENT OF PROTEIN QUALITY IMPORTANT FROM A REGULATORY POINT OF VIEW?

The general consensus is that some segment of our society (government) must take the responsibility for monitoring the protein quality in our foods and for engaging in public rule-making wherever necessary. Regulation may be particularly needed in cases where:

(a) New or substitute food ingredients are introduced or changes are made in proportions of protein-containing ingredients to assure against deteriorations of protein quality.

(b) The dietary protein is from a single source such as infant formula or a few foods.

(c) Because food is increasingly engineered to its possible nutritional detriment or improvement, our diet must be monitored to insure that it does not deteriorate in the quality of protein and other nutrients.

(d) Changes in our economy occur that lead to increased food prices and possible substitutions in manufacturing procedures. These changes may result in a deterioration in the nutritional quality of the dietary protein consumed by individuals.

(e) Unjustified claims are made for protein quality in substituted materials.

IS DIGESTIBILITY AN IMPORTANT FACTOR IN EVALUATING PROTEIN NUTRITIONAL QUALITY?

The answer is yes. Digestibility enters into the equation of protein quality. However, the concept of digestibility as operationally determined may often not merely reflect the ability to break down the ingested protein, but may include the non-availability of certain specific amino acids. Therefore, this issue of digestibility and bioavailability are interconnected and are both of equal importance when new foods and/or new processes are introduced. As far as the "average" American diet is concerned, the problem of digestibility, as a variable in protein nutrition, is of virtually no importance, due mainly to the high levels and generally high quality of the protein consumed. However, the problem is of importance when considering the need for estimating protein quality as discussed in Section 1.

WHAT IS THE DEGREE OF ACCURACY AND PRECISION OBSERVED IN ESTIMATING PROTEIN NUTRITIONAL VALUE IN HUMAN SUBJECTS?

Currently the only accepted technique to measure protein quality in humans is the "long-term" balance study (studies that involve multiple diet periods each lasting from about 10–21 days.) The balance technique must be standardized and carried out with protein intakes close to balance levels. Low intake levels tend to decrease sensitivity. In order to identify with a high degree of statistical confidence and power ($\alpha = 0.05$, $\beta = 0.10$) a protein or mixtures of proteins with a Relative Protein Value of 75% (that of the "average American diet"), about 20 individuals would be required. Experiments with fewer individuals are of course useful, but they are either less sensitive, or their results less sure. This need to use large numbers of subjects makes the use of the long-term procedure as a routine measure almost prohibitive. Other shorter methods are under investigation, and should be actively pursued—including short-term "non-steady state" balance studies, and methods based on measurements of plasma amino acid levels, blood urea nitrogen values, levels of serum proteins synthesized by the liver and having rapid turnover rates (e.g. transferrin, pro-albumin) and, possibly, activity levels of some enzymes. However, to date all of these methods must still be considered entirely in the process of development.

IN VIEW OF YOUR ANSWER TO THE QUESTION ABOVE, WHAT DEGREE OF ACCURACY AND PRECISION IN ESTIMATING PROTEIN NUTRITIONAL VALUE IS SCIENTIFICALLY ACCEPTABLE FOR REGULATORY AND QUALITY CONTROL PROGRAMS?

This question is difficult to answer. Since operationally in humans a coefficient of variation of 15% is about the best we can accomplish, it would appear that this would be the outside limit allowed in any regulation or quality control. However, since it is unlikely that assays which require the use of human subjects will ever become routine, we would suggest that more emphasis be on development of economical non-human methods with low variability and high accuracy (whether ultimately bioassays or chemical methods). As methods are developed and standardized, they should achieve a coefficient of variation of about 15%, or even lower.

ARE ANY OF THE APPROACHES WHICH INVOLVE HUMAN SUBJECTS USEFUL FOR ASSESSMENT OF PROTEIN QUALITY IN RELATION TO REGULATORY PURPOSES?

At present and in the foreseeable future, for routine use, the answer is no. However, since the ultimate use for protein quality in our discussion is for human subjects, the human assay must be the ultimate "reference standard." This reference function of the human assay must remain. Any non-human methods in present use or developed in the future, must highly correlate with the human assay method. Otherwise, they must be rejected for regulatory and scientific assessment. Therefore, periodically either human assays will have to be performed or characterizations of groups of proteins of different biological value determined in humans and used as "standards" for non-human methods.

We wish to emphasize that, in relation to future progress in the development of methods for determining protein quality and for examining factors affecting protein quality, more research on the methodology for determining human requirements of essential amino acids and for estimating the bioavailability of essential amino acids in foods for humans will probably advance this field more rapidly than continued research with whole proteins or even proteins in foods. However, in view of the fact that humans eat foods containing protein, a balance of research (on amino acids and protein, especially food protein) is essential for a complete understanding of bioavailability.

23

Task Force II Report

A.E. Harper

The Task Force II participants were asked to address a series of questions pertaining to the appropriateness of *in vitro* measures of protein quality as the basis for nutritional labelling, regulatory actions and provision of information about the nutritive value of proteins for humans. These issues were addressed by seven groups composed of all of the conference participants. All groups dealt with the same series of questions. This report is based on the answers they prepared.

NUTRITIONAL LABELLING STANDARDS

The first question was: "Have any of the *in vitro* approaches been developed sufficiently to allow serious consideration of their use for nutritional labelling (i.e., is sufficiently precise for practical application; is reproducible between laboratories; is linearly related to nutritive value; measures nutritional value of a protein source when the source is consumed alone or in a mixed diet; detects major processing damage either alone or in combination with a secondary assay, etc.)?"

There was agreement generally that the Protein Efficiency Ratio (PER) method, which is currently used as the basis for nutritional labelling, regulation and providing consumer information about protein, should be replaced by a more appropriate, precise, and easily understood method. The various groups favored a move toward use of a method for protein labelling based on amino acid analyses of food products, which would permit calculation of amino acid scores that could be compared directly with human amino acid requirements. The FAO/WHO amino acid scoring system was considered to be an appropriate model but a need was expressed for critical evaluation of this in relation to other amino acid requirement patterns proposed as standards, such as the NRC scoring pattern. Beyond this, the different groups did not agree on how adequate the various alternative chemical and *in vitro* methods that are currently available for estimating protein quality were as substitutes for the PER method.

The bases for these recommendations were that: 1) The PER method as presently used does not provide values for protein quality that are directly proportional to each other over the entire range of observed values for protein quality; 2) PER values do not provide information that permits calculation of the supplementary or complementary value of food proteins when different proteins are consumed together; 3) the PER method is not based on the use of a uniform standard for regulatory purposes; the standard must be adjusted differently for products of low or high protein quality; 4) values obtained with the PER method are not expressed in units that are readily understood; 5) PER values do not provide information that can be related directly to human nutritional needs. Several groups concluded that PER is being made to serve two purposes: 1) that of providing information about human protein nutriture; and 2) that of providing the basis for regulatory actions and monitoring of the food supply; and that it is not satisfactory for either.

Despite the lack of agreement about the adequacy of currently available alternative methods for providing information about protein quality, all but one of the groups proposed that an amino acid scoring system, based on human amino acid requirements, should be developed and be used together with a measure of the quantity of protein in a food product to give an estimate of utilizable protein. The group that did not propose this procedure still emphasized strongly the need for development of a method based on amino acid analyses. All of the groups that favored development of an amino acid scoring system for nutritional labelling recommended that the score should be adjusted to allow for incomplete digestibility of protein; and, for unavailability of specific amino acids in food products in which there was the possibility of processing damage to the protein. Several groups also emphasized the need for evaluation of food products for the presence of anti-nutritional factors, such as enzyme inhibitors, gossypol, hemagglutinins and other substances that might have deleterious effects on protein utilization or health, either directly or indirectly.

The major differences of opinion among the groups were on the question of the adequacy of the various alternative methods at the present time. Almost without exception, the groups recommended that collaborative studies were needed to assess the precision, reproducibility and accuracy of amino acid analyses of foods and that continued research on methods for measuring digestibility and amino acid availability were needed in order to improve the specificity and reproducibility of these measurements. Most groups emphasized that the ultimate standard for amino acid scoring should be human amino acid requirements. In order to ensure that the standard for the amino acid scoring system would be reliable, and in order to eliminate the need for routine human trials for assessment of the quality of food proteins, continued investigation of human amino acid requirements was recommended.

Some of the groups proposed that a reference bank of standard foods be compiled and that a handbook be prepared which would include information on amino acid composition of foods together with details about methods and conditions of processing.

The view that there would still be a need for simple procedures that can be used to rank proteins even if an amino acid scoring system is instituted, was expressed frequently. In view of this, some of the groups advocated that the PER method, as currently used, be replaced as soon as possible by the relative net protein ratio method. This would permit the use of a single standard for labelling within a short time and would also provide a scale such that values for protein quality would be expressed on a percentage basis. This, it was considered, would improve understanding of the concept by the public generally and would facilitate transition in the future to other more desirable methods for nutritional labelling for protein quality.

Although not all groups discussed the details of amino acid scoring systems, the possibility of scoring for only such critical amino acids as lysine, total sulfur-containing amino acids, and possibly tryptophan was viewed favorably by some. This led to a parallel proposal for establishment of Recommended Dietary Allowances (RDAs), or at least tentative RDAs, for these amino acids to serve as a standard for amino acid labelling of foods.

MORE RESEARCH NEEDED

The second question was: "Is further research on *in vitro* methods warranted and, if so, what specific research should be done?" Some responses to this question were included with those to the first question. As has already been indicated, the responses to this question focused in large measure on the need for collaborative studies of the various methods that might be used in developing a more satisfactory approach to labelling. Emphasis was placed in particular on the need for research on the development of specific chemical and *in vitro* methods for determination of both total and biologically available amino acids, especially available lysine, total sulfur-containing amino acids and tryptophan; and on improvement of enzymatic and chemical methods for digestibility. The need for standardization of all of these methods was emphasized so that values obtained by amino acid scoring systems can be related quantitatively to human amino acid needs and nitrogen utilization.

Question 3 included a group of subquestions concerning scientific considerations pertaining to implementation of the chemical and *in vitro* methods proposed for consideration. These included "How can an allowance for both quantity and quality be made? What kind of "scale"? Can allowances be made for complementation effects? For low quality protein sources when consumed together with a high quality source?"

Most responses to the question concerning development of an expression that would include consideration of both quantity and quality of protein included the suggestion that this be done by indicating the amount of utilizable protein. It was concluded that this could be done most simply by including on the label a value for the product of an amino acid score, based on the pattern of human amino acid requirements, and total protein. Most groups recommended that this value be adjusted by using a factor for digestibility or amino acid availability.

It was recognized that amino acid scoring systems based on human amino acid requirements would not need to take into account the correction factor for average protein quality of U.S. diets that is currently included in estimating the RDA for protein. The amino acid values used as the basis for the score would be comparable to RDAs for other nutrients.

In response to the question about the type of scale that should be used for expressing protein quality measurements, all groups that dealt with this question proposed the use of a percentage scale in which proteins that met completely the accepted scoring standard be given the maximum value of 100. It was noted that if scores for the amino acids that are most likely to be limiting in human diets were included on labels, it would be possible to calculate the extent to which proteins complemented each other in meeting amino acid requirements.

OTHER SUGGESTIONS

Other suggestions were discussed in relation to implementation of the proposed measures of protein quality. Among these were:

1) that, in providing consumer information, proteins ranked by amino acid scores be separated into three groups covering ranges of 70–100, 40–70 and less than 40. The basis for this proposal was to make consumers aware of the importance of combining proteins to improve nutritional quality.

2) that labels include a system of color coding in order to indicate specific limiting amino acids. This would provide consumers with information that would enable them to use amino acid complementation.

3) that a list of some foods that would be complemented by the product be included on the label.

4) that protein information not be included on the label for products that fall below some minimum amount of utilizable protein unless they provide substantial amounts of amino acids that are most likely to be limiting in U.S. diets.

SUMMARY

In summary, the Task Force expressed dissatisfaction with the current method of labelling for protein quality based on PER measurements. All of the groups expressed the need for providing a value for utilizable protein that could be easily understood. The relative net protein ratio method was proposed as an improvement over PER that could be instituted now. The modification most generally accepted as the one to aim for eventually was to substitute for the present method, a procedure based on an amino acid scoring system corrected for digestibility or amino acid availability. This procedure was favored because information on the label could be related directly to human requirements for amino acids. The major unresolved question was over the amount of additional information and testing that was needed before this procedure could be instituted.

24

Conference Overview

Aaron M. Altschul

I have been asked to conclude this meeting by summarizing my impressions of what was intended, what was accomplished, and the direction which seemed most appropriate for the future. Events which led to this meeting and its predecessors are that the dietary of the American consumer is changing; there is a need to manage this change in a way that improves the quality of life for the consumer in its complete and broadest sense. New protein sources are entering the food supply, particularly proteins from vegetable sources, and they are becoming available in a wide variety of processed foods. These would include infant formulas, meat analogs and processed meats, and complete meals.

It is necessary that the new foods be equivalent at least to the old as sources of protein. An added advantage should be that more dietary options are made available by inclusion of these new sources of protein to provide greater flexibility in individual adjustment to the need for changes in intake of other nutrients. There is, therefore, the requirement for reliable methods of estimating protein value of these new foods or combinations that would relate more realistically to human dietary needs. And such methods should be applicable equally for scientific research on human protein requirements, protein metabolisms, and for regulatory purposes.

This particular meeting follows the one held two years ago at Keystone when these same questions were raised. The purpose of this meeting, as was that of the former one, was to assess progress in understanding human protein metabolic needs and in development of methods of measurement of adequacy of dietary protein in fulfilling these needs. The practical objective was to encourage the type of development that would put assessment of protein quality on a more rational basis.

This meeting had its good and bad moments. There were times when it seemed that preoccupations with detail of experimental design and with interpretation of results diverted attention from the objectives. But the meeting pulled itself together and the outcome should exceed the most optimistic projections of its organizers.

I see the following three general conclusions:

1. There was continued and general disaffection with animal surrogates as a means for measuring protein equivalency as it relates to humans and as a means of regulating new protein products in the food supply.

2. Considerable progress has been made in the development of non-invasive techniques for studying protein metabolism in humans. Because of great individual variation and the small number of subjects that can be accommodated, sensitivity in detecting differences between closely related proteins is low. One can distinguish between wheat and egg protein whereas it is difficult to distinguish between egg and soy protein. This could mean that for humans, differences between egg and soy protein are practically unimportant.

But, it never was intended that experiments with humans would be considered as means for practical routine measurement of protein quality. Rather these were intended to provide information on a well-defined protein material or mixtures that would become a standard of performance. This can now be done; and other methods can be applied to describe the properties of the proteins whose nutritional value has been determined in humans.

3. Extraordinary progress has been made in the development of *in vitro* methodology for describing proteins and mixtures of proteins. This includes measures of amino acid content, digestibility, amino acid availability, and levels of physiological inhibitors. It is possible from these observations to develop some sort of score which would rate proteins on the basis of their nutritive value and indicate the limiting factors that are influencing the rating. The score need not necessarily depend on a measure of all of the essential amino acids; it may require just the measure of a few amino acids in these sources.

A word of caution is in order. For the vegetable proteins which enter the protein food supply and for many of the animal proteins that are processed, such as dried milk, processing conditions are most important. They can affect digestibility (reduce it or improve it) or amino acid availability (by binding some of the amino acids such as the epsilon amino group of lysine or the formation of sulfoxides) and they can affect the content of inhibitors such as trypsin inhibitors, hemagglutenins, and gossypol, all of which directly or indirectly affect digestibility and amino acid availability. Hence, *in vitro* methods, in addition to measuring amino acid content, digestibility, and availability, must include some appropriate measure of levels of critical physiological materials peculiar to the protein source being investigated. I cannot conceive, for example, of a chemical assessment of soy protein which does not include assaying for levels of trypsin inhibitor nor can I conceive of a chemical measure of the nutritional value of cottonseed protein which does not contain a measure of gossypol content.

In this connection, I am surprised by the casual way that scientists describe the protein material which they had studied. Those who have been doing extensive work on humans go to considerable trouble in developing experimental design and in carrying out difficult experimental protocols and yet are completely indifferent in their description of the material. It would seem that there ought to be closer discussions between those who conduct experiments in humans and those who do studies on *in vitro* methodology. These two groups of investigators ought to exchange material so that each can better describe or more completely describe the material with which they are working.

It seems likely that the progress described at this meeting will continue to provide information on humans as a standard of protein performance and that parallel progress will be made in describing the proteins or their mixtures by *in vitro* tests. There is, therefore, good reason to believe that PER, now being used traditionally for regulatory purposes in describing protein quality, can be supplanted by something better that can give the consumer greater confidence of its more appropriate relevance to human needs.

There is a momentum clearly observable among the scientists and among those who depend on the scientists for regulatory decisions in moving towards a more rational description of protein quality. This momentum ought to be pre-reserved and fostered by continuing these meetings at appropriate intervals. Aside from the practical objectives that were achieved at this meeting, it was indeed an intellectual experience. It was a good protein meeting that could stand on its own for its scientific merit, aside from the practical objectives achieved.

I take the liberty of speaking for all of us in congratulating the organizers of this conference Drs. Atkins, Bodwell, and Hopkins for their skill in selecting the discussion topics and participants, and their devotion to the details of operation which made for a smooth-functioning meeting. And we thank them for providing, in theory, such agreeable circumstances under which to meet.

I congratulate the representatives of regulatory agencies here for their creative approach to managing such a profound change in the food supply with consequences for both industrialized and less-industrialized societies. This surely is an example of creative partnership between government and science—science from both the public and private sector—in arranging for orderly advance in management of such a critical element of our food supply.

Participants
(Includes contributors who were not able to attend the conference)

Dr. R.A. Abernathy
Dept. of Foods & Nutrition
Purdue University
Lafayette, IN 47907

Dr. J.S. Adkins
Program In Human Nutrition &
    Food
School of Human Ecology
Howard University
2400 6th St., N.W.
Washington, DC 20059

Dr. A. Altschul
Dept. of Community Medicine
Georgetown University
Washington, DC 20007

Dr. R.H. Anderson
General Mills, Inc.
Research Center
9000 Plymouth Ave., N.
Minneapolis, MN 55427

Dr. L.M. Ausman
Dept. of Nutrition
Harvard School of Public Health
665 Huntington Avenue
Boston, MA 02115

Dr. R.J. Bell
Anderson Clayton Foods
3333 N. Central Expressway
Richardson, TX 75080

Dr. Duane Benton
Ross Laboratories
625 Cleveland Ave.
Columbus, OH 43216

Dr. Norma Benton
Corporate Nutrition
Borden, Inc.
990 Kingsmill Parkway
Columbus, OH 43219

Dr. C.E. Bodwell
U.S. Dept. of Agriculture
SEA, Nutrition Center
Nutrition Institute
Room 313, Bldg. 308, BARC-East
Beltsville, MD 20705

Dr. R. Bressani
Instituto de Nutricion de Centro
    America Y Panama
Carretera Roosevelt, Zona 11
Apartado Postal 1180
Guatemala City
Guatemala, Central America

Dr. Mahlon Burnette III
Grocery Manufacturers of America
1425 K Street, N.W.
Washington, DC 20005

Ms. M.I. Cabrera-Santiago
Instituto de Nutricion de Centro
    America Y Panama

Carretera Roosevelt, Zona 11
Apartado Postal 1180
Guatemala City
Guatemala, Central America

Dr. Doris H. Calloway
Dept. of Nutritional Sciences
University of California
Berkeley, CA 94720

Dr. K.J. Carpenter
Dept. of Nutritional Sciences
University of California
Berkeley, CA 94720

Dr. A.H. Cheng
Kraft, Inc.
801 Waukegan Road
Glenview, IL 60025

Dr. Roger Clemens
Carnation Co.
8015 Van Nuys Blvd.
Van Nuys, CA 91412

Dr. Peyton Davis
Nutrition Department
National Livestock & Meat Board
444 North Michigan Ave.
Chicago, IL 60611

Dr. Philip H. Derse
Raltech Scientific Services, Inc.
P.O. Box 7545
Madison, WI 53707

Dr. Norman W. Desrosier
AVI Publishing Co., Inc.
P.O. Box 831
Westport, CT 06880

Dr. Robert Drotman
Proctor & Gamble Co.
Miami Valley Labs
Cincinnati, OH 45247

Dr. C.H. Edwards
School of Human Ecology
Howard University

Washington, DC 20059

Dr. B.O. Eggum
Dept. Animal Physiology
Natl. Inst. of Animal Science
25 Rolighedsvej, DK-1958
Copenhagen, V, Denmark

Dr. John E. Ford
Natl. Inst. for Research in Dairying
Shinfield, Reading
RG2 9AT, England

Dr. M. Friedman
U.S. Dept. of Agriculture
Science & Education
   Administration
Western Regional Research Center
Berkeley, CA 94710

Dr. L.R. Hackler
Dept. of Foods & Nutrition
University of Illinois
Urbana, IL 61801

Dr. R.R. Hahn
A. E. Staley Mfg. Co.
2200 Eldorado
Decatur, IL 62525

Mrs. Muriel Happich
U.S. Dept. of Agriculture
Eastern Regional Research Center
600 East Mermaid Lane
Philadelphia, PA 19118

Dr. A.E. Harper
Dept. of Biochemistry
216 Biochem. Building
University of Wisconsin
Madison, WI 53706

Dr. Mary Heckman
Ralston Purina Co.
3RS, Checkerboard Square
St. Louis, MO 63188

Dr. D. Mark Hegsted
U.S. Dept. of Agriculture

Science & Education
  Administration
Human Nutrition Center
Room 426-A, Administration
  Building
Washington, DC 20250

Dr. D.T. Hopkins
Ralston Purina Co.
Checkerboard Square
St. Louis, MO 63188

Dr. Francis E. Horan
Archer Daniels Midland Co.
P.O. Box 1470
Decatur, IL 62525

Dr. H.T. Huang
Div. of Problem Focus Research
National Science Foundation
1800 G Street, N.W.
Washington, DC 20550

Dr. R. Hurrell
Nestle Products Technical
  Assistance Co., Ltd. (NESTEC)
Case Postale 88, CH-1814
La Tour-De-Peilz, Switzerland

Dr. H. David Hurt
Quaker Oats Co.
617 West Main St.
Barrington, IL 60010

Dr. Seymour Hutner
Haskins Labs, Pace Plaza
41 Park Row at Pace Plaza
New York, NY 10038

Dr. G.R. Jansen
Dept. of Food Science and Nutrition
Colorado State University
Ft. Collins, CO 80523

Ms. Mamie Jenkins
Food and Drug Administration
Division of Nutrition
200 C Street, S.W.
Washington, DC 20204

Dr. J. Kendrick
Dept. of Agricultural Economics
University of Nebraska
Lincoln, NE 68583

Dr. Chor San Khoo
Kellogg Co.
235 Porter St.
Battle Creek, MI 49016

Dr. M.K. Korslund
Dept. of Human Nutrition & Food
Virginia Polytechnic Institute
Blacksburg, VA 24061

Dr. D.M. Larson
Mead Johnson & Co.
Mead Johnson Research Center
Evansville, IN 47721

Dr. Nina Marable
Dept. of Human Nutrition & Food
Virginia Polytechnic Institute
Blacksburg, VA 24061

Dr. S. Margen
School of Public Health
University of California
Berkeley, CA 94720

Ms. W. Martinez
U.S. Dept. of Agriculture
Science & Education
  Administration
Room 137, Bldg. 005 - BARC, East
Beltsville, MD 20705

Dr. D.E. McOsker
Greyhound Corporation
Armour Research Center
15101 N. Scottsdale Road
Scottsdale, AZ 85260

Dr. G.V. Mitchell
Food and Drug Administration
Division of Nutrition
200 C Street, S.W.
Washington, DC 20204

Dr. R.B. Morrissey
Hershey Foods Corp.
1025 Reese Avenue
Hershey, PA 17033

Ms. E.W. Murphy
Food Ingredient Assessment
  Division, FSQS
Room 2148, South Building
U. S. Dept. Agriculture
Washington, DC 20250

Dr. Marshall Myers
Mars, Inc.
12 Emery Ave.
Randolph, NJ 07801

Dr. Robert O. Nesheim
Quaker Oats Co.
617 West Main St.
Barrington, IL 60010

Dr. P. Pellett
Dept. of Food Science & Nutrition
University of Massachusetts
Amherst, MA 01003

Dr. David Peterson
Oat Quality Laboratory
Dept. of Agronomy
University of Wisconsin
Madison, WI 53706

Dr. O. Pineda
Instituto de Nutricion de Centro
  America Y Panama
Carretera Roosevelt, Zona 11
Apartado Postal 1180
Guatemala City
Guatemala, Central America

Dr. W. Rand
Dept. of Nutrition & Food Science
Mass. Institute of Technology
Cambridge, MA 02139

Dr. S.J. Ritchey
Dept. of Nutrition

Va. Polytechnic Institute and State
  University
Blacksburg, VA 24061

Dr. William Van B. Robertson
Div. of Problem Focus Research
National Science Foundation
1800 G Street, N.W.
Washington, DC 20550

Dr. John T. Rotruck
Proctor & Gamble Co.
Miami Valley Labs
Cincinnati, OH 45247

Dr. I.I. Rusoff
NABISCO
Fair Lawn, NJ 07410

Dr. K. Samonds
Dept. of Food Science & Nutrition
University of Massachusetts
Amherst, MA 01002

Dr. G. Sarwar
Food Directorate
Health Protection Branch
Dept. of National Health & Welfare
Tunney's Place
Ottawa 3, Ontario, Canada

Dr. L. Satterlee
University of Nebraska
Dept. of Food Science & Tech.
Room 20, Filley Hall
Lincoln, NE 68583

Dr. N.S. Scrimshaw
Dept. of Nutrition & Food Sciences
Massachusetts Inst. of Technology
Cambridge, MA 02139

Dr. W.H. Seligson
Proctor & Gamble Co.
Winton Hill Technical Center
6071 Center Hill Road
Cincinnati, OH 45224

Dr. N. Shah
General Foods Technical Center
250 North St.
White Plains, NY 10625

Dr. Sandra Skarsaune
Kellogg Co.
235 Porter St.
Battle Creek, MI 49016

Dr. M. Solberg
Dept. of Food Science
Rutgers University
New Brunswick, NJ 08903

Dr. M. Stahmann
Dept. of Biochemistry
University of Wisconsin
Madison, WI 53706

Dr. F.H. Steinke
Ralston Purina Co.
3RS, Checkerboard Square
St. Louis, MO 63188

Dr. M. Swenseid
School of Public Health
UCLA
Los Angeles, CA 90024

Dr. B. Torun
Instituto de Nutricion de Centro
   America Y Panama
Carretera Roosevelt, Zona 11
Apartado Postal 1180
Guatemala City
Guatemala, Central America

Dr. J.E. Vanderveen

Food and Drug Administration
200 C Street, S.W.
Washington, DC 20204

Mr. Emilo Vargus
Instituto de Nutricion de Centro
   America Y Panama
Carretera Roosevelt, Zona 11
Apartado Postal 1180
Guatemala City
Guatemala, Central America

Dr. R.E. Webb
Dept. of Nutrition
Va. Polytechnic Institute and State
   University
Blacksburg, VA 24061

Dr. H.L. Wilcke
Ralston Purina Co.
4 RN, Checkerboard Square
St. Louis, MO 63188

Dr. K.N. Wright
A. E. Staley Mfg. Co.
2200 Eldorado
Decatur, IL 62525

Dr. Allison Yates
Steed College
P.O. Box 3098
Johnson City, TN 37601

Dr. V.R. Young
Laboratory of Human Nutrition
Dept. of Nutrition & Food Sciences
Massachusetts Institute of
   Technology
Cambridge, MA 02139

Index